Microalgae Building Enclosures

Microalgae architecture has gained awareness for its biotechnical potential to achieve net-zero energy architecture while also promoting ecological sustainability and occupant well-being. *Microalgae Building Enclosures: Design and Engineering Principles* aims to provide design, engineering, and biotechnical guidelines for microalgae building enclosures that need to be considered for symbiotic relations among the built environment, humans, and ecosystems.

Part I of the book introduces the theoretical background of microalgae as a bioremediator and future energy system and their potential roles toward sustainable and healthy built environments. Part II exemplifies interventions and multiple benefits of microalgae systems in product, architecture, urban, and infrastructure applications across the globe including Africa, Asia, Australia, Europe, South America, and North America. Part III explains the design and engineering criteria, biotechnical design requirements, and various performance metrics for microalgae architecture. Finally, Part IV investigates potential building applications in low-rise buildings, high-rise buildings, and energy-efficient retrofitting. The book also includes international case studies of microalgae building systems within various building types and climates.

As one of the first books to comprehensively cover this emerging area of microalgae building enclosures, *Microalgae Building Enclosures* is an essential source for professionals and students looking to expand architectural discourse on nature integrated building systems to achieve the triple bottom line of sustainability.

Kyoung Hee Kim, PhD, AIA, NCARB, is a practicing architect, engineer, and entrepreneur. She is an Associate Professor of the School of Architecture and the Director of the Integrated Design Research Lab at the University of North Carolina at Charlotte. Kim, a registered architect in North Carolina, is the founder of Ecoclosure, a university spin-off, and a design principal at Hui Kim Design and Architecture (HK DnA) based in Charlotte, NC. She has developed sustainable, regenerative building technologies toward net zero energy architecture funded by NSF, American Institutes of Architects, and institutional grants. As a facade consultant at Front Inc (2007–2017), Professor Kim engaged in over 50 institutional, cultural, residential, commercial, and governmental projects around the world.

Microalgae Building Enclosures

Design and Engineering Principles

Kyoung Hee Kim

Routledge
Taylor & Francis Group

NEW YORK AND LONDON

Cover image: Mike Basher / www.mikebasher.com

First published 2022
by Routledge
605 Third Avenue, New York, NY 10158

and by Routledge
2 Park Square, Milton Park, Abingdon, Oxon, OX14 4RN

Routledge is an imprint of the Taylor & Francis Group, an informa business

Library of Congress Cataloging-in-Publication Data
Names: Kim, Kyoung Hee, author.
Title: Microalgae building enclosures: design and engineering principles / Kyoung Hee Kim.
Description: New York, NY: Routledge, 2022. | Includes bibliographical references and index. |
Identifiers: LCCN 2021042326 (print) | LCCN 2021042327 (ebook) |
ISBN 9780367410469 (hbk) | ISBN 9780367410452 (pbk) | ISBN 9780367814410 (ebk)
Subjects: LCSH: Facades—Materials. | Microalgae. | Algae as building materials. |
Biofilms. | Sustainable buildings—Materials. | Buildings—Energy conservation. |
Sustainable architecture.
Classification: LCC TH2246 .K56 2022 (print) | LCC TH2246 (ebook) |
DDC 690/.8—dc23/eng/20211115
LC record available at https://lccn.loc.gov/2021042326
LC ebook record available at https://lccn.loc.gov/2021042327

ISBN: 9780367410469 (hbk)
ISBN: 9780367410452 (pbk)
ISBN: 9780367814410 (ebk)

DOI: 10.4324/9780367814410

Typeset in Avenir
by codeMantra

Contents

Figures

Tables

Acknowledgments

The preparation of the book spans a decade, from 2011 to 2021. The book owes its existence to numerous colleagues, research assistants, and organizations.

First, I am deeply grateful to the warm leadership and talented staff at the School of Architecture at UNC Charlotte. They provided not only dedication, support, and professionalism but also friendship, trust, and humor, making the environment encouraging and fun. They are former CoAA Dean Ken Lambla, former SoA director Chris Jarrett, interim SoA director Jose Gamez, current Dean Brook Muller, and current SoA director Blaine Brownell.

I wish to thank research assistants who dedicated long hours and helped with the nuts and bolts work of the book. They are, in alphabetical order, Seyedehhamideh Hosseiniirani, Masoumeh Hosseinzadeh, Daniel Lutarcwych, Bekim Sejdiu, and Kathryn Warren. I am also grateful to those individuals who, though not acknowledged here, directly and indirectly influenced the book.

I would like to thank a particular group of colleagues who successfully undertook a full-scale working prototype funded by the NSF. Chapter 9 introduces some key outcomes from the NSF-funded prototyping, which helps calibrate the gap between the theory and practice. They are Chengde Wu, Garrett Herbst, Yau Shun Hui, Arturo Lujan, Matt Parrow, Ok-Kyun Im, James Hung, Milad Rogha, Alexander Cabral, and Robby Sachs.

I would like to thank the students who participated in my topic studio in 2011 which germinated the concept design of the microalgae façade. The Biofacade seminar in 2013 sponsored by an EPA-P3 award substantiated its nucleus form with small-scale prototyping and an exhibition at the National Sustainable Design Expo in Washington, DC. They are, in alphabetical order, Nathan Aaronson, Martin Andreasson, Chanel Brown, Christine Chlebda, William Cordes, Gina Dematteo, Yiran Hu, Brian Jones, Samaneh Mahabadi, William Mayo, Emily Merill, Jessica Nuts, William Philemon, Angela Scharrer, Patricia Sharp, Nathalie Slobodiuk, Michelle Todd, and Bryan Williams.

I extend my gratitude to the National Science Foundation, the American Institute of Architecture, and the Environmental Protection Agency for their invaluable financial support to carry out proof of concept, prototyping, and associated research.

I would like to thank Assunta Petrone at codeMantra and Christine Bondira at Routledge for their guidance and help on editing, graphic design, and production, Fan Lu and Matt Parrow for reviewing the biological content of the book, and Anne Page for proofreading the draft.

Finally, I want to express my heartfelt thanks to my husband, Yau Shun Hui, and three sons, Anthony, Henry, and Patrick, as well as the families in South Korea and Hong Kong for their support and love, allowing me to devote time to research and writing.

Part I | Microalgae Architecture Introduction

Chapter 1

Sustainable Development

1.1 I = f (P,A,T)

The I = f (PAT) framework, conceived by Ehrlich et al. in the 1970s, is an analytical tool for identifying the forces driving anthropogenic environmental impacts. This framework guides the use of resources with reduced environmental impact. According to the I = f (PAT) framework, there are three major factors affecting environmental impact: (1) population (P), (2) affluence level (A), and (3) technology (T). In other words, environmental impact is expected to rise as a function of increasing population, affluence level as represented by resource consumption, and technological advancement. A leading ecologist Barry Commoner writes,

> Biologically, human beings participate in the environmental system as subsidiary parts of the whole. Yet, human society is designed to exploit the environment as a whole, to produce wealth. The paradoxical role we play in the natural environment—at once participant and exploiter—distorts our perception of it.[1]

According to the 2018 United Nation's population database, two out of three persons will reside in cities by 2050, requiring more construction and infrastructure to absorb the additional 2.3 billion urban residents. The world's population has increased by more than 5 billion over a 70-year period from 2.5 billion in 1950 to 7.8 billion in 2020. It is expected to increase an additional 3 billion by 2050 with a projection of 9.8 billion.[2] Due to the large influx of residents moving from rural areas to cities, the ratio between urban and rural populations has grown rapidly with 30% of world population living in urban areas (0.8 billion out of 1.76 billion total) in 1950 to 56% (4.4 billion out of 7.8 billion total) in 2020 and being expected to reach 68% (6.7 billion of 9.8 billion total) by 2050. With the increasing share of urban population, 90% of the urban population is expected to grow in Asia and Africa by 2050. Just the three countries of India, China, and Nigeria are expected to account for 35% of the urban population growth between 2018 and 2050.[3]

As a result of increased urban influx, we are faced with numerous challenges including problems with creating adequate mechanisms to meet the growing urban population's need for housing, infrastructure, transportation, energy, and employment. A Yale research group projected that new urban development will require nearly three times the urban land area of 2000, expanding to an additional 1.2 million km^2 by 2030 (the size of Manhattan area is 87 km^2 as a comparison).[4]

DOI: 10.4324/9780367814410-2

Accordingly, the new gross floor area construction is projected to increase by approximately 40% (291,700 km^2) in 2030 and 80% (291,700 km^2) in 2030 from circa 2015 (209,900 km^2).[5] These demands have led to an increase in environmental issues such as air pollution, water shortages and contamination, the heat island effect, flooding, and increased energy consumption. According to the International Energy Agency (IEA), global buildings are currently responsible for approximately 40% of all primary energy use and around 30% of global CO_2 emissions annually.[6]

The increasing population and floor area growth from affluent lifestyles have outpaced the energy savings from energy-efficient building technologies, leading to increases in overall energy usage and pollutant emissions from the building sector. In other words, the improvement of energy efficiency is not fast enough to offset the new building construction. The IEA study indicated that the global space-cooling load and the plug load increased by approximately 20% and 18%, respectively, between 2010 and 2017.[7] Air conditioning represents a major cause of electrical end-use and has shown an increase while space heating and lighting loads have been reduced due to improved energy efficiency from equipment and building envelopes.[8] Different climate characteristics also affect the pattern of energy end-users. Warmer winters reduce the heating load, whereas hot summers increase the cooling load. More cooling degree days due to heat waves during summer, for example, would increase global air conditioning use.

Affluence represents income or GDP/capita that reflects the economic power of the population. Increased consumption level per capita has become a significant cause of environmental impact. Although technological innovation can help reduce emissions in some places, thus far technology has not been able to outpace the rapid increases in consumption worldwide. People's lifestyle choices correlate with environmental impacts. According to the U.S. Census Bureau, the average size of a single-family home built in 1950 and 2019 was 983 ft^2 and 2,623 ft^2, respectively, while the average number of family members in a U.S. household decreased from 3.4 in 1950 to 2.5 in 2019.[9] This means that the house floor area per person has more than tripled between 1950 and 2019. With an average of 2.5 people per household and 2,623 ft^2 floor area in 2019, the average U.S. dwelling provided more than 1,000 ft^2 per capita in 2019.[10] Similarly, global house size in many countries has increased in recent decades. For example, in 2018, newly built homes in Japan averaged about 850 ft^2 (down from 947 ft^2 in 1993)[11] and those in the United Kingdom averaged about 970 ft^2 of habitable floor space (up from 945 ft^2 in 2004).[12] In European Union member countries, Luxembourg ranks highest, with 713 ft^2 per capita, but is still well below the U.S. average of 871 ft^2 per capita.[13]

A larger house size means more embodied energy per capita. It also consumes greater operational energy and generates larger environmental impacts from powering the building service equipment such as air conditioner, heater, and artificial lighting. As the house size per capita has increased in the United States, residential energy consumption per capita has more than doubled between 1950 and 2018. In addition, there is an extensive use of furniture and appliances to fill the larger house spaces. A recent study showed that furniture and appliances contribute to approximately 10% and 25%, respectively, of a building's overall environmental impact.[14] In the same study, for office environments, laptops and monitors are the most impacting elements, whereas for dwellings, refrigerators have the greatest impact. In addition to the change in the standard of living, there is a correlation between household income level and energy consumption. The residential energy consumption by a high-income household is higher than that of a low-income household although low-income households spend a larger portion of their income on utility bills.[15] An increased household income and improved living standard result in a bigger carbon footprint and increased municipal wastes.

It has been shown that the combined impact from an increase in urbanization and an improved standard of living has resulted in greater environmental stress, resource depletion, and waste generation. To cope with this situation, low carbon emission per capita is fundamental to energy saving and environmental protection. For example, an average American has a carbon footprint of 15 metric tons of CO_2 compared to the world average per capita of 4.5 metric tons of CO_2.[16] US CO_2 emission per capita, with just a fraction of the world's population, accounts for three time the world's CO_2 emissions per capita. Table 1.1 shows the CO_2 emissions for the four highest emitting countries. As the table demonstrates, although China emitted almost twice as much CO_2 in 2018 as the United States did, per capita emissions for the United States are more than two times higher than those for China. Americans (15 metric tons CO_2) produce about 3.5 times more CO_2 per capita than the global average (4.5 metric tons CO_2).[17] A biologist Walter Howard writes, "The affluent society has become an effluent society. The 6 percent of the world's population in the United States produces 70 percent or more of the world's solid wastes."[18] As 90% of the additional 2.5 billion urban population in 2050 is expected to be concentrated in Asia and Africa, a global climate alliance is necessary to support the developing countries in growing their economies by renewable energy systems and not by fossil fuels. It should be noted that calculating an individual country's carbon footprint is complicated because system boundaries for international trades blur their global carbon footprint. Items manufactured for export, items imported from other countries, and the emissions arising from transporting the items across the globe all contribute to a country's carbon footprint.

Technology can have positive and negative impacts on built environments. Efficiency and innovation of technology can play a key role in reducing environmental impacts while manufacturing production to meet market demand can increase impacts. The climate emergency has motivated net zero energy architectural practices to meet prosperity and environmental goals, engendering a proliferation of sustainable building technologies. Among these, the role of advanced glass window technologies has brought new opportunities for the future of architecture, not only in managing energy transfer between indoor and outdoor spaces but also in harnessing renewable energy resources. Starting with float glass and low-emissivity coating technology in the early 20th century and the 1980s, respectively, energy management techniques for glazed windows range from the integration

▼ Table 1.1

2018 CO_2 emissions for the planet's top emitters[19]

Country	2018 contribution to global CO_2 emissions (%)	2018 per capita CO_2 emissions (t CO_2)	CO_2 emissions growth rate from 2017 to 2018 (%)
China	28	7	2.3
United States	15	16	2.8
EU (28 member states)	9	7	2.1
India	7	2	8
Rest of the World	41	1.3	1.8

of surface treatment techniques (e.g., frit, film, interlayer), to inert gases (e.g., argon, krypton, xenon), to insulative techniques (e.g., aerogel, vacuum), dynamic coatings (e.g., electrochromic, thermochromic, photochromic), or phase change materials (e.g., paraffin, hydrogel). As a building surface serves as the prime place for harnessing solar energy, technologies such as building integrated photovoltaic (BIPV) and solar thermal play an important role in achieving energy savings. For BIPV application in the built environment, solar conversion efficiency and aesthetic appearance are important in increasing wide public acceptance. PV cell technology has advanced significantly since the past decade. A range of PV cells have been developed, including crystalline silicon, thin film amorphous, cadmium telluride (CdTe), copper indium gallium selenide (CIGS), and organic PV cells. Silicon-based PV cells such as single crystalline Si (mono c-Si) and polycrystalline Si (poly c-Si) are first-generation solar cell technology. Second-generation solar cells use thin film technology such as amorphous Si (a-Si), cadmium telluride, and CIGS. Their market shares are 48% (mono c-Si), 35% (poly c-Si), 9% (a-Si), and 1% (CdTe, CIGS, etc.).[20] Crystalline silicon is expected to continue to dominate the global market due to its durability and high conversion efficiency. The embodied energy of a PV cell certainly adds an environmental burden, but the overall net environmental benefit is positive considering that the operation of a PV cell produces clean energy. China is the leader in solar power production with about 300 GW followed by Brazil (90 GW), Canada (80 GW), and India (50 GW).[21]

A green power policy of increasing the supply of electricity from solar cells, however, generates higher waste volumes from solar cells. It is anticipated that approximately 10 million tons of PV waste will be produced by 2050.[22] With rapid technological advancement, cell efficiency will grow, and prices will drop, resulting in the accumulation and acceleration of wastes from frequent replacement of old PV cells with new technologies. The lifecycle of a PV cell includes raw material production, manufacturing, operation, and end-of-life treatment. When it comes to the end-of-life cycle, recycling is a recommended practice. PV cells mainly consist of glass, EVA, and rare metals (e.g., cadmium, selenium, tellurium, gallium, molybdenum, and indium), and the recovery of these materials at the end-of-life cycle is good for a waste-free circular economy. It is important to holistically evaluate the recovery of materials (positive effect) with the energy input and environmental impacts associated with recycling (negative effect). Research was carried out to investigate a method to recycle silicon solar cells by separating metals from coated silicon wafers and turning them into marketable products such as poly-aluminum-hydroxide-chloride or turning thin film photovoltaic wastes into float glass production.[23] It is also important to inform the public and other stakeholders of the extent to which recycling of PV cells is an economically and environmentally viable practice.

Energy consumption and construction material use have intensified. Secretary-General Petteri Taalas of the World Meteorological Organization said,

> If we do not take urgent climate action now, then we are heading for a temperature increase of more than 3°C by the end of the century, with ever more harmful impacts on human wellbeing. We are nowhere near on track to meet the Paris Agreement target.[24]

Growing construction activities to accommodate urbanization along with affluent lifestyles have turned the building sector into a major contributor to environmental impact and non-renewable resource use. The global construction market is expected to grow by 85%, reaching $15.5 trillion worldwide by 2030, and three countries lead the way– China, United States, and India – accounting for 57% of all global growth.[25] Accordingly, building materials production, especially steel manufacturing and cement production, will continue to grow. Steel production is the highest industrial energy consumer due to the use of blast furnaces typically powered by coal power plants. The cement industry is the second largest consumer of industrial energy use. Carbon dioxide emissions due to fossil fuel use and cement production have increased by 60% since 1990. Substantial cement manufacturing has continued to rise steadily with surging economic growth, especially in emerging developing nations. A reduced ratio of clinker-to-cement content and an increase in clinker substitutes could mitigate environmental impacts from cement production. The use of cleaner fossil fuels, shifting from coal to renewable energy sources, is another way to mitigate impact. Cement production in Europe, which can reduce the clinker-to-cement ratio to 70%, results in a further CO_2 savings of 4%, whereas cement production in the United States uses a 95% clinker-to-cement ratio.[26]

To improve ecological performance of concretes, eco-friendly solutions have been introduced that replace cement with the by-product of other materials or recycled waste such as fly ash from power plants, glass powder, and demolished waste from construction sites. Researchers have experimented with living organisms such as cyanobacteria that use photosynthesis and produce calcium carbonate similar in attributes to cement. This bio concrete pulls CO_2 out of the air for biomass increase and when the concrete cracks, the organism provides active growth of hard calcium carbonate to fill in the crack.[27] Another study was done using marine algae in the concrete mix to create an economical and environmentally friendly concrete with increases of 20% in compression strength, 20% in split tensile strength, and 25% in flexural strength with an 8% addition of microalgae.[28]

In addition to sustainable construction, the energy efficiency of energy end-users plays an important role in reducing environmental impact. Air conditioning and refrigeration are responsible for approximately 17% of the world's total electricity consumption.[29] Heat wave increases in the urban environment have caused a dramatic surge in cooling energy consumption. Air conditioner technologies have been advancing in function and energy efficiency over the past century. Three-quarters of U.S. homes use air conditioners resulting in the consumption of 6% of U.S. electricity, a $29 billion annual energy cost, and 117 million tons of CO_2 emissions.[30] Despite an overall building energy efficiency improvement of 2.5% in 2015, energy efficiency has fallen in 2018 by 0.5%, in part due to a lack of energy policy implementation and economic incentives.[31] The energy performance of buildings has significantly improved with the use of highly insulated materials and energy-efficient HVAC (heating, ventilation, and air conditioning) systems. Traditional architecture has been successful in achieving climate-responsive design, along with natural heating, cooling, and ventilation without relying on external energy equipment. However, the improvement of living standards and economic growth have favored the hermetically sealed built environment operated by active building service systems.

At the resource consumption level, various anthropogenic activities affect water consumption, land use, and building operations. Buildings consume 70% of electricity generation in the United States, and a majority of power plants use fossil fuels such as coal and natural gas that contribute a significant amount of air pollutants, including arsenic, mercury, SO_2, and NO_x. These pollutants have been recognized as a threat to human and ecosystem health. As the population and construction increase, the reliance on fossil fuels is inevitable. Scientific data indicate that anthropogenic greenhouse gas (GHG) emissions have drastically increased compared to the pre-industrial era, which has pushed us to a new geological era called the Anthropocene. Of the greenhouse gases, anthropogenic CO_2 accounts for 65% of the global warming trend, methane: 16%, nature-produced CO_2: 11%, nitrous oxide: 6%, and fluorinated gas: 2% in 2010.[32] Different greenhouse gases have different climate effects. The current CO_2 level, the principal driver of global warming, has increased by 40% compared to the pre-industrial era. It was produced at around 280 parts per million (ppm), (10 million tons CO_2) per year before the pre-industrial era in 1750, increased to an annual 290 ppm (2 gigatons [Gt] CO_2) in 1900, and is now produced at over 410 ppm (36 $GtCO_2$) per year in 2019.[33]

Methane, a secondary anthropogenic greenhouse gas, is emitted from activities such as burning fossil fuels, agriculture (e.g., livestock and rice paddies), and biogas from solid waste landfills. Atmospheric methane has more than doubled from 700 ppb (parts per billion) in 1750 to 1,900 ppb in 2019.[34] Nitrous oxide (N_2O) known as "laughing gas" is mostly caused by agricultural activities (e.g., nitrogenous fertilizer) as well as during combustion and human waste disposal.[35] The CO_2 equivalent is a way to normalize the effect of trapping heat for a specific length of period compared with CO_2 as a reference gas. One Methane, for example, has an approximate 30 CO_2 equivalent over a 100-year timescale, meaning it has 30 times more warming potential than CO_2. Nitrous oxide has around a 300 CO_2 equivalent for a 100-year timescale. The measurement unit for methane and nitrous oxide is ppb, one-thousandth of ppm, but small increases in emissions from such gases have detrimental environmental impacts.

Unbalance between anthropogenic emissions and natural sinks calls for global citizenship in climate change abatement. The global carbon cycle describes carbon storage and fluxes between primary components of the Earth's systems—atmosphere, lithosphere, biosphere, and hydrosphere. These primary systems serve as carbon reservoirs as well as carbon exchange fluxes. According to the 2019 United Nations Emissions Gap Report,

> GHG emissions have risen at a rate of 1.5 per cent per year in the last decade, stabilizing only briefly between 2014 and 2016. Total GHG emissions, including from land-use change, reached a record high of 55.3 $GtCO_2e$ in 2018.... Fossil CO_2 emissions from energy use and industry, which dominate total GHG emissions, grew 2% in 2018, reaching a record 37.5 $GtCO_2$ per year.[36]

According to Intergovernmental Panel on Climate Change (IPCC's) 2013 Climate Change Report, fossil fuel burning and cement production contribute to around 7.8 GtC (gigatons of carbon) per year and land use changes such as deforestation result in annual 1.1 GtC emissions.[37] While anthropogenic carbon production averaged 8.9 GtC annually, nature has been absorbing only about 4.9 GtC per year into ocean and land sinks.[38] As a result, around 4 GtC of anthropogenic carbon accumulates in the atmosphere every year, causing a global warming trend. Using a conversion ratio of 1 GtC to 3.667 $GtCO_2$, the annual CO_2 accumulation in the atmosphere is approximately 15 $GtCO_2$.

Scientific studies show the correlation between the CO_2 level and Earth's average global surface temperature. Global warming is driven by heat-trapping greenhouse gases absorbing infrared heat radiated by Earth. As a result, climate change impacts such as flooding, drought, and sea-level changes are evident. Globally, these natural disasters cause the loss of human lives as well as economic damage. Desertification is a result of climate change, but it also has an impact on climate change by altering regional surface temperatures and water availability.[39] Desertification threatens plant and animal biodiversity and results in agricultural and socioeconomic impacts. Drought and desertification can impact human health in myriad ways. For example, it is estimated that land degradation reduced agricultural incomes in Ghana by $4.2 billion between 2006 and 2015, increasing the national poverty rate by 5.4% in 2015.[40] Land use and land cover changes have already been shown to increase the frequency and intensity of dust storms, generating particulate matter in the atmosphere and causing severe cardiorespiratory effects. Dust storms impact energy generation and infrastructure. The IPCC reports,

> There is robust evidence and high agreement that dust storms negatively affect the operational potential of solar and wind power harvesting equipment through dust deposition, reduced reach of solar radiation and increasing blade-surface roughness, and can also reduce effective electricity distribution in high-voltage transmission lines.[41]

Other impactful climate events are heat waves that are intensified by the heat island effect, impacting energy use and socioeconomic consequences. According to the National Oceanic and Atmospheric Administration (NOAA), "the total cost of U.S. billion-dollar disasters over the last five years (2015–2019) exceeds $525 billion, with a five-year annual cost average of $106.3 billion (Consumer Price Index [CPI]–adjusted), both of which are records."[42] The U.S. Global Change Research Program's 2018 report emphasizes that "the evidence of human-caused climate change is overwhelming and continues to strengthen, that the impacts of climate change are intensifying across the country, and that climate-related threats to Americans' physical, social, and economic well-being are rising."[43]

The World Meteorological Organization reports on the extreme weather events of the past 50 years:

> From 1970 to 2012, 8,835 weather-, climate- and water-related disasters were reported globally. Together they caused the loss of 1.94 million lives and economic damages of US$ 2.4 trillion. The 10 worst reported disasters in terms of human lives lost represented only 0.1% of the total number of events, but accounted for 69% (1.34 million) of the total deaths. The 10 most costly disasters accounted for 19% (US$ 443.6 billion) of overall economic losses. Storms, droughts, floods and extreme temperatures all figure on both lists of the worst disasters.[44]

The urban heat island (UHI) effect leads to higher energy use in cities. UHI is typically caused by urban structures such as tall buildings and hardscape that trap heat during the daytime and emit heat during the night. Research has shown that Central Park in NYC has indicated UHI phenomona from 2.0°C to 2.5°C over the century despite being a vegetated park because of reduced wind circulation due to an increase in tall buildings surrounding the park.[45] A UHI in Nanjing, China creates a 12.0–24.0% higher cooling demand in residential buildings. The urban areas in Antwerp, Belgium show an increase in the cooling demand of residential buildings by 60.8–90% compared to the rural areas.[46] In Rome, Italy, urban heat islands increase cooling energy demands by 18% to 24%.[47] The UHI effect increases the demand on air conditioning systems which can mitigate some of the health and safety impacts from heat waves.

1.2 Carbon-Neutral Built Environment

The world's population has more than doubled since 1970. The global consumption of non-renewable energy resources, such as coal, oil, and natural gas, has increased accordingly. Current urban settlements take up only a small fraction of Earth's surface, but they contribute to a series of major environmental problems globally. The amount of impervious surface in urban lands causes not only UHI but also modifications to the hydrological, air, and terrestrial systems. Impervious surfaces have been a source of urban flooding, excessive runoff and groundwater reduction, and contaminating pollutants in downstream water systems. Urban settlements also increase anthropogenic emissions from smokestacks in the operation of buildings and the support of manufacturing activities, and from the tailpipes of vehicles.

According to IPCC's Greenhouse Gas Emissions by Economic Sectors, the main contributor of GHG emissions comes from electricity and heat production (25%), forestry and other land use (24%), industry (21%), transport (14%), building (6.4%), and others (9.8%).[48] This anthropogenic climate change has pushed us into a new geological epoch—the Anthropocene. The global environmental threat from humanity necessitates redefining the widely accepted notion of the triple pillars of sustainability—social, economic, and environmental. We cannot provide social and economic sustainability if we keep increasing environmental pressure on our planet. In other words, sustainable development supports social and economic sustainability in the nest of Earth's life support system, not as three pillars.[49] This paradigm shift to safeguard the planet calls for a new definition of sustainable development by reframing the Brundtland Commission's definition—that "Sustainable development is the kind of development that meets the needs of the present without compromising the ability of future generations to meet their own needs (the Brundtland Commission, 1987)"—as "development that meets the needs of the present while safeguarding Earth's life-support system, on which the welfare of current and future generations depends."[50]

One of the primary sustainable development goals is to change human activities in ways that keep the global temperature increase at or below 1.5°C relative to the pre-industrial era. It has been proven that anthropogenic emissions are causing rising Earth temperatures, increasing global sea levels, frequency of heat waves, and food insecurity. There are also increases in the frequency and intensity of heavy precipitation, droughts, and wildfires in different regions. The recent Australian and Californian wildfires exemplify the evidence of global warming. Therefore, worldwide CO_2 emission cuts are necessary to keep the global temperature rise to below 1.5°C relative to the pre-industrial level as outlined in the IPCC Special Report on Global Warming of 1.5°C in 2018. According to the IPCC report, zero carbon emission should be achieved by 2050 to meet the global temperature target through system and behavioral changes in energy, building and infrastructure, transportation, and industry sectors.

The 1.5°C special report released in 2018 sparked environmental initiatives from many cities. Government policy and building energy codes are playing a key role in mitigating environmental impacts from the building sector. At least 88 countries including the top ten CO_2 emitters—China, the United States, India, Russia, Japan, Germany, South Korea, Canada, Iran, and the United Kingdom—have implemented mandatory building codes.[51] These ten countries contribute two-thirds of the global CO_2 emissions. The 1970 oil crisis prompted the government-led energy conservation efforts of adopting building energy codes and implementing energy-efficient building technologies and renewable energy systems. Mandatory building energy codes have been effective in reducing carbon intensity and energy use while maintaining suitable indoor environments for the occupants' well-being. In addition, in recent decades, government policy and incentives have been enacted to foster affordable implementation of renewable energy production. Despite energy-efficient improvements and renewable energy systems, resource depletion and carbon emission from the building sector have increased by 5% annually. In other words, an improved lifestyle and growing construction outpace building energy efficiency.

Buildings are responsible for more than 40% of the global energy consumption and carbon emissions. Therefore, buildings offer the greatest opportunity for reducing anthropogenic emissions. To that end, net zero energy and/or net zero carbon building has attracted stakeholders such as policymakers, developers, A/E/C (architecture/engineers/construction) professionals, building users, and researchers. Net zero energy means building energy consumption is offset by renewable energy sources such as photovoltaic panels connected to the utility grid called net metering. Net zero carbon means that emissions produced by powering the building for its operation are offset by carbon storage or sequestration technologies. Despite significant initiatives from multiple stakeholders, net zero energy and/or net zero carbon practices have been slow in garnering a substantial number of users due to their complexity and challenges in implementation. The definition of net zero carbon building varies and is often used as a synonym for high-performance building, green/sustainable building, or low-carbon low-energy building. In the United States, it is categorized into four concepts depending on the system boundary and metrics that measure the zero energy performance: net zero site energy, net zero source energy, net zero energy costs, and net zero energy emissions.[52] Typically, net zero carbon buildings are more challenging to achieve than net zero energy buildings as different energy systems to operate the building are associated with different levels of carbon emissions.

The energy efficiency of new construction and existing buildings is fundamental and important in reducing energy consumption and environmental impacts. Energy use intensity (EUI) is a building energy efficiency metric that is determined by dividing annual total energy use by total floor area. It allows different buildings with various sizes to compare their energy efficiency. A lower EUI indicates higher building efficiency or lower energy use to operate the building. Global constructed floor area in 2018 was 240 billion m^2 and has increased by 60% since 2010 while the EUI, energy use normalized by unit floor area in ft^2 or m^2, has declined by 25% only in the same period. This means that building energy efficiency is not keeping pace with construction area growth and the subsequent increasing global energy use and carbon emissions from the building sector. This is in part due to the lack of financial incentives and high efficiency building code reinforcement or poor efficiency requirements for both existing and new construction, if available, in many countries. For example, two-thirds of countries lack building energy codes, and, as a result, 5 billion m^2 of building area (approximately 55 billion ft^2) was constructed without energy standards in 2019.[53]

In addition to energy code compliance, voluntary commitment to net zero energy practice is important to outpace the energy carbon intensity from growing construction. Therefore, it is recommended that all countries adopt building energy codes to mandate high-efficiency building construction and adopt net zero energy architecture practices for both new construction and existing buildings. Besides net zero energy buildings, it is crucial that power plants limit pollutant emissions and increase renewable energy production from solar, wind, and hydro power. In addition to renewable energy sources, smart grid systems reduce wasted energy through a modernized grid technology with more energy efficiency and reliability in transmission, distribution, and transformer use. The smart grid technology also allows easy coordination with renewable energy production. Additional technology for energy conservation is the smart meter, which monitors energy consumption by buildings and transmits usage information to the power plant for improved operation and communication to better manage power delivery and provide restoration capability after an outage.

As of 2019, 25 states and territories within the United States had joined the U.S. Climate Alliance. This group started in 2017 as a partnership between the states of California, New York, and Washington and expanded to the current participants. The alliance overall aims to reduce GHG emissions and increase renewable energy production with the goal to "reduce their collective greenhouse gas (GHG) emissions by at least 26 to 28 percent below 2005 levels by 2025, with many adopting substantially more ambitious emission reduction targets."[54] The participating states and territories have adopted policy strategies to adhere to GHG reduction targets through building energy efficiency, clean transportation systems, renewable power generation, and carbon sequestration. They are implementing legislation committed to net zero or nearly net zero GHG emissions and a 100% renewable energy target by no later than 2050. Three states, Virginia, Washington, and Wisconsin, and the territory of Puerto Rico signed legislation committing to 100% carbon-free electricity by no later than 2050.[55] In addition to environmental impact, the relation between the green economy and green jobs is clear, and the promotion of clean technology is expected to increase the number of jobs in renewable energy and the energy efficiency market. Climate change policies and capital allocations in the green industry affect the implementation of net zero carbon buildings.

According to the U.S. Climate Alliance, California's Senate Bill 32 (2016) mandates GHG reductions of 40% below 1990 levels by 2030, and a 2018 executive order targets carbon neutrality and 100% zero carbon electricity sales by 2045. The state also aims to achieve net zero energy residential construction by 2020 and commercial buildings by 2030. California is the leading state in curtailing GHG emissions, supplying a 100% renewable energy supply by 2045, and creating over 450,000 green jobs. New York's Climate Leadership and Communities Protection Act (CLCPA) mandates carbon neutrality by 2050 and calls for a building energy efficiency improvement of 23% above 2012 by 2030. The new Clean Energy Standard aims to supply 100% clean renewable electricity by 2040. Effective November 15, 2019, Local Laws 92 and 94 in New York City require that all new buildings, building expansions, and structural roofs cover all available roof space with green roofs, solar, or both. Local Law 95 requires buildings over 25,000 ft^2 to post their energy grades. Local Law 96 (Property-Assessed Clean Energy) established financing guidelines to encourage projects for energy efficiency and renewable energy. Local Law 97 sets GHG limits for all buildings larger than 25,000 ft^2 and targets city GHG reductions of 40% by 2025 and 50% by 2030. It also requires NYC Housing Authority properties to meet GHG reductions of 40% by 2030 and sets emissions reporting requirements.[56]

The American Council for an Energy Efficient Economy ranked Boston as the number 1 energy-efficient city in the country from 2013 to 2019.[57] In 2011, Boston set carbon reduction goals below 2005 levels of 25% by 2020 and 80% by 2050. By 2017, the city had reduced emissions from municipal buildings and fleets by more than 40% below 2005 levels, with an updated goal to reduce emissions by 60% by 2030.[58] Boston has pledged to be carbon neutral by 2050, which involves "transitioning to zero-net carbon new construction."[59] Article 37 of the Boston Building Code targets green buildings and climate resiliency. All projects must "achieve at a minimum the 'certifiable' level utilizing the most appropriate U.S. Green Building Council Leadership (USGBC) in Environmental and Energy Design (LEED) Rating System(s)."[60] The article provides specific Boston Green Building Credits that can be substituted for up to four points toward LEED certification. In April 2013, Boston enacted the Building Energy Reporting and Disclosure Ordinance (BERDO), which requires that the energy and water use of all commercial and residential buildings that are 35,000 ft^2 or have 35 units or more be reported to the city each year. In addition to this reporting requirement, it must be shown that action has been taken to reduce the buildings' energy use or emissions by 15% every five years or that "a detailed assessment of options to reduce their energy use"[61] has been conducted.

In 2019, the District of Columbia committed to 100% renewable electricity by the year 2032.[62] The plan includes doubling the amount of solar energy, improving the energy efficiency of existing buildings, and requiring all public transportation and privately owned fleet vehicles be emission-free by the year 2045, among other provisions.[63] Washington, DC had 31,810,018 ft^2 of LEED-certified buildings (52.86 ft^2 per capita), 143 completed projects, and 2,597 LEED-certified professionals.[64] Although the USGBC does not include DC in its rankings because it is not a state, DC outranks Colorado (2019s #1 LEED state) in every metric except number of LEED professionals. (CO had 23,962,344 ft^2 of LEED building in 2019, 4.76 ft^2 per capita, 102 finished projects, and 6,339 LEED professionals.[65])

Hawaii's Act 15 (2018) targets carbon neutrality by 2045, and the 100% renewable portfolio standard requires 100% clean renewable electricity by 2045. As the second largest solar-installed state, North Carolina's Executive Order (EO 80) targets a GHG reduction of 40% below 2005 and a building energy efficiency of 21% below 2009 by 2025. "Washington passed a landmark clean energy package in 2019, ensuring clean electricity by 2030, phasing down hydrofluorocarbons (HFCs), implementing a first-of-its-kind energy standard for commercial buildings and incentivizing electric vehicles (EVs)."[66] Washington also "leads the nation" in carbon-free energy, with over 75% of its electricity coming from hydro and other renewable sources.[67] Washington is one of only three U.S. states (along with New Jersey, California, and the District of Columbia) to require energy use disclosure upon the sale or lease of commercial buildings.[68] With Senate Bill 5116, the state of Washington committed to 100% carbon-neutral electricity by 2030 and 100% carbon-free electricity by 2045 (Table 1.2).[69]

U.S. Climate Alliance's GHG emissions reduction targets, renewable/clean energy targets, and green jobs in the "green" sector[70]

State	GHG emissions reduction target	Renewable energy target	Number of green jobs	Number of green jobs per 1,000
California	1990 levels by 2020 40% below 1990 levels by 2030 Carbon neutral by 2045	100% for retail sales by 2045	461,998	11.69
Colorado	26% below 2005 levels by 2025 50% by 2030; 90% by 2050	30% by 2020 (10% or 20% for smaller utilities)	50,414	8.75
Connecticut	10% below 1990 levels by 2020 45% below 2001 levels by 2030 80% below 2001 levels by 2050	40% by 2030	38,499	10.79
Delaware	26–28% below 2005 levels by 2025	25% by 2025	13,243	13.59
Hawaii	Carbon neutral by 2045	100% by 2045	10,190	7.19
Illinois	26–28% below 2005 levels by 2025	25% by 2025	123,247	9.72
Maine	45% below 1990 levels by 2030 At least 80% by 2050	80% by 2030 (target) 100% by 2050 (goal)	10,845	8.06
Maryland	40% below 2006 levels by 2030 80–95% by 2050	50% by 2030	77,981	12.89
Massachusetts	25% below 1990 levels by 2020 At least 80% by 2050	35% renewable energy by 2030 with a 1% annual increase each year thereafter; 40% clean energy by 2030 (includes 35% renewable requirement) with a 2% increase each year thereafter, totaling 80% by 2050	110,767	16.06
Michigan	20% below 2005 levels by 2020 80% by 2050	15% renewables by 2021 (target) 35% combined renewables and energy efficiency by 2025 (goal)	126,081	12.62
Minnesota	30% below 2005 levels by 2025 80% by 2050	25% by 2025 (Xcel Energy 31.5% by 2020; other investor-owned utilities 26.5% by 2025)	54,279	9.62

(Continued)

State	GHG emissions reduction target	Renewable energy target	Number of green jobs	Number of green jobs per 1,000
Montana	Net zero emission for average annual electric loads by 2035	15% by 2015	9,991	9.34
Nevada	28% below 2005 levels by 2025 45% by 2030; net zero by 2050	50% renewable by 2030 (target) 100% carbon free by 2050 (goal)	21,045	6.83
Nevada	28% below 2005 levels by 2025 45% by 2030; net zero by 2050	50% renewable by 2030 (target) 100% carbon free by 2050 (goal)	21,045	6.83
New Jersey	80% below 2006 levels by 2050 (the goal of reaching 1990 emission levels by 2020 has already been achieved)	50% by 2030 100% clean energy by 2050 (goal)	46,410	5.22
New Mexico	45% below 2005 levels by 2030	40% clean energy by 2045 and 50% by 2030 for investor-owned utilities and rural electric cooperatives; 80% clean energy by 2040 for investor-owned utilities and by 2050 for rural electric cooperatives	10,161	4.84
New York	Net zero by 2050	70% renewable by 2030; zero carbon by 2040	143,873	7.39
North Carolina	40% below 2005 levels by 2025	12.5% renewable and/or energy efficient by 2021	96,957	9.24
Oregon	10% below 1990 levels by 2020 75% by 2050	50% by 2040	51,205	12.13
Pennsylvania	26–27% below 2005 levels by 2025 80% by 2050	8% from Tier I sources and 10% from Tier II sources by 2021	76,780	5.99
Puerto Rico	50% by 2025	40% by 2025; 60% by 2040 100% by 2050	N/A	N/A
Rhode Island	10% below 1990 levels by 2020 45% by 2035 80% by 2050	38.5% by 2035	16,021	15.12
Vermont	40% below 1990 levels by 2030 80–95% by 2050	55% Tier I renewables by 2017 75% by 2032	13,497	21.63

(Continued)

State	GHG emissions reduction target	Renewable energy target	Number of green jobs	Number of green jobs per 1,000
Virginia	N/A	30% by 2030 100% carbon free by 2050	85,126	9.97
Washington	Return to 1990 levels by 2020 25% below by 2035 50% below by 2050	100% carbon neutral by 2030 100% clean energy by 2045	74,719	9.81
Wisconsin	N/A	10% renewable by 2015 100% zero carbon electricity by 2050 (executive goal)	68,710	11.80

Source: United States Climate Alliance. 2019 State Fact Sheets: Climate Leadership across the Alliance. PDF file. https://static1.squarespace.com/static/5a4cfbfe18b27d4da21c9361 /t/5db99b0347f95045e051d262/1572444936157/USCA_2019+State+Factsheets_ 20191011_compressed.pdf

1.3 Building Lifecycle

Lifecycle analysis (LCA) is an effective research technique for quantifying and comparing environmental impacts of various building materials, systems integration and operation, thus facilitating the decision-making process of various scenarios. From the building lifecycle perspective, both the construction and operation of a building significantly contribute to resource depletion and environmental impacts. A challenge for the building designers, construction industries, and building owners is to provide healthy built environments with resource conservation, pollutant reduction, and global warming mitigation. In addition, reducing building energy consumption during pre-use (manufacturing, construction), use (operation), and post-use of a building is a design imperative, and innovative design strategies integrated with sustainable building technology are critical for tackling future climate change and non-renewable energy depletion while promoting the satisfaction and well-being of occupants.

LCA evaluates the environmental implications of building products or processes depending on different stages of the lifecycle. It allows for the estimation of material flows, input resources, and output pollutants. The first LCA studies were carried out for product use phases in the 1960s. In 1970, the Midwest Research Institute analyzed resources, emissions, and wastes for Coca Cola beverage containers.[71] LCA gained popularity in the mid-1980s as it holistically evaluates different environmental impact categories. Significant research on building-focused LCA has been carried out during the past two decades. According to the ISO 14044 standard, LCA requires four steps: (1) goal and scope definition, (2) lifecycle inventory (LCI) analysis, (3) lifecycle impact assessment (LCIA), and (4) lifecycle interpretation.[72] For the scope definition, it is important to determine the system boundary and functional unit. The functional unit describes a building systems' functionality, durability, or ease of maintenance for the studied alternatives. For example, a functional unit for light bulb options would be lighting a room with 500 lux for three years of operation, and the environmental impacts would vary depending on the choice of light bulb type to supply the illumination for the target period. Typical examples of system boundaries for a building system include cradle to gate, cradle to grave, and cradle to cradle. Using the light bulb example, the cradle to gate deals with the environmental impacts associated with the material extraction and product manufacturing while the cradle to grave includes transportation, installation, use, maintenance, and landfill burial. The cradle to cradle includes the environmental impact associated with the recycling process instead of landfill burial.

Using a microalgae building, the following section exemplifies LCA procedures to quantify environmental impacts associated with a microalgae building. In accordance with the ISO 14040 series of standards, an LCA procedure starts with the goal and scope definition and continues to inventory analysis, impact assessment, and finally, interpretation of the results. The first phase of the LCA framework under ISO 14040 starts with defining the scope of the work using the functional unit, system boundaries, and assumptions and limitations of the study. Using a microalgae window, the goal of the LCA is to evaluate the environmental benefits a microalgae building yields compared with a counterpart building with conventional windows. The functional unit of the microalgae window is to cover a 5 feet wide by 10 feet tall area for an operational period of 30 years. Depending on the project goal, the microalgae window can be examined for environmental impacts at a cradle to gate, cradle to grave, or cradle to cradle system boundary. The study can also focus on the use phase only, excluding the pre-use and post-use phase of the microalgae window.

The second phase of the LCA framework under ISO 14040 is the LCI analysis, which quantifies material use, energy input, and pollutant emissions during a specified lifecycle phase. The inventory data typically rely on published databases and manufacturers directly engaged in product fabrication and processing activities. The LCA for the microalgae window can use published databases. There are around a dozen LCA databases available for construction materials, and when selecting an LCA database, transparency and suitability of the database are important.[73] Material use and primary energy consumption are calculated in the forms of kg/functional unit and MJ/functional unit, respectively, and pollutant emissions are expressed in terms of kg/functional unit.

The third phase of the LCA framework is the LCIA. ISO 14040 defines the impact assessment phase as consisting of classification, characterization, and weighting. Classification assigns inventory results to impact categories while the characterization process involves defining characterization factors to convert each pollutant emission into equivalent potentials represented by a reference substance (e.g., CO_2 equivalent). The weighting process combines the impact categories into a single score (e.g., ecopoint, SimaPro eco indicator method). For the microalgae window case, the impact assessment can focus on the global warming potential specifically caused by four major pollutants such as CO_2, CH_4, CF_4, and C_2F_6.

Based on these four major pollutants, the last phase of an LCA interprets the degree of environmental impact. The inventory analysis and the impact assessment are combined in order to reach a conclusion and make recommendations on the studied system's performance contrasts with a counterpart in the areas of impact categories discussed later in this section. Additional sensitivity analysis can be integrated to understand how different functional units, system boundaries, or input variables change the LCA results or how critical uncertain parameters impact environmental performance. As construction expands in countries such as China and India, absolute emission and energy use figures will rise. Building construction accounts for significant energy use and CO_2 emission due to rising production demand on construction materials. For example, according to data presented in Architecture 2030, the embodied carbon from building structure and enclosures accounts for 11% of global GHG emissions and 28% of global building sector emissions. In addition, the cement industry alone accounts for 7% of the total worldwide CO_2 emissions from fossil fuel burning, equivalent to approximately 2.2 billion tons of CO_2 emission annually.[74] The global cement production in 2017 was reported to be at 4 billion tons and is expected to increase. The global iron and steel production supplied approximately 1.7 billion tons of steel in 2017 and generated an annual 2 billion tons of CO_2 emission. Manufacturing industries have been adopting low-carbon technologies in addition to using green energy sources. They are also maximizing the recycling of scraps and employing alternatives to energy-intensive materials (e.g., clinker content in cement production). Government policy regulating CO_2 emissions could improve the carbon and energy intensity of major construction materials. Eliminating these emissions is key to addressing climate change and meeting Paris Climate Agreement targets.

In addition to embodied carbon and energy from the manufacture of building materials and construction of buildings, building operations account for one-third of the global final energy, making it essential to tackle these environmental challenges. Building energy efficiency serves as a metric to discuss energy performance and environmental impacts from a building. For example, EUI indicates the degree of building energy efficiency. A lower EUI indicates higher building efficiency or lower environmental impact. A target carbon footprint per capita or unit floor area serves as another performance metric. According to the 2000-watt society, Americans currently consume an average of 12,000 watts per person per day and contribute 20 tons of CO_2 emissions annually. They advocate that to meet sustainable development or 1.5°C imperatives, individuals should consume 500 watts from non-renewable energy sources equivalent to 1 ton of CO_2 emissions and use 1,500 watts from renewable energy sources, totaling 2,000 watts and 1 ton of CO_2 emissions.[75] Another performance metric used in buildings is carbon footprint. In 2018, NYC passed the Climate Mobilization Act, which set new targets for GHG emission reductions by 2050. The act set emission standards for different building categories, for example, spaces in occupancy group A must not exceed CO_2 emissions of 0.01074 tCO_2e/sf, and buildings in occupancy group B must not exceed CO_2 emissions of 0.00846 tCO_2e/sf.[76] Similar limits are set for electricity and the combustion of natural gas and fuel oil.

For a product lifecycle, five stages are typically considered: extraction of raw materials, design and production, packaging and distribution, use and maintenance, and end of the product lifecycle. The cradle to gate system boundary refers to the lifecycle from the extraction of raw materials through the use and maintenance of the system, whereas the cradle to grave system boundary covers the entire lifecycle of the extraction through the end of the lifecycle. After the functional unit and system boundary are defined, the LCI is then followed with the collection of input and output data of the study subject. During the LCI, natural resources and energy input as well as output emissions are quantified depending on system boundaries. All construction activities and building operations require raw materials and energy and generate pollutants into the air, water, and soil.

After input and output data are collected, the LCIA focuses on characterization and weighting of environmental impacts. The U.S. Environmental Protection Agency's (EPA) Tool for the Reduction and Assessment of Chemical and Other Environmental Impacts (TRACI) is a spreadsheet impact assessment tool that can be used to quantify the potential impacts of processes, products, facilities, companies, and communities. It can be used to assess sustainability, lifecycle impact, process design, and pollution prevention. According to TRACI, impact categories include ozone depletion, global warming, acidification, eutrophication, smog formation, human health (cancer and noncancer), ecotoxicity, and fossil fuel use. Environmental impacts are then weighted and categorized into human health measured in DALY (Disability Adjusted Life Years), ecosystem quality or ecotoxicity measured in PDF m^2/year (Potentially Disappeared Fraction), and resources measured in MJ. Table 1.3 describes the impact categories and scope of impacts included in the tool.

▼ Table 1.3

Cause–effect chain selection for TRACI, used to calculate environmental impacts[77]

Impact category	Midpoint level selected	Level of site specificity selected	Possible endpoints
Ozone depletion	Potential to destroy ozone based on chemical's reactivity and lifetime	Global	Skin cancer, cataracts, material damage, immune system suppression, crop damage, other plant and animal effects
Global warming	Potential global warming based on chemical's radiative forcing and lifetime	Global	Malaria, coastal area damage, agricultural effects, forest damage, plant and animal effects
Acidification	Potential to cause wet or dry acid deposition	United States, east or west of the Mississippi River, U.S. census regions, states	Plant, animal, and ecosystem effects, damage to buildings
Eutrophication	Potential to cause eutrophication	United States, east or west of the Mississippi River, U.S. census regions, states	Plant, animal, and ecosystem effects, odors and recreational effects, human health impacts
Photochemical smog	Potential to cause photochemical smog	United States, east or west of the Mississippi River, U.S. census regions, states	Human mortality, asthma effects, plant effects
Ecotoxicity	Potential of a chemical released into an evaluative environment to cause ecological harm	United States	Plant, animal, and ecosystem effects
Human health: criteria air pollutants	Exposure to elevated particulate matter less than 2.5µ	United States, east or west of the Mississippi River, U.S. census regions, states	Disability-adjusted life-years (DALYs), toxicological human health effects

(Continued)

Impact category	Midpoint level selected	Level of site specificity selected	Possible endpoints
Human health: cancer	Potential of a chemical released into an evaluative environment to cause human cancer effects	United States	Variety of specific human cancer effects
Human health: non-cancer	Potential of a chemical released into an evaluative environment to cause human non-cancer effects	United States	Variety of specific human toxicological noncancer effects
Fossil fuel	Potential to lead to the reduction of the availability of low-cost energy/fossil fuel supplies	Global	Fossil fuel shortages leading to use of other energy sources, which may lead to other environmental or economic effects
Land use	Proxy indicator expressing potential damage to threatened and endangered species	United States, east or west of the Mississippi River, U.S. census regions, states, counties	Effects on threatened and endangered species (as defined by proxy indicator)

Summary

Anthropogenic activities cause climate changes and extreme weather events that affect human health, ecosystems, and built environments. The world population growth has resulted in increased construction and operation of more buildings with greater floor area per capita. As the urban population and affluence continue to grow, the environmental impact from the building sector is of increasing concern. Due to high embodied energy and energy consumption from buildings, there needs to be a growing effort to promote net zero energy carbon-neutral practices. Lifecycle assessment perspective is important when making decisions for material selection, design development, construction, and end of building life. In the lifecycle of a building, non-renewable resources as well as natural resources and energy are consumed, and environmental pollutants are generated as a result of design, engineering, and construction activities followed by building operations and demolition. Reducing material input and pollutant output at every lifecycle stage of the building is indispensable to carbon neutrality. The U.S. Climate Alliance aims toward achieving carbon neutrality and economic enhancement in the clean energy sector. An increasing number of energy efficiency and clean energy legislation, polices, initiatives, and leadership have been deployed at the federal, state, and local levels.

Notes

1   Commoner, Barry, *The Closing Circle: Nature, Man, and Technology* (Courier Dover Publications, 2020), 12.
2   United Nation, "Population Division," accessed August 22, 2020, https://population.un.org/wup/DataQuery/
3   United Nation, "World Urbanization Prospects 2018," accessed August 23, 2020, https://population.un.org/wup/Publications/Files/WUP2018-Highlights.pdf.
4   Karen C. Seto, Burak Güneralp, and Lucy R. Hutyra, "Global Forecasts of Urban Expansion to 2030 and Direct Impacts on Biodiversity and Carbon Pools," Proceedings of the National Academy of Science of the United States of America 109, no. 40 (2012): 16083–16088, https://doi.org/10.1073/pnas.1211658109
5   Statista, "Global Building Floor Area Growth Forecast by Region 2015–2050," accessed August 23, 2020, https://www.statista.com/statistics/731858/projected-global-building-floor-area-growth-by-region/
6   Internal Energy Agency (IEA), "Data and Statistics," accessed August 23, 2020, https://www.iea.org
7   Ibid.
8   Ibid.
9   "Highlights of Annual 2019 Characteristics of New Housing," United States Census Bureau, revised June 1, 2020, http://www.census.gov/construction/chars/highlights.html?
10 "QuickFacts: United States," United States Census Bureau, accessed August 25, 2020, http://www.census.gov/quickfacts/fact/table/US#.
11 "10-7 Dwelling Construction Started and Floor Area by Type of Construction (2010 to 2018)," Statistics Bureau of Japan, accessed August 25, 2020, https://www.stat.go.jp/english/data/nenkan/69nenkan/1431-10.html;
12 Matt Jennings and Rhys Lewis, "House Price Per Square Metre and House Price Per Room, England and Wales: 2004 to 2016," Office for National Statistics, updated October 11, 2017, https://www.ons.gov.uk/economy/inflationandpriceindices/articles/housepricepersquaremetreandhousepriceperroomenglandandwales/2004to2016.
13 Kees Dol and Marietta Haffner, *Housing Statistics in the European Union 2010.* Delft University of Technology, 2010, 51, https://www.researchgate.net/publication/334030779_housing-statistics-in-the-european-union-2010.
14 Endrit Hoxha and Thomas Jusselme, "On the Necessity of Improving the Environmental Impacts of Furniture and Appliances in Net-zero Energy Buildings," *Journal of Science of the Total Environment* 596–597(2017), https://doi.org/10.1016/j.scitotenv.2017.03.107
15 André Schaffrin and Nadine Reibling, "Household Energy and Climate Mitigation Policies: Investigating Energy Practices in the Housing Sector," *Journal of Energy Policy* 77(2015): 1–10, https://doi.org/10.1016/j.enpol.2014.12.002
16 The World Bank Data, "CO2 Emissions (metric tons per capita)," accessed August 27, 2020, https://data.worldbank.org/indicator/EN.ATM.CO2E.PC
17 Pierre Friedlingstein, et al., "Global Carbon Budget 2019," 1810.
18 Commoner, "The Closing Circle," 5.
19 Pierre Friedlingstein, et al., "Global Carbon Budget 2019," 1810.
20 Grand View Research, "Solar Cell Market Analysis by Product," accessed August 27, https://www.grandviewresearch.com/industry-analysis/solar-cell-market
21 REN21, *Renewables 2020 Global Status Report* (Paris: REN21 Secretariat, 2020). ISBN 978-3-948393-00-7
22 Véronique Monier and Mathieu Hestin, "Study on Photovoltaic Panels Supplementing the Impact Assessment for a Recast of the WEEE Directive," *Final Report* 6 (2011).

23   Wolfram Palitzsch and Ulrich Loser, "Economic PV Waste Recycling Solutions—Results from R&D and Practice," in *2012 38th IEEE Photovoltaic Specialists Conference* (IEEE, 2012), 000628–000631.

24   "2019 Concludes a Decade of Exceptional Global Heat and High-impact Weather," World Meteorological Organization, Press Release #03122019, published December 3, 2019, https://public.wmo.int/en/media/press-release/2019-concludes-decade-of-exceptional-global-heat-and-high-impact-weather.

25   Graham Robinson, "Global Construction Market to grow $8 Trillion by 2030: Driven by China, US and India."

26   IEA, "Energy Efficiency 2019," accessed August 24, 2020, https://www.iea.org/reports/cement

27   Christer Jansson and Trent Northen, "Calcifying Cyanobacteria—The Potential of Biomineralization for Carbon Capture and Storage," *Current Opinion in Biotechnology* 21, no. 3 (2010): 365–371.

28   R. Ramasubramani, R. Praveen, and K.S. Sathyanarayanan, "Study on the Strength Properties of Marine Algae Concrete," *Rasayan Journal of Chemistry* 4 (2016): 706–715.

29   UN Energy Performance, "2018 TOC Refrigeration, A/C and Heat Pumps Assessment Report," accessed August 23, 2020, https://ozone.unep.org/sites/default/files/2019-04/RTOC-assessment-report-2018_0.pdf

30   Department of Energy, "Air Conditioning," accessed August 23, 2020, https://www.energy.gov/energysaver/home-cooling-systems/air-conditioning

31   IEA, "Energy Efficiency 2019," accessed August 24, 2020, https://www.iea.org/reports/energy-efficiency-2019

32   Intergovernmental Panel on Climate Change (IPCC), "Climate Change 2007: The Physical Science Basis," *IPCC's Fourth Assessment Report*, 2007.

33   Hannah Ritchie and Max Roser, "CO_2 and Greenhouse Gas Emissions," Published online at OurWorldInData.org, 2020. https://ourworldindata.org/co2-and-other-greenhouse-gas-emissions [Online Resource].

34   Ibid.

35   Ibid.

36   "Executive Summary," in *2019 Emissions Gap Report* (United Nations Environment Programme, 2019), XIV.

37   IPCC, *CLIMATE CHANGE 2013 The Physical Science Basis*, 2013, www.ipcc.ch/site/assets/uploads/2018/02/WG1AR5_all_final.pdf.

38   Ibid.

39   Alisher Mirzabaev, et al., "Desertification," 268.

40   Ibid., 272.

41   Ibid., 275.

42   "National Climate Report—Annual 2019," NOAA National Centers for Environmental Information, published January 14, 2020, https://www.ncdc.noaa.gov/sotc/national/201913.

43   Alexa Jay, et al., "Overview," in *Impacts, Risks, and Adaptation in the United States: Fourth National Climate Assessment, Volume II*, ed. D.R. Reidmiller, C.W. Avery, D.R. Easterling, K.E. Kunkel, K.L.M. Lewis, T.K. Maycock, and B.C. Stewart (Washington, DC: U.S. Global Change Research Program, 2018), 36.

44   World Meteorological Organization, *Atlas of Mortality and Economic Losses from Weather, Climate and Water Extremes (1970–2012)*, WMO-No. 1123 (Geneva, Switzerland: WMO, 2014), 6.

45   Gaffin, S.R., C. Rosenzweig, R. Khanbilvardi, L. Parshall, S. Mahani, H. Glickman, R. Goldberg, R. Blake, R.B. Slosberg, and D. Hillel, "Variations in New York City's Urban Heat Island Strength Over Time and Space," *Theoretical and Applied Climatology* 94, no. 1 (2008): 1–11.

46   Xiaohai Zhou, et al., "Energy-efficient Mitigation Measures," 2.

47   Ibid.

48 Ibid., p. 9. The direct greenhouse gas emission from the building sector from burning fuel on site accounts for 6.4%, whereas the indirect emission from the building sector supplied by the power plants accounts for 12%.

49 D. Griggs, M. Stafford-Smith, O. Gaffney, et al., "Sustainable Development Goals for People and Planet," *Nature* 495 (2013): 305–307, https://doi.org/10.1038/495305a

50 Ibid.

51 IEA, *Global CO$_2$ Emissions in 2019* (IEA, Paris, 2020), https://www.iea.org/articles/global-co2-emissions-in-2019

52 P. Torcellini, S. Pless, M. Deru, and D. Thu Crawley, "Zero Energy Buildings: A Critical Look at the Definition; Preprint," United States, https://www.osti.gov/servlets/purl/883663.

53 IEA, *Building Envelopes* (IEA, Paris, 2020), https://www.iea.org/reports/building-envelopes

54 *Strength in Numbers: American Leadership in Climate*, United States Climate Alliance 2019 Annual Report (U.S. Climate Alliance, 2019), https://static1.squarespace.com/static/5a4cfbfe18b27d4da21c9361/t/5df78938e7c320168ad2e19a/1576503687285/USCA_2019+Annual+Report_final.pdf.

55 Ibid., 12.

56 "NYC Climate Mobilization Act," *Mayor's Office of Sustainability*, PowerPoint presentation.

57 City of Boston, *Climate Action Plan 2019 Update*, by Martin J. Walsh (Boston, MA: City of Boston, 2019), 8. https://www.boston.gov/sites/default/files/embed/file/2019-10/city_of_boston_2019_climate_action_plan_update_4.pdf.

58 Ibid., 10.

59 Katherine Eshel, "Reducing Emissions," *City of Boston*, updated October 8, 2019, https://www.boston.gov/environment-and-energy/reducing-emissions.

60 "Article 37 Green Building and Climate Resiliency Guidelines," *Boston Planning & Development Agency*, accessed July 31, 2020, http://www.bostonplans.org/planning/planning-initiatives/article-37-green-building-guidelines.

61 City of Boston, *Climate Action Plan 2019 Update*, by Martin J. Walsh (Boston, MA: City of Boston, 2019), 22.

62 "Mayor Bowser Signs Historic Clean Energy Bill, Calling for 100% Renewable Electricity by 2032," DC Department of Energy & Environment, January 18, 2019, https://doee.dc.gov/release/mayor-bowser-signs-historic-clean-energy-bill-calling-100-renewable-electricity-2032.

63 "Mayor Bowser Signs Historic Clean Energy Bill, Calling for 100% Renewable Electricity by 2032," DC Department of Energy & Environment.

64 Sarah Stanley, "USGBC Releases the Top 10 States for LEED."

65 Ibid.

66 "Washington," 2019 State Factsheets, *United States Climate Alliance*, http://www.usclimatealliance.org/state-climate-energy-policies.

67 Ibid.

68 Chris Perry, "Building Energy Efficiency Policies," 78–79.

69 Washington State Legislature, Senate, SB 5116, 66th Legis., 2019 regular sess., passed April 22, 2019, http://lawfilesext.leg.wa.gov/biennium/2019-20/Pdf/Bills/Senate%20Passed%20Legislature/5116-S2.PL.pdf?q=20200722114807.

70 "Inventory of Climate and Clean Energy Policies," United States Climate Alliance, accessed July 22, 2020, http://www.usclimatealliance.org/state-climate-energy-policies.

71 A. Darnay and G. Nuss, "Environmental Impacts of Coca-Cola Beverage Containers," *Midwest Research Institute for Coca-Cola USA* (1971).

72 ISO Technical Committee, "ISO 14044:2006 Environmental Management—Life Cycle Assessment—Requirements and Guidelines."

73   A. Martínez-Rocamora, Jaime Solís-Guzmán, and Madelyn Marrero, "LCA Databases Focused on Construction Materials: A Review," *Renewable and Sustainable Energy Reviews* 58 (2016): 565–573.

74   D.J. Barker et al., "CO_2 Capture in the Cement Industry," *Energy Procedia* 1, no. 1 (2009): 87–94.

75   2000-watt Society. What." Accessed May, 2021. https://www.2000-watt-society.org/what.

76   New York City Council, Commitment to achieve certain reductions in greenhouse gas emission by 2050 (Climate Mobilization Act), Int 1253-2018, introduced November 28, 2018, 9.

77   Jane C. Bare, et al., "TRACI: The Tool for the Reduction and Assessment of Chemical and Other Environmental Impacts," *Journal of Industrial Ecology* 6, no. 3–4 (2003): 55, https://doi.org/10.1162/108819802766269539.

Chapter 2

Why Microalgae?

Microalgae consist of phytoplankton and terrestrial algae. Phytoplankton microorganisms live in aquatic environments while terrestrial algae can grow on surfaces of soil, rock, and so on. These photosynthetic microorganisms use light energy to convert water and carbon dioxide (CO_2) into organic compounds such as lipids, carbohydrates, proteins, pigments, and other valuable products. Algae, one of the first photosynthetic organisms on Earth over 3 billion years ago, have been the lungs for the planet, fixing CO_2 and generating oxygen through their highly efficient photosynthetic process. Humans have harvested seaweed algae for millions of years. Since being promoted as a food resource in the 1950s and discussed as a biofuel candidate during the oil crisis of the 1970s, the industrial-scale production of microalgae has reached a level of thousands of tons per year since the 2000s.[1] It is estimated that hundreds of thousands of microalgae species exist, and over the past few decades, many microalgae-based bioproducts have emerged on the market. Microalgae have been cultivated for various uses in bioenergy, agricultural, aquacultural, pharmaceutical, and food industries due to their fast growth rate and a potential high content of nutritional and bioactive compounds. The lipid and carbohydrate content in microalgae are important criteria for selecting species for biofuel production. The percentage of content of organic compounds varies depending on species, strain type, culture conditions (e.g., pH, temperature), light intensity, CO_2 concentration, and nutrients. The world's annual production of dry algae biomass is estimated at approximately 15,000 tons, primarily consisting of *Spirulina* (10,000 tons), *Chlorella* (4,000 tons), and other species such as *Dunaliella*, Haematococcus, and *Schizochytrium*.[2]

Microalgae live everywhere from fresh, brackish, and salty water to desert, arctic, stone, soils, and on plants. Their sizes range from the few micrometers of a single cell to the giant seaweed kelp at over 100 ft long. Algae can be grouped into macroalgae (i.e., seaweed) and microalgae (e.g., phytoplankton) depending on their physiology. Microalgae in the ocean are an important food source for many aquatic organisms. Macroalgae are multicellular aquatic plants, whereas microalgae are unicellular microscopic organisms. The symbiotic relationship between coral and microalgae is a well-known example of supplying resources for each other. Another categorization can be based on the abundance of types: (1) diatoms (Bacillariophyceae; over 100,000 species), (2) Red algae (Rhodophyta; ~6,000 species), (3) Green algae (Chlorophyceae; ~4,000 species), (4) Brown algae (Phaeophyceae; ~3,000 species), (5) Blue-green algae (Cyanophyceae; ~2,000 species), and (6) Golden algae (Chrysophyceae; ~1,000 species). [3]

Although algal biomass is about one-tenth of all the terrestrial plants, they contribute to approximately half of all the oxygen produced on Earth.[4] Microalgae have a higher photosynthetic efficiency at 3–9%, meaning they convert 3–9% of the sunlight they receive into chemical bond energy, compared to averages of 2.4% for fuel crops with a C3 photosynthetic pathway (e.g., wheat) or 3.7% for fuel crops with a C4 photosynthetic pathway (e.g., corn).[5] These efficiencies mean that algae can grow more quickly than land-based fuel crops, with some species doubling their biomass 1 to 3 times in one day.[6] Owing to their high photosynthetic efficiency and full canopy absorption, microalgae growth rate is reported to be up to 12 times more productive than terrestrial plants.[7] Due to the simpler unicellular structure, microalgae growth rate is much faster than terrestrial plants which have reproductive structures like branches or roots. Microalgae's high photosynthetic efficiency and fast growth rate could effectively offset anthropogenic CO_2 emissions. Microalgae are known to mitigate carbon due to their remarkable CO_2 sequestration by uptaking 1.8 kg of CO_2 per 1 kg biomass. Chemical compounds generated by photosynthesis include lipids, carbohydrates, and proteins that constituent the algal cell mass.

Phytoremediation for fossil fuel-based power plant and wastewater treatment is another benefit that microalgae offer. A quick growth rate favors microalgae as an optimal candidate for biofuel, and they are able to grow on non-arable land that presents less risk of competing with food crops. Microalgae-based biofuel is still more expensive than petroleum, but one way to make biofuel comparable to its counterpart is to integrate with sequestering flue gas and treating wastewater. Microalgae's ability to utilize poor quality water and flue gas for nutrients and CO_2 requirements gives them another advantage over other fuel crops. Microalgae have been incorporated into wastewater treatment where essential nutrients including nitrogen and phosphorous are retained to help the cultivation process become a more economical and sustainable option. Many research studies have demonstrated that microalgae could bioremediate municipal wastewater (contaminated with nitrates and phosphates) and the flue gas of combustible power plants (containing sulfates and CO_2) as potential waste streams.[8] As an economic solution for biofuel production, microalgae can use nutrients from wastewater and flue gas for lipid production before producing biofuel in its lifespan and the residues after lipid extraction can be turned into other bioproducts such as feed and fertilizer.

Microalgae offer various agricultural advantages in their land and water use. Microalgae can be cultivated without competing for arable lands and freshwater throughout the year. They can grow with substantial alkaline, saline non-potable water or rainwater and use nutrients from wastewater or flue gas from power plants. Microalgae can grow more quickly than land-based fuel crops, and their fast growth rate coupled with their ability to grow on non-arable land further reduces land use. For this reason, they are a more flexible fuel crop than their land-based counterparts and present little to no risk of competing with food crops for land space. Researchers estimate that the United States could meet its entire gasoline fuel demand (10 million barrels per day) with algae grown on approximately 30 million acres (the size of North Carolina).[9] This sustainable agriculture is critical because crop land in the world has declined by more than 20% in the past decade, with the loss of 20 million hectares per year due to a combination of soil salination and soil erosion.[10]

As a sustainable biological system, microalgae do not rely on agricultural resources and crop land that could be otherwise allocated for food production. They can grow in salty or brackish water thus alleviating pressure on freshwater. Microalgae likely do not cause soil erosion and water pollution because they can use recycled culture medium or wastewater. Microalgae can also grow over a greater range of sunlight and temperature environments. The decarbonization and oxygenation by superior photosynthesis could be a cost-effective and sustainable means to address global environmental issues. Microalgae have several factors that make them a promising alternative to fossil fuels. Ongoing R&D and investment will make the substantial promise of microalgae commercially viable.

Microalgae biotechnology research has dramatically increased over the past decade due to their multifunctional benefits. Although thousands of microalgae strains exist, only a few have been investigated in depth. *Chlorella*, spherical green microalgae of 2–10 μm in diameter, is a popular candidate for biofuel, wastewater remediation, and flue gas sequestration. They are one of the few microalgae that have been widely studied, have a tolerance for high CO_2 concentrations like flue gas, and have a high lipid content for biofuel production. As a fast-growing and robust strain, they can double their biomass in four days.[11] Optimum pH for growth is 7.5–8.0 with some Chlorella strains growing in a pH range of 9–12, but a pH over 11 inhibits growth.[12] Due to their high assimilation rates, *Chlorella* is one of the best microalgae for wastewater treatment, reducing nitrates and phosphates in municipal wastewater by 90% and 83%, respectively.[13] In addition, they have a high tolerance for high CO_2 concentrations and are a good candidate to remediate flue gas emissions. The lipid content of *Chlorella* varies depending on growing conditions such as temperature, light intensity, and nutrients. Studies reveal a lipid content of 30–60% of their dry weight.[14] The round shape of 2–10 micrometers makes *Chlorella* a challenge in dewatering and drying for biofuel production. The annual global production of *Chlorella* is over 2,500 tons as a food supplement, and it contains over 50% protein, 10% fat, 15% carbohydrates, 5% chlorophyll, and various minerals.[15]

Spirulina is a filamentous cyanobacterium largely cultivated in freshwater. It accounts for the majority of the annual global production with a commercial market size of $90–230 million per year.[16] Spirulina exhibits resistance to high temperatures, light intensity, salinity, and alkalinity and grows in different environments. The growth rate of Spirulina has a high productivity of 12–15 g/sf/day with an average production of 4g/sf/day when tested in lab environments.[17] Optimum pH 9.0 showed the highest growth rate with highest chlorophyll a and carotenoids at pH 8.5, and the highest phenolic content at pH 9.5.[18] The main chemical compounds of dry Spirulina are 60–70% protein and the remaining are lipid, carbohydrates, and other natural antioxidants.[19]

Haematococcus is a unicellular green microalga that lives in freshwater and is the only strains that produce valuable astaxanthin, which is a red-colored carotenoid. They live at 25–28°C and around pH 7.0. It has a relatively slow growth rate and therefore is often cultivated in photobioreactors rather than open ponds to prevent contamination by other microorganisms. A maximum growth of around 4 g/l/day and 1 mg/g astaxanthin were produced using wastewater from a bioethanol plant with the removal of 90% of the total nitrogen and 100% of the total phosphorous from the wastewater.[20] Astaxanthin, an important antioxidant for human nutrition and natural pigment, comprises 3–5% of the biomass. The commercial market for *Haematococcus* is estimated at $20 million/year with an approximate annual production of 300 tons. *Haematococcus* grows in two phases with biomass accumulation in the first stage, and astaxanthin production in the second stage. The introduction of stress conditions such as nutrient deficiency, especially reduced nitrogen and high irradiance, promotes astaxanthin production.

Dunaliella is a marine-dwelling green microalga, and its mass production is used mainly for β-carotene for the food industry due to a high content of over 10%.[21] β-carotene is produced to overcome excess solar radiation. An optimum light intensity is 20–50 μmol/photons/μmol m^{-2} s^{-1} and when UVA is supplemented along with photosynthetic photon flux density (PPFD), total carotene content increases by up to 300%.[22] Lower temperatures increase biomass production. Nitrate and sulfate starvation encourages the production of carotene to withstand starvation stress. Due to low-volume production from microalgae, the majority of commercial β-carotene is produced through chemical synthesis. The residue remaining from extraction can be utilized for biofuel production such as hydrothermal liquefaction by converting wet microalgae into low-molecular-weight liquid fuels. The global market size of *Dunaliella* was approximately $100 million in 2021.[23]

Chlorococcum is a green microalga that lives in freshwater. It is a good biofuel candidate due to their high lipid content. Biomass production depends on light, temperature, pH, nutrients, and salinity. The lipid amounts to 56% for dry weight at optimum pH 8.0–8.5,[24] which is a higher lipid content compared to *Nannochloropsis* (reported at 25.0 wt% to dry biomass) and *Botryococcus braunii* (reported at 28.6 wt% to dry biomass).[25]

Bioluminescent algae are dinoflagellates that produce blue sparkling light at night to deter grazers. Like nearly all marine bioluminescence, dinoflagellates emit light at a peak wavelength near 490 nm.[26] The marine dinoflagellates have generic characteristics of containing unique pigments showing a golden-brown color.[27] Shear stress from fluid affects the intensity and decay rate of bioluminescence where the optimum stress ranges of 0.02 and 0.3 N/m for a typical flash duration of 100 ms.[28] *Pyrocystis lunula* emits spontaneous bioluminescence without generating significant glow, whereas *Pyrocystis noctiluca* emits light of 10^9 photons per cell.[29]

2.1 Aerospace Applications

Aerospace researchers have explored microalgae as a biological life support system (BLSS) that provides a long-term high level of regeneration and restoration for far-distance space exploration. The BLSS is a promising multifunctional and closed-loop system that does not rely on external resources in outer space. The first BLSS was conceived in 1883 and published in 1903 by K.E. Tsiolkovsky, and subsequent physical experiments were carried out in 1915.[30] In the 1950s, American scientists carried out the first theoretical calculations of BLSS using *Chlorella* photosynthesis and animal interaction followed by *Chlorella* photosynthesis and human interaction experiments in 1960–1961.[31]

In 1979, researchers conducted a series of "man-chlorella-microorganisms" experiments and confirmed the feasibility of a closed-loop BLSS system for generating clean air, water, and food. The study, however, did not include the operational details of the BLSS necessary to determine size and reliability in the spaceship. Built on the prior study, additional research on BLSS was carried out and identified major research issues such as long-term culture stability and reliability under long-term space environment, the optimal design of algal growth reactors considering microgravity, and post-growth harvesting and processing. Research found that space cabin environments are favorable for *Chlorella vulgaris* in temperature, pH of the growth medium, and high concentrations of CO_2 while light intensity from the spacecraft was not adequate for efficient photosynthesis.[32] *Chlorella* needs optimum growing environments with a light spectrum of 400–700 nm, light intensity of 150–350 μmol m^{-2} s^{-1}, and a 12/12 day/night cycle.[33] For maximum utilization of the light spectrum, stacking of green algae (e.g., *Chlorella*) and blue-green algae (e.g., *Spirulina*) was proposed as they absorb light in the range of 650–675 nm and 500–650 nm, respectively.

In 2003, additional NASA studies were carried out to understand how simulated microgravity affects cell morphology, physiological activity, and protein production. One NASA study concluded that the microgravity reduces cell growth and protein production while it had little effect on photosynthetic capacity and CO_2 fixation.[34] Another study indicated that algae-based BLSS can function in space for up to 12 months, although additional data are required to fully understand the most suitable algal strains and dynamic culture behaviors, especially the effect of radiation.[35]

The primary function of the BLSS is to provide astronauts with the basic needs of nutrition, water, oxygen, and waste treatment. Astaxanthin is one of the important dietary supplements to support astronaut health in outer space, and *Haematococcus pluvialis* has been tested in the International Space Station (ISS) to understand how gravitation stress could change its growth characteristics.[36] In 2018, four samples of *Haematococcus pluvialis* were tested in the ISS for 25 days to investigate color changes of *Haematococcus pluvialis* and the production of astaxanthin as a health supplement for future astronauts.[37] A recent study included the outer-space feasibility of algal photobioreactors containing *Chlorella vulgaris* used for air revitalization, waste water management, food production, radiation shielding, and thermal control.[38] This study allows the comparison of a photobioreactor environmental control and life support system (ECLSS) to the ECLSS currently in use on the ISS. *Chlorella vulgaris* is another species that has been investigated for future food and oxygen production. Microalgae as a BLSS requires further development to scale up the operation to produce significant levels of food and oxygen and possibly treat wastewater.

2.2 Biofuel Applications

Global energy consumption is projected to increase due to growing populations and GDP. Microalgae have received global attention as a renewable energy resource. There are several benefits to using microalgae as a fuel crop, including agricultural land use advantages, no freshwater requirements, fast growth rate, and the physical and chemical properties of the resultant biofuel. It is projected that biofuel could supply 27% of global transportation fuel by 2050 with 2.1 Gtons of CO_2 reduction annually.[39]

Algae could produce renewable fuels in the form of gaseous compounds (e.g., hydrogen, methane), liquids (e.g., biodiesel, viscous oil, alcohols, hydrocarbons), and carbon solids (e.g., coke).[40] The earlier microalgae biofuel research in the 1940s through 1970 focused on methane production via anaerobic digestion, followed by biodiesel in the early 1980s.[41] Microalgae biodiesel is considered a third-generation biodiesel, following first-generation food crops such as corn and soy, and second-generation nonedible seed crops such as jatropha and karanja, as well as waste cooking oil and animal fats. To make microalgae biofuel commercially viable and cost-competitive with petroleum, it is important to optimize biotechnology (e.g., algae strains, cultivation, lipid variations, genetic modifications) and production processes (e.g., harvesting, dewatering, oil extraction and purification, product).[42] Coproduction potentials integrated with power plant and/or wastewater treatment could favor biofuel production. Finally, investments and support from the government and private companies are vital to growing the microalgae biofuel industry.

High photosynthesis rates and rapid growth yield a greater fuel conversion than terrestrial crops for the same arable land, making microalgae biofuel a sustainable clean energy source. Tremendous research has made substantial improvements in microalgae biofuel technology, but commercialization in real-world application has remained challenging. Market analysis revealed microalgae biofuel prices of roughly $1.1–1.9/L ($4–4.5/gallon) as achievable by 2020 with $0.54/kg for dry microalgae biomass.[43] An important market barrier is related to the high cost associated with cultivation and harvesting of fuel stocks compared to conventional fuel crops. Dewatering microalgae and lipid extraction have been recognized as hindrances to the scale-up of biofuel production. The harvesting and drying process of microalgae is known to be the most energy intensive with a high operation cost. The limited extraction techniques for microalgae-specific biofuel production further challenge energy efficiency and cost-effectiveness. Another economic hurdle could be related to the Energy Independence and Security Act (EISA), which requires the lifecycle greenhouse gas (GHG) emission of biofuels meet minimums of 50% less than that of petroleum-based transportation fuels.[44]

There are several benefits to using microalgae as a fuel crop, including agricultural land use advantages, no freshwater requirements, and the physical and chemical properties of the resultant biodiesel. Biofuel production can be more economically incentivized if it accompanies the coproduction of high-value products. Co-products include pharmaceuticals, nutritional supplements for humans and livestock, petrochemical replacements, animal feeds, fertilizers, industrial enzymes, bioplastics, and surfactants.[45] Coupling the microalgae biofuel production with flue gas sequestration and/or wastewater treatment could be another way to achieve a low-cost process (Table 2.1).

▼ Table 2.1

Biofuel feedstock comparison[46]

Crop	Oil yield (gal/acre/year)
Soybean	48
Camelina	59.8
Sunflower	101.9
Jatropha	201.7
Oil palm	634.0
Algae	1,500 (FY14)
	2,500 (FY18)
	3,700 (FY20)
	5,000 (FY 22)

Of over hundreds of thousands strains, biofuel researchers narrowed down the 300 most promising strains consisting mainly of green algae and diatoms.[47] A promising species needs fast growth and high lipid content to make microalgae-based biofuel economical. *Chlorella vulgaris* and *Nannochloropsis* species were identified as two top lipid producers and additional species such as *Haematococcus pluvialis* and *Chlamydomonas reinhardtii* are being studied for lipid production improvement.[48] *Chlorella* sp., a freshwater green microalga, grows in a high-pH solution that provides resistance to contamination by bacteria. With up to 8% photosynthetic efficiency, *Chlorella's* environmental tolerance and high lipid rates make them a good candidate for biofuel production, achieving a range of 10–48% lipid content in their biomass.[49] *Haematococcus pluvialis,* a freshwater microalgae, can reach a lipid content of up to 40% of cell dry mass.[50] *Nannochloropsis* sp. is another top candidate for algal biodiesel with 60% lipid content in their biomass.[51] One of the most widely cultivated algae in the world are *Spirulina* (*Arthrospira platensis* and *Arthrospira maxima*). *Spirulina*, a blue-green freshwater microalga, is largely used as a dietary supplement in humans, agriculture, and aquaculture due to its high vitamin and nutrient content. Various studies have shown the lipid content of *Spirulina* to range between 5% and 19.5%, which could be used for biofuel production.[52] With a combined bioethanol and methane process, *Spirulina* achieved an 82% starch conversion efficiency in the saccharification process to produce bioethanol, followed by an 83% efficiency during fermentation to produce methane.[53] *Spirulina* is also a more attractive option for membrane filtration than either *Chlorella* or *Nannochloropsis*, which are both circular in shape, because *Spirulina's* spiral filamentous structure makes it easier to mechanically separate them from the growth culture.[54]

The Renewable Fuels Standard (RFS) mandates the blending of 36 billion gallons (approximately 25% of annual transportation fuel consumption) of renewable fuels by 2022, including up to 21 billion gallons from biofuel other than corn-based ethanol.[55] Algal biofuel could be a viable source to meet this standard due to its biological diversity, fast growth, and high lipid and carbohydrate content. In addition to the mandate of increasing the blending portion of biofuel, policies and regulations on reducing carbon emissions could promote the economic viability of microalgae biofuel.

2.3 Bioelectricity Applications

According to the 2016 U.S. Department of Energy's (DOE) biotechnology report, microalgae bioenergy has multiple levels of technical hurdles in three primary areas: (1) feedstock (e.g., algae biology, algal cultivation, harvesting, and dewatering), (2) conversion (e.g., extraction, conversion, and co-products), and (3) infrastructure (e.g., distribution, utilization, siting). To make scale-up applications more economical and appealing to stakeholders, microalgae systems should promote synergistic product operation integrated with wastewater treatment, flu gas sequestration, and biofuel production simultaneously. Taking the same principle of offering multiple functionalities, microalgae-assisted microbial fuel cells (mMFC) is another area that has gained attention in the clean electricity research field.

The use of microorganisms in producing electricity was first conceived during the late 19th and early 20th century.[56] The mMFC uses microalgae to convert solar energy into electricity, and over the past 20 years, considerable research has been carried out on biophotovoltaic generators coupled with wastewater treatment, desalination, and biodiesel production.[57] The mMFC is similar to a hydrogen fuel cell and consists of an anode chamber and cathode chamber separated by an electrolyte membrane. In principle, the anode chamber contains heterotrophic microorganisms, and the cathode chamber is filled with phototrophic microalgae. Microbial metabolism in the anode chamber produces electrons, photons, and CO_2 in the process of wastewater treatment. The microalgae in the cathode chamber produce O_2 as a result of photosynthesis. Electrons travel through an external circuit and generate electricity while protons traveling through the electrolyte membrane combine with O_2 and produce water. Carbon dioxide generated from the anode chamber is used to grow microalgae at the cathode chamber. Oxygen generation from photosynthesis in the cathode chamber facilitates the reaction in the cathode chamber.

Another arrangement of microalgae serve as a substrate for exoelectrogenic bacteria that produce electrons and protons.[58] Microalgae used in the anode with maximum power density include *Chlorella vulgaris* at 2,000 mW/m^2, *Scenedesmus obliquus* at 150 mW/m^2, and *Spirulina maxima* at 20 mW/m^2 while microalgae at the cathode are *Scenedesmus obliquus* at maximum power density of 153 mW/m^2 and *Chlorella vulgaris* at 20 mW/m^2.[59] The mMFC is still in academic research scale and therefore requires additional development to improve energy efficiency for commercial scale. Other key challenges include the selection of the microalgae strain, hazards in the wastewater that could hurt the algae, light penetration for algae, and algal biofilm inhibiting the reaction by blocking light.

2.3 High-Value Bioproducts

Algae coproduction integrated with biofuel production can make biofuel more economic and commercially viable. If biofuel infrastructure is in place, adding facilities to produce co-products helps minimize cost implications.[60] The global market of microalgae-based products was estimated at approximately $34 billion in 2017 and is projected to reach around $57 billion by 2026.[61] Although microalgae have considerable potential in different industries, aquaculture and human consumption are the dominant fields today, and are the only microalgae industries that exist on an industrial scale, which represent over 75% of the market for algal products.[62] *Spirulina* is the most common commercially produced algae, with *Chlorella*, *Dunaliella*, and *Haematococcus* following next in common usage. Major high-value products from microalgae include proteins, lipids, carbohydrates, carotenoids, fatty acids, and polysaccharides, among others, which are used in different sectors such as pharmaceuticals, nutraceuticals, food and beverage, agriculture, and aquaculture.[63]

Microalgae have also been used for nutritional remedies and as enhancements for human health and well-being. Studies have also shown that extracts from microalgae can help treat cancer and tumors by inhibiting angiogenesis, the process of developing new blood vessels from preexisting blood vessels.[64] Their medicinal attributes include "cardiovascular health, antioxidant, anti-cancer, anti-inflammatory, antimicrobial, anti-aging and skin protective and other medicinal properties".[65] *Chlorella*, for example, helps antitumor and stomach ulcer healing while their β-1,3-glucan lowers blood lipid and strengthens the immune system.[66] *Chlorella* also helps control obesity by enhancing blood hemoglobin level and reducing blood sugar levels.[67] *Spirulina* has attributes to improve the immune system and sulfated polysaccharides with antiviral properties.[68] *Dunaliella* is used in the production of antihypertensive, analgesic, and broncholytic drugs, and administering *Spirulina*

Microalgae applications in food, bioproducts, and biofuel

Microalgae biomass	Microalgae applications
Direct use	Human food, animal feeds, food supplements
Bioactive compounds	Poly unsaturated fatty acid, proteins, antioxidants, astaxanthin, β-Carotene, vitamins
Biofuel	Solid (e.g., biochar)
	Liquid (e.g., bioethanol, biodiesel, vegetable oil)
	Gaseous (e.g., biohydrogen and biosyngas)

maxima to runners helped reduce their blood triglyceride levels.[69] *Haematococcus pluvialis* has been shown to decrease the chances of developing Alzheimer's and Parkinson's, as well as enhance cardiovascular health. Some researchers are exploring the potential for microalgae products in fighting obesity, since some strains produce compounds that can limit fat accumulation.[70] Table 2.2 summarizes various applications of microalgae in food, bioproducts, and biofuel industry.

Summary

A considerable amount of research on microalgae has been conducted due to their multifunctional environmental benefits. A fast growth rate makes microalgae an attractive option for biofuel production as a sustainable power alternative to reduce our dependence on petroleum fuel. In addition, algal ability to utilize wastewater and flue gas for nutrients and CO_2 requirements gives them another advantage over other fuel crops. Microalgae-assisted microbial fuel cells (mMFC) have been studied as means of cost-effective renewable electricity production coupled with wastewater treatment. In processes where the lipids are extracted from microalgae cells to produce biodiesel, the excess biomass can be used in other high-value products. The commercialization of co-products in nutraceutical or cosmetic industries may help improve the economics of algal biofuel because the expense of extending into the fuel market is less than it would be if starting from scratch. Microalgae have been used in numerous pharmacological studies researching their bioactive compounds related to medicinal attributes such as anticancer drugs, obesity, tumor treatments, Alzheimer's and Parkinson's. Despite multiple values, commercialization of a microalgae-based energy system is still not economically viable due to the high investment and operational costs in real-world applications on a large scale. Therefore, energy-efficient, cost-effective harvesting and dewatering processes are an important research area for the widespread development of microalgae for broader societal impact. Growth enhancement and genetic engineering can also be used to improve microalgae productivity.

Notes

1   Pauline Spolaore, Claire Joannis-Cassan, Elie Duran, and Arsène Isambert, "Commercial Applications of Microalgae," *Journal of Bioscience and Bioengineering* 101, no. 2 (2006): 87–96.

2   L.M.L. Laurens, "State of Technology Review—Algae Bioenergy," *Golden: IEA Bioenergy* (2017), 35.

3   Russell Leonard Chapman, "Algae: The World's Most Important 'Plants'—An Introduction," *Mitigation and Adaptation Strategies for Global Change* 18, no. 1 (2013): 5–12.

4   Ibid., 5.

5   Carla S. Jones and Stephen P. Mayfield, *Our Energy Future: Introduction to Renewable Energy and Biofuels* (University of California Press, 2016).

6   Ibid.

7   Ibid.

8   Michael Hannon, Javier Gimpel, Miller Tran, Beth Rasala, and Stephen Mayfield, "Biofuels from Algae: Challenges and Potential," *Biofuels* 1, no. 5 (2010): 763–784.

9   Ibid.

10 David Pimentel, "Biofuels versus Food Resources and the Environment," *Review (Fernand Braudel Center)* 33, no. 2/3 (2010): 178, https://www.jstor.org/stable/23346881.

11 J.R. Malapascua, et al., "Photosynthesis and Growth Kinetics of *Chlorella vulgaris* R-117 Cultured in an Internally LED-illuminated Photobioreactor," *Photosynthetica* 57, no. 1 (2019): 103, http://doi.org/10.32615/ps.2019.031.

12 Virthie Bhola, et al., "Effects of Parameters Affecting Biomass Yield and Thermal Behavior of *Chlorella vulgaris*," *Journal of Bioscience and Bioengineering* 111, no. 3 (2011): 380, http://doi.org/10.1016/j.jbiosc.2010.11.006.

13 Jyoti Sharma, et al., "Microalgal Consortia for Municipal Wastewater Treatment—Lipid Augmentation and Fatty Acid Profiling for Biodiesel Production," *Journal of Photochemistry & Photobiology* 202, no. 111638 (2020): 2, https://doi.org/10.1016/j.jphotobiol.2019.111638.

14 Jin Liu, et al., "Heterotrophic Production of Algal Oils," 130.

15 West M. Bishop and Heidi M. Zubeck, "Evaluation of Microalgae for Use as Nutraceuticals and Nutritional Supplements," *Journal of Nutrition and Food Sciences* 2, no. 5 (2012): 4, http://doi.org/10.4172/2155-9600.1000147.

16 L.M.L. Laurens, "State of Technology Review—Algae Bioenergy," *Golden: IEA Bioenergy* (2017), 35.

17 Junfeng Wang, et al., "Field Study on Attached Cultivation of *Arthrospira* (*Spirulina*) with Carbon Dioxide as Carbon Source," *Bioresource Technology* 283 (2019): 270, https://doi.org/10.1016/j.biortech.2019.03.099.

18 Mostafa Mahmoud Sami Ismaiel, et al., "Role of pH on Antioxidants Production by *Spirulina*," 298.

19 Mohammad A. Alshuniaber, et al., "Antimicrobial Activity of Polyphenolic Compounds from *Spirulina*," 459.

20 Fatima Haque, et al., "Integrated *Haematococcus pluvialis* Biomass Production and Nutrient Removal Using Bioethanol Plant Waste Effluent," *Process Safety and Environmental Protection* 111 (2017): 131, http://dx.doi.org/10.1016/j.psep.2017.06.013.

21 Aharon Oren, "A Hundred Years of Dunaliella Research: 1905–2005," *Saline Systems* 1, no. 1 (2005): 1–14.

22 R. Raja, S. Hemaiswarya, and R. Rengasamy, "Exploitation of Dunaliella for β-carotene Production," *Applied Microbiology and Biotechnology* 74, no. 3 (2007): 517–523.

23 Research and Markets. "Dunaliella Salina Market by End User and Geography – Global Forecast to 2028." Accessed November, 2021. https://www.researchandmarkets.com/reports/5438307/dunaliella-salina-market-by-end-user-and.

24 Theresia Umi Harwati, Thomas Willke, and Klaus D. Vorlop, "Characterization of the Lipid Accumulation in a Tropical Freshwater Microalgae *Chlorococcum* sp.," *Bioresource Technology* 121 (2012): 54–60.

25 A. Kirrolia, N.R. Bishnoi, and Rajesh Singh, "Effect of Shaking, Incubation Temperature, Salinity and Media Composition on Growth Traits of Green Microalgae *Chlorococcum* sp.," *Journal of Algal Biomass Utilization* 3, no. 3 (2012): 46–53.

26 Jeremiah D. Hackett et al., "Dinoflagellates: A Remarkable Evolutionary Experiment," *American Journal of Botany* 91, no. 10 (2004): 1524, http://doi.org/10.3732/ajb.91.10.1523.

27 Ibid., 1523.

28 Charlotte L.J. Marcinko, Stuart C. Painter, Adrian P. Martin, and John T. Allen, "A Review of the Measurement and Modelling of Dinoflagellate Bioluminescence," *Progress in Oceanography* 109 (2013): 117–129.

29 Martha Valiadi and Debora Iglesias-Rodriguez, "Understanding Bioluminescence in Dinoflagellates—How Far Have We Come?" *Microorganisms* 1, no. 1 (2013): 3–25.

30 Y.Y. Shepelev, "Biological Systems for Human Life Support: Review of the Research in the USSR" (1979).

31 Ibid., 2.

32 Tobias Niederwieser, Patrick Kociolek, and David Klaus, "Spacecraft Cabin Environment Effects on the Growth and Behavior of Chlorella Vulgaris for Life Support Applications," *Life Sciences in Space Research* 16 (2018): 8–17.

33 Ibid., 15.

34 W. Ronald Mills, "Growth and Metabolism of the Green Alga, Chlorella Pyrenoidosa, in Simulated Microgravity" (2003).

35 Tobias Niederwieser, Patrick Kociolek, and David Klaus, "A Review of Algal Research in Space," *Acta Astronautica* 146 (2018): 359–367.

36 NASA, "About Microalgae," December 19, 2019, https://www.nasa.gov/audience/foreducators/stem_on_station/ncas_microalgae/about_microalgae/index.html.

37 Sandra May, "About Microalgae," NASA, accessed November 8, 2020. https://www.nasa.gov/audience/foreducators/stem_on_station/ncas_microalgae/about_microalgae/index.html

38 Michael T. Flynn, Marc Cohen, Renée L. Matossian, Sherwin Gormly, Rocco Mancinelli, Jack Miller, Jurek Parodi, and Elyssee Grossi, "Water Walls Architecture: Massively Redundant and Highly Reliable Life Support for Long Duration Exploration Missions" (2018).

39 L.M.L. Laurens, "State of Technology Review—Algae Bioenergy," *Golden: IEA Bioenergy* (2017), 20.

40 Amanda Barry, Alexis Wolfe, Christine English, Colleen Ruddick, and Devinn Lambert, "2016 National Algal Biofuels Technology Review" (2016).

41 Ibid., 4.

42 Faizal Bux, ed. *Biotechnological Applications of Microalgae: Biodiesel and Value-added Products* (CRC Press, 2013).

43 Laurens, "State of Technology Review—Algae Bioenergy," 76.

44 A. Barry, A. Wolfe, C. English, C. Ruddick, and D. Lambert, "National Algal Biofuels Technology Review," *US Department of Energy, Office of Energy Efficiency and Renewable Energy, Bioenergy Technologies Office* (2016).

45 Ibid., 13.

46 Darzins et al. 2020. Current Status and Potential of Algal Biofuels Production. IEA Bioenergy Task 39. Report T39-T2. http://www.fao.org/uploads/media/1008_IEA_Bioenergy_-_Current_status_and_potential_for_algal_biofuels_production.pdf, cited in Barry et al. "2016 National Algal Biofuels Technology Review" (2016), 2.

47 Barry et al., "2016 National Algal Biofuel," 4.

48 Ibid., 4.

49 Luisa Gouveia, "Microalgae as a Feedstock for Biofuels," in *Microalgae as a Feedstock for Biofuels* (Springer, Berlin, Heidelberg, 2011), 1–69.

50 Mirash Zhekisheva, Sammy Boussiba, Inna Khozin-Goldberg, Aliza Zarka, and Zvi Cohen, "Accumulation of Oleic Acid in Haematococcus pluvialis (chlorophyceae) under

Nitrogen Starvation or High Light Is Correlated with That of Astaxanthin Esters1," *Journal of Phycology* 38, no. 2 (2002): 325–331.

51 Liliana Rodolfi, Graziella Chini Zittelli, Niccolò Bassi, Giulia Padovani, Natascia Biondi, Gimena Bonini, and Mario R. Tredici, "Microalgae for Oil: Strain Selection, Induction of Lipid Synthesis and Outdoor Mass Cultivation in a Low-cost Photobioreactor," *Biotechnology and Bioengineering* 102, no. 1 (2009): 100–112.

52 Soha S.M. Mostafa and Nour Sh. El-Gendy, "Evaluation of Fuel Properties for Microalgae *Spirulina platensis* Bio-diesel and Its Blends with Egyptian Petro-diesel," *Arabian Journal of Chemistry* 10 (2013): S2043, https://doi.org/10.1016/j.arabjc.2013.07.034.

53 Alan Rempel, et al., "Bioethanol from *Spirulina platensis* Biomass and the Use of Residuals to Produce Biomethane: An Energy Efficient Approach," *Bioresource Technology* 288 (2019): 4–5, https://doi.org/10.1016/j.biortech.2019.121588.

54 Jorge Alberto Vieira Costa, et al., "Operational and Economic Aspects of *Spirulina*-based Biorefinery," *Bioresource Technology* 292 (2019): 3, https://doi.org/10.1016/j.biortech.2019.121946.

55 Barry et al., "2016 National Algal Biofuel," 2.

56 Michael C. Potter, "Electrical Effects Accompanying the Decomposition of Organic Compounds," *Proceedings of the Royal Society of London. Series b, Containing Papers of a Biological Character* 84, no. 571 (1911): 260–276.

57 Mostafa E. Elshobary, Hossain M. Zabed, Junhua Yun, Guoyan Zhang, and Xianghui Qi, "Recent Insights into Microalgae-assisted Microbial Fuel Cells for Generating Sustainable Bioelectricity," *International Journal of Hydrogen Energy* (2020).

58 Z. Baicha, M.J. Salar-García, V.M. Ortiz-Martínez, F.J. Hernández-Fernández, A.P. De los Ríos, N. Labjar, E. Lotfi, and M. Elmahi, "A Critical Review on Microalgae as an Alternative Source for Bioenergy Production: A Promising Low Cost Substrate for Microbial Fuel Cells," *Fuel Processing Technology* 154 (2016): 104–116.

59 Ibid.

60 Michael Hannon, Javier Gimpel, Miller Tran, Beth Rasala, and Stephen Mayfield, "Biofuels from Algae: Challenges and Potential," *Biofuels* 1, no. 5 (2010): 763–784.

61 Credence Research (2016) Algae Products Market by Application, accessed November 11, 2020, http://www.credenceresearch.com/report/algae-products-market ().

62 F.G. Acién, E. Molina, J.M. Fernández-Sevilla, M. Barbosa, L. Gouveia, C. Sepúlveda, J. Bazaes, and Z. Arbib, "Economics of Microalgae Production," in *Microalgae-based Biofuels and Bioproducts* (Woodhead Publishing, 2017), 485–503.

63 Md Asraful Alam, Jing-Liang Xu, and Zhongming Wang, *Microalgae Biotechnology for Food, Health and High Value Products* (Springer, 2020).

64 M.I. Khan, J.H. Shin, and Kim, "The Promising Future of Microalgae: Current Status, Challenges, and Optimization of a Sustainable and Renewable Industry for Biofuels, Feed, and Other Products," *Microbial Cell Factories* 17, no. 1 (2018), 1–21.

65 Sajid Basheer, Shuhao Huo, Feifei Zhu, Jingya Qian, Ling Xu, Fengjie Cui, and Bin Zou, "Microalgae in Human Health and Medicine," in *Microalgae Biotechnology for Food, Health and High Value Products* (Springer, Singapore, 2020), 149–174.

66 Ibid., 121.

67 Ibid., 121.

68 Ibid., 121.

69 Ibid., 121.

70 Muhammad Imran Khan, Jin Hyuk Shin, and Jong Deog Kim, "The Promising Future of Microalgae: Current Status, Challenges, and Optimization of a Sustainable and Renewable Industry for Biofuels, Feed, and Other Products," *Microbial Cell Factories* 17, no. 1 (2018): 36.

Chapter 3

The New Symbiotic Architecture

Recently, microalgae-integrated building systems have also drawn the attention of architects and designers in the field of net zero carbon, net zero energy buildings due to their effective role in enhancing building energy efficiency, good air quality, and renewable power production. Bibliographic analysis reveals that microalgae-integrated building systems occurred as niche experimentation in the 2000s and are an incrementally active and fast-growing area in research and practice. One theoretical promise of biology-integrated architecture is the possible symbiotic nature of microalgae and the built environment with environmental benefits such as energy savings, pollutant reduction, and depollution for air, water, and soil quality. In a symbiotic system, different organisms live together by benefiting each other and developing a synergetic relationship.

Microalgae grow in various aquatic habitats and can tolerate a wide range of temperatures, sunlight, salinities, and pH levels. In a microalgae-integrated built environment, photosynthetic microalgae become a part of the building materials or service systems. The biocatalytic outputs of microalgae benefit the building and occupants by sequestrating CO_2, generating O_2, and producing renewable energy potentials. Their superior photosynthesis provides effective summer shading and winter solar heating, and year-round daylighting penetration, thus reducing building energy consumption. Microalgae take life-sustaining resources of CO_2 and nutrients from the occupants and building operations. High concentrations of CO_2 created by the occupants' respiration increase biomass productivity. Nitrates and phosphate from domestic wastewater maximize biomass growth. This symbiotic relationship results in CO_2 fixation, wastewater treatment, contaminant reduction, and biomass production for various end-uses such as foods (human, animal), bioproducts (e.g., antioxidants, astaxanthin, beta carotene, anti-cancer, anti-Alzheimer), and biofuels (e.g., solid, liquid, and gaseous).

Integration of microalgae within the built environment can be an open cultivation system, a closed cultivation system, or a hybrid of both. The open cultivation system includes open ponds, tanks, and raceway ponds, and the closed system is a photobioreactor (PBR). While the open cultivation system has advantages of low upfront and operational cost and low operation energy, it requires larger ground areas for light availability at shallower depths. It is susceptible to contamination, water evaporation, and adverse weather such as extreme temperatures. On the other hand, the closed cultivation system overcomes the disadvantages of the open system by minimizing land use and controlling growing environments (i.e., control of contamination, light availability, and temperature). However, it incurs

DOI: 10.4324/9780367814410-4

Microalgae's promising solutions for anthropogenic pollution

Anthropogenic activities	Environmental benefits	Bioremediation
Manufacturing, power plant, traffic, building operation	Atmospheric decarbonation	CO_2 capture
Municipal/domestic wastewater, agriculture run-off, industrial/domestic effluents	Wastewater treatment	Effluent removal
Urbanization, agriculture	Soil decontamination	Heavy metal removal

higher costs and energy for installation, operation, and maintenance. The majority of the microalgae-growing industry adopts open ponds.

For microalgae-integrated system feasibility, it is vital to understand key culturing technologies and environmental growth factors. High biomass productivity means higher carbon sequestration, wastewater treatment, and bioproducts. Culture productivity is dependent on strains and environmental factors such as nutrients, temperature, light intensity, pH, salinity, inorganic carbon concentrations, and so on. Production conditions also affect the overall yield including aeration, mixing, harvest frequency, and dilution rate. Aeration is one of the essential operations for mixing nutrients and CO_2, circulating microalgae cells, and allowing light penetration to the culture. Harvesting schedules are another important criterion for increased biomass yields. Batch cultivation keeps algae growing in a storage condition. Semi-continuous production harvests a partial volume of the culture at a certain interval, and the harvested culture is replaced by nutrient-rich media.

With increasing environmental impacts from the built environment and global responsibilities, in addition to providing applications from biomass, microalgae's promise lies in its ability to mitigate and bioremediate anthropogenic pollution to air, soil, and water. One of the bioremediation techniques using microalgae is wastewater treatment. Microalgae take nutrients and organic wastes from the wastewater and help clean water by removing nutrients. Photosynthetic carbon capture is another efficient microalgae technology. Each ton of biomass production consumes 1.8 tons of CO_2, exemplifying the commercial viability of installing microalgae reactors next to CO_2 emitters such as power plants, factories, and breweries. Microalgae's biosorption and bioaccumulation contribute to decontaminating anthropogenic pollution. Table 3.1 provides an overview of bioremediation roles in air, soil, and water environments.

3.1 Microalgae and Air Quality

Outdoor air quality: The World Health Organization (WHO) estimates that 9 out of 10 urban dwellers are exposed to air quality levels that do not meet WHO limits for healthy air.[1] For outdoor air, there are primary and secondary air pollutants. Primary air pollutants are ones that are directly emitted to the atmosphere from sources such as power plants, vehicles, and building heating equipment. They include sulfur dioxide, nitrogen oxides, carbon monoxide, volatile organic compounds (VOC), and particulate matter (PM).[2] Secondary air pollutants are formed within the atmosphere itself and arise from chemical reactions such as photochemical smog from nitrogen oxides, oxides of nitrogen, and secondary PM.[3]

Fossil-based power plants are the primary source of sulfur dioxide. Cities near heavily industrialized areas still show high concentrations. Research has shown a positive association between sulfur dioxide and health impacts especially related to respiratory, cardiovascular, all-cause mortality, and morbidity risks.[4] It was also found that children's asthma and bronchial hyper-responsiveness were associated with sulfur dioxide. WHO sulfur dioxide air quality guidelines are 20 μg/m³ for a 24-hour average and 500 μg/m³ for a 10 minute average.[5]

The major source of anthropogenic nitrogen dioxide is rush-hour traffic emissions outdoors. Research has shown positive correlation between daily concentrations of nitrogen dioxide with respiratory, cardiovascular, and overall mortality risks. Some studies included nitrogen dioxide's effect on children's hospital admissions, lung function, and asthma attacks.[6] It also shows association with fetal intrauterine growth retardation (exposure to nitrogen dioxide during the first month of pregnancy) and preterm birth (exposure to nitrogen dioxide during the last month of pregnancy).[7] WHO nitrogen dioxide air quality guidelines are below 40 μg/m³ for an annual average and 200 μg/m³ for a 1 hour concentration.[8]

Rising global temperatures help facilitate ozone formation, especially in sunny urban and industrial environments such as Los Angeles, California.[9] In sunlight, nitrogen dioxide from rush-hour traffic emissions turns into nitric oxide and atomic oxygen, and atomic oxygen combines with molecular oxygen to form ozone. According to WHO, exposure to ozone is responsible for about 150,000 deaths globally each year from respiratory conditions.[10] Ozone can impact human health by triggering coughing, throat irritation, and airway inflammation, and can also have long-term effects such as permanently reducing lung function and worsening emphysema and asthma, as well as causing children to develop asthma in the first place.[11] Nitric oxide generated by traffic causes ozone formation exacerbates respiratory symptoms.[12]

PM can be directly emitted from sources or formed by chemical reactions. PM with a 10 μm aerodynamic diameter, $PM_{10,}$ is typically generated from the combustion of fossil fuels and biomass burning. PM with a 2.5 μm aerodynamic diameter, $PM_{2.5,}$ is emitted from residual oil and diesel fuel combustion. PM has a great impact on human health with adverse effects on the respiratory and cardiac systems, leading to increased hospitalizations for issues related to the lungs and heart and higher death rates in cities on days with PM pollution spikes.[13] PM from coal burning has a particularly damaging effect on the cardiovascular system, with an ischemic mortality risk 5 times higher than with the average $PM_{2.5}$ particle.[14] According to data from 2009, the Centers for Disease Control and Prevention (CDC) estimates that a 10% reduction in $PM_{2.5}$ pollution in the air could prevent over 13,000 deaths per year in the United States.[15] Ozone and black carbon are known to have adverse effects on not only human health but also reduced crop yields, making them a serious threat to food security in some parts of the world.[16] Ozone is toxic to plants, and black carbon can block sunlight required for photosynthesis.

Outdoor pollutants and their health impacts[17]

WHO guideline levels at average time periods for each pollutant ($\mu g/m3$)			Anthropogenic major source	Health impact
Sulfur dioxide, SO_2	24 hours	20	Fossil power plants	Respiratory, cardiovascular, all-cause mortality and morbidity risks, children's asthma and bronchial hyper-responsiveness
	10 minute	500		
Nitrogen dioxide, NO_2	1 year	40	Congestion traffic emission	Children's hospital admissions, lung function, and asthma attacks; fetal and reproductive effect
	1 hour	200		
O_3	8 hours, daily maximum	100	Photochemical reactions with NO_2	Pulmonary, cardiovascular, reduced lung function, chronic asthma
PM_{10}	1 year	20	Fossil fuel combustion, biomass burning	Cardiovascular, heart, ischemic heart, stroke, respiratory, pneumonia, asthma
	24 hours (99th percentile)	50		
$PM_{2.5}$	1 year	10	Emissions from residual oil and diesel fuel combustion	
	24 hours (99th percentile)	25		

Indoor air quality: Indoor environments also present their own challenges, such as the off-gassing of building materials becoming trapped indoors. Typical indoor air pollutants include PM, NO_2, CO, O_3, SO_2, and VOCs.[18] Common sources of indoor air pollution include fuel-burning appliances, tobacco smoking, building materials and furnishings, cleaning products, improper maintenance of heating, ventilation, and air conditioning (HVAC) devices, excess moisture, and outdoor sources such as radon, pesticides, and outdoor air pollution.[19] Incomplete combustion of solid fuel burning and lack of ventilation result in high concentrations of indoor air pollution. In addition, insufficient ventilation and lack of indoor air quality monitoring systems often result in poor indoor air quality. Multiple studies have shown that improving indoor air quality can improve workplace productivity. Research data demonstrated that good indoor air quality greatly reduces asthma risk at home by a factor of more than 2, increases workers' productivity in offices by 10%, and accelerates students' learning in schools by 15% compared to conventional counterparts.[20]

In the late 1990s, Fisk and Rosenfeld estimated that a reduction in sick building syndrome (SBS) in the United States could yield around $10–20 billion in increased productivity.[21] A different study found that increasing ventilation rates from 5 l/s (10 cfm) per person to 10 l/s (20 cfm) per person would change building energy use by just 10%, but provide benefits in productivity yielding at least ten times the energy and maintenance costs.[22] A study in Norway estimated that improving indoor climate conditions could increase productivity by 10–100 times the operational and maintenance costs for the building and systems.[23] According to the World Trade Organization (WTO), Sick Building Syndrome (SBS) is defined as "a collection of nonspecific symptoms including eye, nose and throat irritation, mental fatigue, headaches, nausea, dizziness and skin irritations, which seem to be linked with occupancy of certain workplaces."[24] The U.S. Environmental Protection

Sick building syndrome sources[25]

Causes of sick building syndrome	Examples
Inadequate ventilation	ASHRAE 1973: 10 cfm/p (classroom), 15 cfm/p (office) ASHRAE 62 1981: 5 cfm/p (classroom), 5 cfm/p (office) (due to 1973 oil embargo) ASHRAE 62 1989: 15 cfm/p (classroom), 20 cfm/p (office) ASHRAE 62 1999: 15 cfm/p (classroom), 20 cfm/p (office) ASHRAE 62 2001: 15 cfm/p (classroom), 20 cfm/p (office) ASHRAE 62 2004: 10 cfm/p (classroom), 5 cfm/p (office) ASHRAE 62 2007: 10 cfm/p (classroom), 5 cfm/p (office) ASRHAE 62 2013: 10 cfm/p (classroom), 5 cfm/ p (office) ASHRAE 62 2019: 10 cfm/p (classroom), 5 cfm/p (office)
Chemical contaminants from indoor sources	Adhesives, carpeting, upholstery, manufactured wood products, copy machines, pesticides, cleaning agents, gas space heaters, woodstoves, fireplaces, and gas stoves
Chemical contaminants from outdoor sources	Motor vehicle exhaust, plumbing vents, building exhausts, nearby garage
Biological contaminants	Bacteria, molds, pollen, and viruses from stagnant water that has accumulated in ducts, humidifiers, drain pans, ceiling tiles, carpeting, or insulation

Agency (EPA) defines it as a method "to describe situations in which building occupants experience acute health and comfort effects that appear to be linked to time spent in a building, but no specific illness or cause can be identified."[26] Table 3.2 explains multiple factors causing SBS such as biological, ergonomic, lack of ventilation, physical, chemical, and building material selections (Table 3.3).

Indoor Air Quality Regulation: In the late 19th century, the first building ventilation standards called for a minimum of 30 cubic feet per minute (cfm) of outside air exchange for each building occupant.[27] The first ventilation standard, ASA A53.1-1943, Light and Ventilation, required a minimum ventilation rate of 20 cfm per person.[28] In the American Society of Heating, Refrigerating, and Air-Conditioning Engineers–established ASHARE 62-1973, the ventilation rate minimum was 10 cfm/person for classrooms and 15 cfm/person for offices. However, ASHRAE 62-1981 reduced the minimum requirement to 5 cfm in response to the energy crisis in the 1970s to save ventilation energy. This inadequate ventilation rate turned out to cause negative impacts on occupant health and comfort, and the ASHRAE standard 62-1989 was updated with a minimum of 15 cfm/p for classrooms and 20 cfm/p in offices. However, since ASHRAE 62-2004, and today's ASHRAE 62-2019, the standard specifies between 5 and 10 cfm of outdoor air flow per occupant and additional air exchange depending on the program of the space.

Mitigation of indoor air pollution is typically achieved by removing the pollutant source, increasing outdoor air ventilation, and incorporating air cleaning technology such as carbon filters, UV filtration, bipolar ionization, photocatalysis, and so on. The EPA provides a number of ways to combat SBS and indoor air pollution. One such method is not to use high VOCs in carpets or wallpaper, and the removal of contaminating products that may be stored in the building, such as paint cans and cleaning products.[29] The EPA also recommends increasing ventilation rates and regular maintenance of HVAC systems with appropriate air filters. They acknowledge that other nontechnical approaches can also help, such as increasing education and communication about indoor air pollution and developing standard investigation procedures.[30]

In the 1970s, NASA's life science researchers investigated the effectiveness of foliage plants to improve air quality in spacecraft and found that plant phytoremediation can reduce the concentration of VOCs specifically focusing on formaldehyde, benzene, and trichlorethylene.[31] Subsequent research has continued to verify plant effects on VOC reduction in varying environmental factors, focusing on the most dangerous VOCs such as benzene, toluene, ethylbenzene, formaldehyde, and trichloroethylene.[32] In 2000, an active phytosystem was introduced to enhance removal rates by forcing air to pass through active biofiltration.[33] Environmental factors such as light intensity, pH, and nutrients, as well as plant species and plant density, affect VOC removal. The lab experiments showed that plants can reduce some VOCs, but their removal rate has not been verified within the constraints of real-world applications. No studies have verified VOC removal by microalgae.

Experiments on microalgae with flue gas emissions have been performed to create green power plants. When microalgae growth is coupled with nonrenewable combustion flue gas, the flue gas directly supplies high CO_2 levels for photosynthesis and NO_x or SO_x as nutrients.[34] This symbiotic process does not require a separate pollutant scrubbing system for power plants, and the microalgae biomass can be converted into high-value products or biofuel, which can offset upfront and operational costs with less environmental impact. However, only a few microalgae species can adapt to rigorous flue gas conditions such as high concentrations of CO_2 and other pollutants, high temperatures, high aeration rates, and low pH. Although microalgae's ability to mitigate outdoor air pollution is well documented, the scientific literature lacks information about the role microalgae can play in indoor air pollution mitigation. Some research suggests that microalgae may be able to absorb radon, a toxic gas that is usually released indoors from "the potable water contaminated with dissolved radon supplied through different distribution systems."[35] Research from the 1960s shows that strains of *Chlorella* sp. and *Scenedesmus* sp. have the ability to absorb and use carbon monoxide.[36] Another study showed limited success with a microalgae air filter that helped produce oxygen and reduce PM in indoor air.[37] Potted and outdoor plants have demonstrated an ability to absorb air pollutants including PM and VOCs such as benzene, toluene, *n*-hexane, and formaldehyde.[38] These results suggest that it may be possible for microalgae to improve indoor air quality, but further specific research is needed. However, Wood et al. cite sources that consider bacteria to be a more effective microorganism for VOC removal than "fungi, cyanobacteria or algae."[39]

3.2 Microalgae and Water Quality

Water consumption is attributed to ecosystems, agriculture, manufacturing, human settlement, and energy production. As population, urbanization, and lifestyle consumption increase, there is expected to be a substantial rise in water demand. Water consumption in the United States consists of thermoelectric power (41%), irrigation (37%), and public supply (12%). The remaining 10% of water use supplies domestic (1%), industrial (5%), aquaculture (2%), mining (1%), and livestock (2%) requirements.[40] The building industry consumes water resources for construction and operation. The building sector discharges 80 gallons per capita of wastewater from daily activities such as flushing toilets, personal washing, laundry, food prep, and kitchen uses.[41]

The use of traditional systems for the treatment of wastewater is very expensive and requires intensive energy in terms of operation and investment. Surface runoff exceeds treatment capacity resulting in sanitary sewer overflows. Examples of sanitary sewer overflow causes are blockage or broken sewer lines or excessive storm water into sewer lines. Wastewater from the building sector contains large concentrations of water contaminants such as organic matter, nitrogen, phosphorus, and other trace elements including carbon, calcium, potassium, and iron. The wastewater needs to be treated before being discharged into waterbodies to avoid damaging the water's ecosystem. Wastewater treatment technology includes chemical treatment, biological treatment, and electrocoagulation.

Conventional wastewater treatment separates solids from wastewater and then goes through biological and chemical processes. Microalgae have been investigated as a bioremediator for wastewater treatment because they are abundant in nature, overall highly cost-effective, and effective against a broad range of pollutants.[42] Microalgae cultivation requires a substantial amount of nitrogenous nutrients, and 8–16 tons of nitrogen is required for one hectare of microalgae production.[43] They can assimilate nutrients, metals, and organic matter and can remove pathogens from wastewater using only solar energy as fuel.[44] Microalgae wastewater treatment was studied in the 1950s, and research has substantially advanced in the past decade as a result of growing interest in microalgae's potential as a fuel stock.[45] Nitrates and phosphates in wastewater and sulfates and CO_2 contained in the flue gas of combustible power plants can serve as living resources for biomass growth.[46] In addition, microalgae show specific promise in remediating heavy metals and eliminating pathogens in contaminated water. Microalgae cultivation using wastewater far surpasses all other terrestrial remediators with 40–50% higher biomass productivity, 80–100% pollution removal rates, and 1.8 kg of CO_2 capture per kg of microalgae biomass.[47] Microalgae have been researched for wastewater treatment for other manufacturing industries such as brewery wastewater, domestic wastewater, textile wastewater, pharmaceutical waste streams, slaughterhouse waste, heavy metal–containing wastewater, palm oil mill effluents, and agro-industrial wastewater.[48] Research indicated that microalgae-based wastewater treatment is cost-effective and yields higher biomass with ample nutrients. Table 3.4 shows the nutrient removal rates from different experiments growing algae in wastewater by uptaking these nutrients directly into their cells.

Microalgae bioremediation is mainly processed by (1) nutrient assimilation through photosynthesis and heterotrophic metabolism, (2) bioaccumulation of heavy metals through active and passive uptake, and (3) pathogen removal through raising the pH, temperature, and dissolved oxygen concentration of their growth slurry.[49] Once the wastewater is separated from the solid sludge, heterotrophic microalgae break down the biodegradable organic pollutants. Microalgae photosynthesis raises the pH level which inhibits pathogens with additional disinfection provided by sunlight penetration into the wastewater.

Precedent studies showing nutrient removal by microalgal species grown in different wastewater streams (Adapted from Shahid, et al., 2020)[50]

Microalgae	Wastewater	% Nitrogen removal	% Phosphorus removal	Biomass productivity (mg/L-day)	Source
Ascochloris sp.	Raw dairy	80	97	94–98	Kumar, et al., 2019
Ascochloris sp. (ADW007)	Raw dairy	78	98	34	Kumar, et al., 2018
Asterarcys quadricellulare	Municipal wastewater treatment	48	50	21.9	Odjadjare, et al., 2018
Chlamydomonas sp.	Palm oil mill effluent	65	34	–	Hazman, et al., 2018
Chlorella	Seafood processing	94.5	68.3	77.7	Gao, et al., 2018
Chlorella sp.	Soybean processing (diluted)	85	97	–	Qui, et al., 2019
Chlorella sp. MM3	Diluted winery and piggery (80:20)	89.3	56.5	$4.4×10^6$ cells/ mL	Ganeshkumar, et al., 2018
Chlorella sorokiniana	Diluted municipal & industrial	84.2	47	1524	De Francisci, et al., 2018
Chlorella vulgaris	Swine manure	71	54	9.1	Deng, et al., 2017
Chlorella vulgaris	Aquaculture and pulp	76.5	92.7	187	Daneshvar, et al., 2018
Coelastrum sp.	Cattle farm	80	100	281	Mousavi, et al., 2018
Desmodesmus sp.	Synthetic industrial effluent	85	>90	0.05	Rugnini, et al., 2018
Dunaliella salina	Tertiary treated municipal	57.5	69	28.25	Liu and Yildiz, 2018
Dunaliella tertiolecta	Diluted food leachate	>80	80	200	Wu, et al., 2018
Ecuadorian Chlorella sp.	Secondary effluent	52–93	67	0.6–1.8	Benítez, et al., 2019
Scenedesmus acuminatus	Paper and pulp	100	>97	685	Tao, et al., 2017
Scenedesmus Obliquus	Brewery effluent	88	30	–	Ferreira, et al., 2019
Scenedesmus sp.	Meat market	90	85	98.5	Apandi, et al., 2019
Tetraselmis suecica	Dairy	83	100	–	Daneshvar, et al., 2019
Tribonema minus	Tofu-whey	92.8	72	431.6	Wang, et al., 2019
Parachlorella kessleri	Agro-waste	>98	59	62	Koutra, et al., 2018

Nutrient Assimilation: Microalgae are able to assimilate nutrients from wastewater through the process of photosynthesis and heterotrophic metabolism. In other words, nutrient assimilation is mainly attributed to light and inorganic carbon availability. For example, microalgae assimilate nitrates and phosphates in municipal wastewater and the flue gas of sulfates and CO_2 in combustible power plants. Nutrient assimilation rates vary depending on the culture environment, and nitrogen and phosphorus removal efficiency averages 75% and 64%, respectively, as shown in Table 3.4. *Chlorella* sp., *Chlorella vulgaris*, and *Scenedesmus* sp. have shown promise in the removal of nutrients from wastewater.

Biosorption, Bioaccumulation, and Biodegradation of Heavy Metals: Biosorption is defined as passive heavy metal uptake and relies on pollutant attachment on nonliving algal cell walls.[51] Exposure to heavy metals causes detrimental health impacts and damages the function of internal organs. Common heavy metals found in building materials include lead, arsenic, cadmium, chromium, and mercury used for paint, insulation, metal coating, batteries, and power plants. Bioaccumulation is called active biosorption and has a two-step active process: (1) Active mechanisms transport metal ions into the cells and (2) heavy metals become attached to the surface of the algae.[52] Uptake of heavy metals occurs through metabolism of metal ions absorbed and transported inside the cells. Phytochelatins are chemical compounds in microalgae that can help bind and detoxify heavy metals.[53]

In addition to active biosorption using metabolism, dead microalgae biomass shows promise for wastewater remediation of heavy metals by absorption of heavy metal ions. Dead algae do not require nutrients or a growth medium and are immune to the toxic effects that high concentrations of heavy metals can have on living algae.[54] Dead biomass has shown a higher absorption capacity than live cells. One study indicated that the dead biomass of *Chlamydomonas reinhardtii* exhibited five times higher biosorption rates of heavy metals than bioaccumulation by living cells[55] and non-active cells of *Chlorella kessleri* showed greater biosorption rates of copper than active cells.[56] Single-cell algae typically have higher biosorption capacities than filamentous algae due to their higher surface-to-volume ratio.[57]

Microalgae degrade and transform contaminants into benign organic matter. Different environmental factors affecting the bioremediation process include biotic factor (e.g., microalgae strains) and abiotic factors (e.g., light, pH, temperature, pollutant concentration, contact time, and ionization). Studies have progressed to increase microalgae resistance to high heavy metal concentrations by genetically manipulating species, adapting cells to progressively higher pollutant concentrations, and by isolating strains from heavily contaminated sites.[58] Microalgae from the genera *Chlorella—Chlamydomonas* and *Scenedesmus*—are able to absorb 30–200 mg metal/g microalgae for metals such as copper, zinc, and lead. *Spirulina* has achieved absorption capacities of 240–420 mg metal/g microalgae for the same metals.[59] *Chlorella variabilis* exhibits the ability to remediate 100% of the nickel, aluminum, and iron in textile effluents with a biomass productivity of 74 g/m^2/day and a lipid yield of 20%.[60] *Chlamydomonas Ehrenberg* and *Chlorella beijerinck* were equally effective at removing zinc from wastewater. Both algae removed 100% of the zinc from a culture with a 10 mg/L concentration of zinc and 90% from a 100 mg/L zinc solution.[61] *Chlamydomonas reinhardtii* has gained attention for its ability to remove heavy metals from aqueous solutions. It shows wide heavy metal tolerance, sequestering metals such as lead, cadmium, and mercury.[62] Table 3.5 shows prior research that has investigated microalgae absorption of heavy metals in aqueous environments.

Microalgae's bioabsorption capacity to remove heavy metals in water solutions

Microalgae	Toxic elements	Absorption capacity	References
Ceramium virgatum	Cd(II)	39.7 mg/g	Sari et al., 2008
Chlamydomonas reinhardtii	Hg(II)	89.5 mg/g	Bayramŏglu et al., 2006
	Cd(II)	66.5 mg/g	
	Pb(III)	253.6 mg/g	
Chlorella vulgaris (dead)	Cd(II)	96.8%	Cheng et al., 2017
Chlorella vulgaris (live)	Cd(II)	95.2%	
Durvillaea potatorum	Cd(II)	90%	Matheickal et al., 1999
Sargassum sp.	Cd(II)	72.5 mg/g	Karthikeyan et al., 2007
Ulva fasciata	Cd(II)	73.5 mg/g	
Padina sp.	Cd(II)	90%	Kaewsarn et al., 2001
Scenedesmus quadricauda	Cd	66%	Mirghaffari et al., 2015
	Pb	82%	
Spirogyra sp.	Cr(VI)	4.7×10^3 mg metal/kg	Gupta et al., 2001
Ulva lactuca	Cd(II)	85%	Lupea et al., 2012
Ulva lactuca	Cd(II)	29.2 mg/g	Sarı et al., 2008
	Pb(II)	34.7 mg/g	
Ulva sp.	Zn(II)	29.63 mg/g	Badescu, 2017

Pathogen Removal: Microalgae can help remove pathogens in wastewater by changing abiotic factors such as increasing the pH, temperature, and dissolved oxygen concentration. They also compete with pathogens for nutrients, release algal toxins that inhibit pathogen growth, and adhere to certain pathogens.[63] The increase in pH and dissolved oxygen content during microalgal photosynthesis can result in the deactivation of pathogens. Microalgae like *Chlorella vulgaris* produce toxins of long-chain fatty acids when they are under stress; these toxins can destroy pathogens. *Synechocystis* sp. also produce toxins that are harmful to fecal bacteria.[64] Pathogens may also attach to the sediment that settles on algae cells, attracted by the negative charge of the algal cells.[65] *Scenedesmus* sp. showed a 100% removal of *Salmonella enterica* from swine wastewater after two days.[66] In the 1960s, scientists observed that the environmental factors that were favorable for algal growth were unfavorable for the survival of coliform bacteria such as *E. coli*.[67] Other experiments demonstrated 88.8% and 99.6% reductions in the number of coliform bacteria in raceway ponds.[68] Table 3.6 summarizes microalgae strains that are effective in removing heavy metals, nutrients, and pathogens.

End-Uses for Algae: Microalgae grown in wastewater can be used for continuing bioremediation or biofuel production. Wastewater-cultured microalgae, *Chlamydomonas reinhardtii*, show fast growth and exhibit significant lipid production.[69] However, it is not recommended to use the same microalgae used in the remediation of heavy metals for nutritional supplements. Agricultural waste and fish processing waste have been used to feed algal cultures that were then used to produce microalgal oil, methane energy, and fertilizer.[70] Microalgae that were grown to remediate nutrients in wastewater may be used in the pharmaceutical and nutraceutical industries and contain valuable nutrients such as omega-3 fatty

Microalgae in removing heavy metals, nutrients, and pathogens in wastewater

Effective Microalgae	Pollutant	Source
Chlamydomonas, Chaetophora, Chlorella, Oedogonium, Scenedesmus, Spirulina, Ulva	Heavy metals	Bilal, et al., 2018, p. 9 Bulgariu and Gavrilescu 2015, 462 Flouty and Estephane, 2012, p. 107 Shahid et al., 2020, p. 135306
Chlorella, Chroococcus, Dunaliella, Scenedesmus, Spirulina	Nutrients	Abdel-Raouf, et al., 2012, p. 262 Chowdhary, et al., 2015, p. 487 Dar, et al., 2019, 250 Shahid, et al., 2020, p. 135306
Chlorella, Scenedesmus, Synechocystis	Pathogens	Mezzari et al., 2017, p. 2 Rouf Ahmad Dar, et al., 2019, p. 254

acids, pigments, and amino acids.[71] Many researchers see wastewater remediation as a valuable co-process that can reduce costs and increase benefits to algae production for biofuel.[72]

3.3 Microalgae and Soil Quality

Increased urbanization, traffic pollution, industry manufacturing, as well as agricultural pest treatment and synthetic fertilizers become sources of soil contamination. Urbanization causes environmental and social stress primarily due to industrial production and inadequate disposal of solid wastes, affecting water quality, biodiversity, and filtering of pollutants. Traffic pollution such as gasoline spills and polyaromatic hydrocarbons from asphalt, tars, and incomplete combustion of fuels introduce urban soil contaminants. Manufacturing sites, smelting, and machine/auto repair shops with harmful processes and products can easily contaminate urban soils. Brownfields are typically contaminated with hazardous heavy metals such as lead, cadmium, and mercury in quantities above the ecological baseline or regulatory levels. Brownfield sites are estimated at over 5 million (totaling over 50 million acres) globally[73] with nearly half-a-million brownfields in the United States.[74] Superfund sites, estimated at 664 potential sites, require long-term management and clean-up plans.[75] Typical contaminants at brownfield sites include lead, petroleum, asbestos, and other heavy metals discharged from industry/manufacturing/agriculture processes, metal fabrication, and mining. Potential health effects include brain damage, various cancers, and organ damage. Table 3.7 summarizes past activities producing contaminants in brownfield sites and potential effects on human health.

As of 2016, phytoremediation has treated over 100 brownfield sites worldwide.[76] Three primary remediation techniques for contaminated soils include (1) containment, (2) extraction removal, and (3) solidification/stabilization. The containment method functions to keep polluted soil within a contaminated site boundary and prevent the spread of contaminants. Typically, waterproofing membrane is used to cap the containments preventing it from permeating the nearby ecosystem. This method consists of surface capping and encapsulation. Extraction removal adsorbs

Eight most common pollutants found at brownfield sites in the United States[77]

Contaminant	Examples of past uses	Potential health effects
Lead (Pb)	Mining, fuel, paint, inks, piping, batteries, ammunition	Damage to brain, nerves, organs, and bone; cancer
Arsenic (As)	Pesticides, agriculture, manufacturing, wood preservative	Nausea, vomiting, and stomach pain; blood disorders; nerve damage; skin disease; lung and skin cancer
Other metals	Metal fabrication, plating, mining, industry/manufacturing	Immune, cardiovascular developmental, gastrointestinal, neurological, reproductive, respiratory, and kidney damage; cancer
Petroleum	Drill and refining; fuel, chemical, and plastic production	Headache; nervous system, immune, liver, kidney, and respiratory damage; cancer
Asbestos (fiber in rock)	Mining and processing, piping, insulation, fireproofing, brakes	Lung scarring, mesothelioma, and lung cancer
Polycyclic aromatic hydrocarbons (PAHs) Hydrocarbon compounds, combustion byproducts	Coal tar, creosote, soot, fire, industry/manufacturing by-products	Liver disorders; cancer
Volatile organic compounds	Industry and commercial product solvents, degreasers, paint strippers, dry cleaning	Eye irritation; nausea; liver, kidney, and nervous system damage; birth defects; cancer
Polychlorinated biphenyls (PCBs)	Heat and electrical transfer fluids, lubricants, paint and caulk, manufacturing, power plant	Disruption or damage to the immune, hormone, and neurological systems; liver and skin disease

contaminants electronically, chemically, or naturally. It actively removes contaminants by using electrokinetic extraction, landfilling, soil flushing, soil washing, verification, and phytoremediation. Solidification serves to solidify the contaminated area with binding agents, but the downside to this method is that it blocks the phytoremediation process and therefore should be considered as a last option.

Containment method: Surface capping covers the contaminated site with waterproof material to stop contaminants from bleaching into the water table and causing human exposure. Typical use of this surface capping treatment is for parking lots or other city infrastructure use. The method is beneficial for small sites under 2,000 m^2 and costs range from $20/m^2 to $90/m^2 in the United States.[78]

Extraction removal method: This method involves active removal and decomposition of contaminants. The excavation method involves digging up the contaminated soil and transportation, disposal, and refilling of the site; this is more costly yet quicker and easier to employ.[79] Electrokinetic extraction uses low electric conductivity to decontaminate heavy metals in saturated or partially saturated permeable soils with an average cost of $117/m^3 soil.[80] Soil flushing is an in-situ process that passes fluid through the contaminated soil with high permeability and costs range from $20/m^3 to $104/m^3 soil depending on soil permeability and water table locations.[81]

Solidification uses chemical methods to trap and consolidate contaminants and reduce movement and solubility of contaminants in soil and water. These chemical elements not only trap heavy metals but also turn them into less soluble end-products, which is good for a small-scale project and needs continuous monitoring.[82] Costs for solidification average $520/m^3 soil with costs as high as $1,500/m^3 soil depending on the chemicals used and the costs for drilling and mixing.[83]

These detoxifying methods are costly and have little to no impact on revitalization of ecosystems. Phytoremediation, on the other hand, uses plants to remove and/or stabilize heavy metals and other contaminants, promoting the physical and biological quality of the site and supporting ecosystem. It is an affordable and non-invasive process of uptaking and immobilizing contaminants with aesthetically pleasing ecosystems. The efficacy of phytoremediation techniques is dependent upon biotic factors, such as plant species used, and abiotic factors such as the nature of the contaminant, soil conditions (e.g., pH, texture, and organic content), site climate, and geography. In general, phytoremediation can be categorized as phytoextraction and phytostabilization: (1) Phytoextraction depends on the shoots and leaves of plants to accumulate contaminants and (2) phytostabilization uses plant roots to immobilize contaminants in the soil.[84] Similar to phytoremediation, bioremediation uses living and nonliving organisms to decontaminate soils. These organisms vary from algae, lichens, mushrooms, bacteria to biowaste such as rice husks, wood fibers, and saw dust and decontaminate through operations such as valence transformation, biosorption, extracellular chemical precipitation, and volatilization.[85]

Research suggests that microalgae can fixate pesticides, heavy metals, and toxic pollutants in soil. With the same bio-assimilation and bio-adsorption techniques discussed in the wastewater treatment section, microalgae can serve as a bioremediator in contaminated soil in which uptake and storage capability vary depending on algae type, environmental conditions, and contaminant types.[86] For good soil quality, microalgae can be used as biofertilizer which improves the nutrients in agriculture and support the activity of living microorganisms. Another potential use of microalgae is to act as biostimulants by adding inactive microalgae and adsorbing soil pollution. Microalgae can control bioagents and protect them against pathogens and pests because they can regulate water permeability and improve the physical property of soil.[87] By increasing nutrients, organic matter, and soil crust, microalgae play a role in fertilizing soil and bio-controlling pests although excessive pesticides can cause detrimental effects on microalgae populations affecting the rate of mutation.[88] Research data indicated that green algae are more effective in removing heavy metals (e.g., zinc, lead, copper) than blue-green algae and diatoms.

Summary

Microalgae are microscopic photosynthetic organisms that have gained attention as effective bioremediators for anthropogenic pollutants combined with wastewater treatment along with bioproducts production. Compared to other chemical bioremediators, microalgae are natural in origin, overall low cost, and effective against a broad range of contaminants. Considering wastewater contains nutrients such as phosphorous and nitrogen, microalgae cultivation can be integrated with wastewater treatment. Carbon dioxide and flue gas from factories and power plants can be used for microalgae photosynthesis. With assimilation and the biosorption process, microalgae offer a cost-effective, ecofriendly, biophilic, recyclable remediation method. Assimilation occurs when pollutants accumulate in the interior of the cell. Biosorption processes use nonliving or living microalgae to uptake and neutralize pollutants. Microalgae act as biofertilizers, biostabilizers, and bioagents that are useful for soil enhancement and remediation. Using photosynthesis and heterotrophic metabolism, microalgae can assimilate heavy metals through active and passive mechanisms and destroy pathogens by raising the pH, temperature, and dissolved oxygen concentrations of their growth slurry. The challenge lies in the bioengineering of algal strains that tolerate and depollute toxicity while achieving high growth rate and high-quality chemical compounds.

Notes

1 WHO, *Air Quality Guidelines: Global Update 2005: Particulate Matter, Ozone, Nitrogen Dioxide, And Sulfur Dioxide* (WHO, 2006).
2 Ibid., 4.
3 Ibid., 4.
4 Ibid., 415.
5 Ibid., 415.
6 Ibid., 365.
7 Ibid., 366.
8 Ibid., 455.
9 *State of the Air 2020*, 6–7.
10 WHO, *Reducing Global Health Risks*, 1.
11 George D. Thurston, John R. Balmes, Erika Garcia, Frank D. Gilliland, Mary B. Rice, Tamara Schikowski, Laura S. Van Winkle et al., "Outdoor Air Pollution and New-onset Airway Disease. An Official American Thoracic Society Workshop Report," *Annals of the American Thoracic Society* 17, no. 4 (2020): 387–398.
12 Ibid., 389.
13 Ibid., 391.
14 Ibid., 391.
15 "Outdoor Air," Centers for Disease Control and Prevention, updated April 13, 2017, https://ephtracking.cdc.gov/showAirData.
16 WHO, *Reducing Global Health Risks through Mitigation of Short-lived Climate Pollutants: Scoping Report for Policymakers*, by Noah Scovronick (WHO, 2015), 2.
17 WHO, *Air Quality Guidelines: Global Update 2005: Particulate Matter, Ozone, Nitrogen Dioxide, and Sulfur Dioxide* (WHO, 2006).
18 "Introduction to Indoor Air Quality," U.S. Environmental Protection Agency, accessed June 2, 2020, https://www.epa.gov/indoor-air-quality-iaq/introduction-indoor-air-quality.
19 Ibid.
20 P.O. Fanger, "What Is IAQ," Indoor Air 16: 328–334, https://doi.org/10.1146/annurev.energy.25.1.537.
21 R. Kosonen and F. Tan, "The Effect of Perceived Indoor Air Quality on Productivity Loss," *Energy and Buildings* 36 (2004): 981, http://doi.org/10.1016/j.enbuild.2004.06.005.
22 Ibid., 982.
23 Ibid., 981.
24 WHO, "World Health Organization. Indoor Air Pollutants: Exposure and Health Effects," *EURO Reports and Studies* 78 (1983): 1–42. Quoted in Amirhosein

Ghaffarianhoseini, Husam AlWaer, Hossein Omrany, Ali Ghaffarianhoseini, Chaham Alalouch, Derek Clements-Croome, and John Tookey, "Sick Building Syndrome: Are We Doing Enough?" *Architectural Science Review* 61, no. 3 (2018): 99–121.

25   Quah, Stella. *International Encyclopedia of Public Health.* Academic Press, 2016, 502.

26   U.S. Environmental Protection Agency (EPA). 1991. "Indoor Air Facts No. 4 (Revised) Sick Building Syndrome," Environmental Protection, accessed June 10, 2018, https://www.epa.gov/sites/production/files/2014-08/documents/sick_building_factsheet.pdf.

27   Andrew Persily, "What We Think We Know about Ventilation." *International Journal of Ventilation* 5, no. 3 (2006): 275–290.

28   Ibid., 275.

29   EPA, *Indoor Air Facts No. 4 (revised) Sick Building Syndrome*, 3.

30   Ibid., 2–3.

31   Bill C. Wolverton and John D. Wolverton, "Plants and Soil Microorganisms: Removal of Formaldehyde, Xylene, and Ammonia from the Indoor Environment," *Journal of the Mississippi Academy of Sciences* 38, no. 2 (1993): 11–15.

32   A. King, "Plant A/C," *Chemistry & Industry* 83, no. 3 (2019): 2629.

33   A. Darlington, M. Chan, D. Malloch, C. Pilger, and M. Dixon, "The Biofiltration of Indoor Air: Implications for Air Quality." *Indoor Air* 10, no. 1 (2000): 3946.

34   Xing Zhang, "Microalgae removal of CO2 from Flue Gas," *IEA Clean Coal Centre, UK* (2015).

35   Debabrata Pradhan and Lala Behari Sukla, "Removal of Radon from Radionuclide-Contaminated Water Using Microalgae," in *The Role of Microalgae in Wastewater Treatment*, ed. L.B. Sukla, V. Subudhi, and D. Pradhan (Singapore: Springer, 2018), 75.

36   Emmett W. Chappelle, "Carbon Monoxide Oxidation by Algae," *Biochimica et Biophysica Acta* 62, no. 1 (1962): 50, http://doi.org/10.1016/0006-3002(62)90491-2.

37   Qian Lu, et al., "Application of a Novel Microalgae-Film Based Air Purifier to Improve Air Quality through Oxygen Production and Fine Particulates Removal: Microalgae-film Based Air Purification and Study on Its Mechanism," abstract, *Journal of Chemical Technology & Biotechnology* 94, no. 4 (2018), http://doi.org/10.1002/jctb.5852.

38   R.A. Wood, et al., "Potted-plant/Growth Media Interactions and Capacities for Removal of Volatiles," 120–121.

39   Ibid., 121.

40   Liandong Zhu and Tarja Ketola. "Microalgae Production as a Biofuel Feedstock: Risks and Challenges," *International Journal of Sustainable Development & World Ecology* 19, no. 3 (2012): 268–274, 269.

41   Wayne B. Solley, Robert R. Pierce, and Howard A. Perlman, *Estimated Use of Water in the United States in 1995*, vol. 1200 (US Geological Survey, 1998).

42   Muhammad Bilal, et al., "Biosorption," 1.

43   Peter J. le B Williams. "Biofuel: Microalgae Cut the Social and Ecological Costs," *Nature* 450, no. 7169 (2007): 478–478.

44   E. Posadas, et al., "Microalgae Cultivation in Wastewater," in *Microalgae-based Biofuels and Bioproducts: From Feedstock Cultivation to End-products* (Cambridge, MA: Woodhead Publishing Series in Energy, 2017): 67–68; O.B. Akpor, et al., "Pollutants in Wastewater Effluents," 57.

45   Ibid., 67–68.

46   Michael Hannon, et al., "Biofuels from Algae," 3.

47   Ayesha Shahid, Sana Malik, Hui Zhu, Jianren Xu, Muhammad Zohaib Nawaz, Shahid Nawaz, Md Asraful Alam, and Muhammad Aamer Mehmood. "Cultivating Microalgae in Wastewater for Biomass Production, Pollutant Removal, and Atmospheric Carbon Mitigation; A Review," *Science of the Total Environment* 704 (2020): 135303.

48   Ibid., 7.

49   Cynthia Alcántara, et al., "Microalgae-based Wastewater Treatment," 444; E. Posadas, et al., "Microalgae Cultivation in Wastewater," 73, 74.

50 Shahid et al., "Cultivating Microalgae in Wastewater," 7.

51 Roula Flouty and Georgette Estephane, "Bioaccumulation and Biosorption of Copper and Lead by a Unicellular Algae *Chlamydomonas Reinhardtii* in a Single and Binary Metal Systems: A Comparative Study," *Journal of Environmental Management* 111 (2012): 106, http://dx.doi.org/10.1016/j.jenvman.2012.06.042.

52 Muhammad Bilal, et al., "Biosorption," 4.

53 Sergio Balzano, Angela Sardo, Martina Blasio, Tamara Bou Chahine, Filippo Dell'Anno, Clementina Sansone, and Christophe Brunet. "Microalgae Metallothioneins and Phytochelatins and Their Potential Use in Bioremediation." *Frontiers in Microbiology* 11 (2020): 517.

54 Muhammad Bilal, et al., "Biosorption," 5.

55 Roula Flouty and Georgette Estephane, "Bioaccumulation and Biosorption," 112.

56 Ibid., 113.

57 Azam Heidarpour, et al., "Bio-removal of Zn from Contaminated Water," 2.

58 E. Posadas, et al., "Microalgae Cultivation in Wastewater," 69.

59 Cynthia Alcántara, et al., "Microalgae-based Wastewater Treatment," 446; E. Posadas, et al., "Microalgae Cultivation in Wastewater," 73.

60 Ayesha Shahid, et al., "Cultivating Microalgae in Wastewater," 135306.

61 Azam Heidarpour, et al., "Bio-removal of Zn from Contaminated Water," 8.

62 Roula Flouty and Georgette Estephane, "Bioaccumulation and Biosorption," 107.

63 Cynthia Alcántara, et al., "Microalgae-based Wastewater Treatment," 444; Rouf Ahmad Dar, et al., "Feasibility of Microalgal Technologies in Pathogen Removal from Wastewater," in *Application of Microalgae in Wastewater Treatment* (Springer International Publishing, 2019), 251–253.

64 Rouf Ahmad Dar, et al., "Feasibility of Microalgal Technologies in Pathogen Removal from Wastewater," 253.

65 Ibid., 254.

66 Melissa Mezzari, et al., "Elimination of Antibiotic Multi-resistant *Salmonella typhimurium* from Swine Wastewater by Microalgae-induced Antibacterial Mechanisms," *Journal of Bioremediation and Biodegradation* 8, no. 1 (2017): 2, http://doi.org/10.4172/2155-6199.1000379.

67 N. Abdel-Raouf, A.A. Al-Homaidan, and I.B.M. Ibraheem. "Microalgae and Wastewater Treatment," *Saudi Journal of Biological Sciences* 19, no. 3 (2012): 257–275.

68 N. Abdel-Raouf, et al., "Microalgae and Wastewater Treatment," *Saudi Journal of Biological Sciences* 19 (2012): 263, http://dx.doi.org/10.1016/j.sjbs.2012.04.005.

69 J. Paniagua-Michel, "Bioremediation with Microalgae," 473.

70 Ibid.; Beatriz Molinuevo-Salces, et al., "From Piggery Wastewater Nutrients to Biogas," 1103.

71 Ayesha Shahid, et al., "Cultivating Microalgae in Wastewater," 135303.

72 Poonam Choudhary, et al., "Phycoremediation-Coupled Biomethanation of Microalgal Biomass," in *Handbook of Marine Microalgae: Biotechnology Advances* (Academic Press, 2015), 483; J. Paniagua-Michel, "Bioremediation with Microalgae," 478.

73 Lianwen Liu, et al., "Remediation Techniques for Heavy Metal-contaminated Soils," 206–207.

74 "Overview of EPA's Brownfields Program," (EPA, April 7, 2020), https://www.epa.gov/brownfields/overview-epas-brownfields-program.

75 EPA, *Superfund: Transforming Communities, Accomplishments Report FY 2018*, 8, 18, https://semspub.epa.gov/work/HQ/100001884.pdf.

76 Lianwen Liu, Wei Li, Weiping Song, and Mingxin Guo, "Remediation Techniques for Heavy Metal-contaminated Soils: Principles and Applicability," *Science of the Total Environment* 633 (2018): 206–219, 213.

77 United States, *Environmental Contaminants Often Found at Brownfield Sites*, 2.

78 Ibid., 209.

79   Ibid., 214.
80   Ibid., 210.
81   Ibid., 210, 211.
82   Ibid., 211.
83   Ibid., 211.
84   Ibid., 212.
85   Ibid., 213.
86   Krishna Kumar Yadav, Neha Gupta, Vinit Kumar, and Jitendra Kumar Singh, "Bioremediation of Heavy Metals from Contaminated Sites Using Potential Species: A Review," *Indian Journal of Environmental Protection* 37, no. 1 (2017): 65.
87   Neveen Abdel-Raouf, A.A. Al-Homaidan, and I.B.M. Ibraheem, "Agricultural Importance of Algae," *African Journal of Biotechnology* 11, no. 54 (2012): 11648.
88   Ibid., 11649.

Part II | Microalgae Architecture Case Studies

Microalgae Infrastructure Intervention

Microalgae have been recognized as an attractive biological system that can be integrated with built environments to mitigate global challenges due to anthropogenic activities. Microalgae contribute to effectively offsetting carbon footprints and processing wastewater while increasing in biomass. These environmental attributes give microalgae great potential for integrating urban infrastructure that are related to food, energy, air, soil, and water use. Symbiotic efficiency from the biological system and the built environment can be developed into a closed-loop circular system where anthropogenic wastewater and carbon dioxide are recycled for microalgae growth which in return supplies valuable bioproducts. Biomass production as a result of photosynthesis is commercially available for essential foods and nutrients. Waste valorization can use microalgae as a biocatalyst to provide wastewater treatment and sequester flue gas carbon dioxide from power plants or transportation sources. Besides their role in cleaning air and water, they can also uptake and bioremediate pollutants such as heavy metals from contaminated soil. Their agricultural and environmental benefits make microalgae sustainable fuel stock for a renewable future energy system.

Microalgae are a diverse group of microorganisms living in various environmental conditions. The use of microalgae as a food source and ailment treatment dates back thousands of years and an active application of bioproducts was promoted in the middle of the previous century. For centuries, microalgae served as a primary source in the human food chain. In 900 CE, people discovered *Spirulina* as a food source, and the Aztecs in 1300–1521 CE harvested *Spirulina* from Lake Texcoco.[1] In response to population growth and food insufficiency, microalgae as a food source and antibiotic began in the 1950s. Beginning with the publication of "*Algae Culture. From Laboratory to Pilot Plant*" in the Algae Mass Culture Symposium at Stanford University in 1952, the first commercial culture of *Chlorella* was produced in Japan in the 1960s followed by *Spirulina* production in Mexico City in the 1970s.[2] The energy crisis accelerated research on microalgae as a renewable energy source in the 1970s.[3] In 1980, the global annual production of over 10 tons was produced in Asia, consisting of *Dunaliella salina*, as a source of β-carotene and *Haematococcus pluvialis* as a source of astaxanthin.[4] Nowadays, for human food use, the global market produces 5,000 tons of dry mass with *Spirulina* (3,000 tons), *Chlorella* (2,000 tons), *Dunaliella* (1,200 tons), and *Haematococcus* (300 tons) with future market growth at approximately $2.5 billion.[5] Numerous commercial applications of microalgae have been used for human health (e.g., food supplements,

DOI: 10.4324/9780367814410-6

pharmaceutical valued molecules, cosmetics, pigments) and agri-aquaculture (e.g., animal feed, fertilizer, aquaculture feed).

Microalgae have high photosynthetic performance and high lipid content compared to terrestrial plants and promise a new sustainable energy system as the next generation of biofuel sources. Besides biofuel production, microalgae cultivation becomes an active biological agency in which nutrients and CO_2 are supplied by wastewater treatment facilities and flue gas power plants, resulting in good water quality, clean air quality, and healthy soil quality. Microalgae are playing an active role in enhancing the health of the environment through decarbonation, oxygen generation, and phytoremediation against contaminants. These multi-functionalities and manifold benefits have drawn attention from architects, designers, and engineers and inspired them on to initiate and pursue speculative projects, scientific testing, prototypes, and technology demonstrations. Chapters 4–7 introduce various approaches to integrating microalgae biological systems in different scales of the built environment, and their environmental and societal benefits. The precedent projects in the chapters act as a reference of the feasibility of a microalgae system in various applications in four main areas: (1) infrastructure intervention, (2) urban intervention, (3) building intervention, and (4) product intervention, which consist of conceptual design or built examples (Figure 4.1; Table 4.1).

AlgoMed (2000)[6] is a *Chlorella* food production farm located in Klötze, Germany. Discovered in the 19th century, *Chlorella* is one of the most studied microalgae—with over 5,000 scientific publications that include the search term "chlorella" in PubMed, one of the largest online scientific databases—primarily due to its superior physiological properties. AlgoMed's farm consists of 500 km glass tubing that is housed in a protected greenhouse with optimum sunlight (Figure 4.2). The patented cultivation method further increases biomass production. With a fully integrated facility, from the first algae seed to the finished production, the closed bioreactor system prevents the growing environment from external contaminants. The bioreactor also utilizes mineralized pure water from an artesian well, resulting in high-quality *Chlorella* as a food source. When the *Chlorella* is harvested, a centrifuging system separates the microalgae from water and moderately dries the *Chlorella* for immediate consumption or further processing for food supplements. Optimizing the cultivation and production process allows a reduction in electricity use and water consumption.

X SEA TY[7,8] (2009) project is a concept proposal of an algae-producing, floating city that aims to offer clean energy autonomy, depollution, biodiversity, and biophilic life quality. The project uptakes pollutants produced by city centers and produces oxygen to improve air quality and ocean ecosystems (Figure 4.3). Building surfaces are enclosed with microalgae and living walls to maximize biofuel production and encourage biodiversity. As a living island city, it produces O_2, biomass energy, and other bioproducts such as bioplastics while treating wastewater and filtering sea water. The primary floating structure consists of a honeycomb geometry made of porous concrete. The modularity of the honeycomb concrete structure allows fast construction and cellular growth of the floating city to accommodate increasing demand. This floating structure supports microalgae-enclosed buildings while harnessing ocean current and protecting ecosystems. The ocean, the site of this microalgae city, serves as a heat sink to regulate heat transfer for

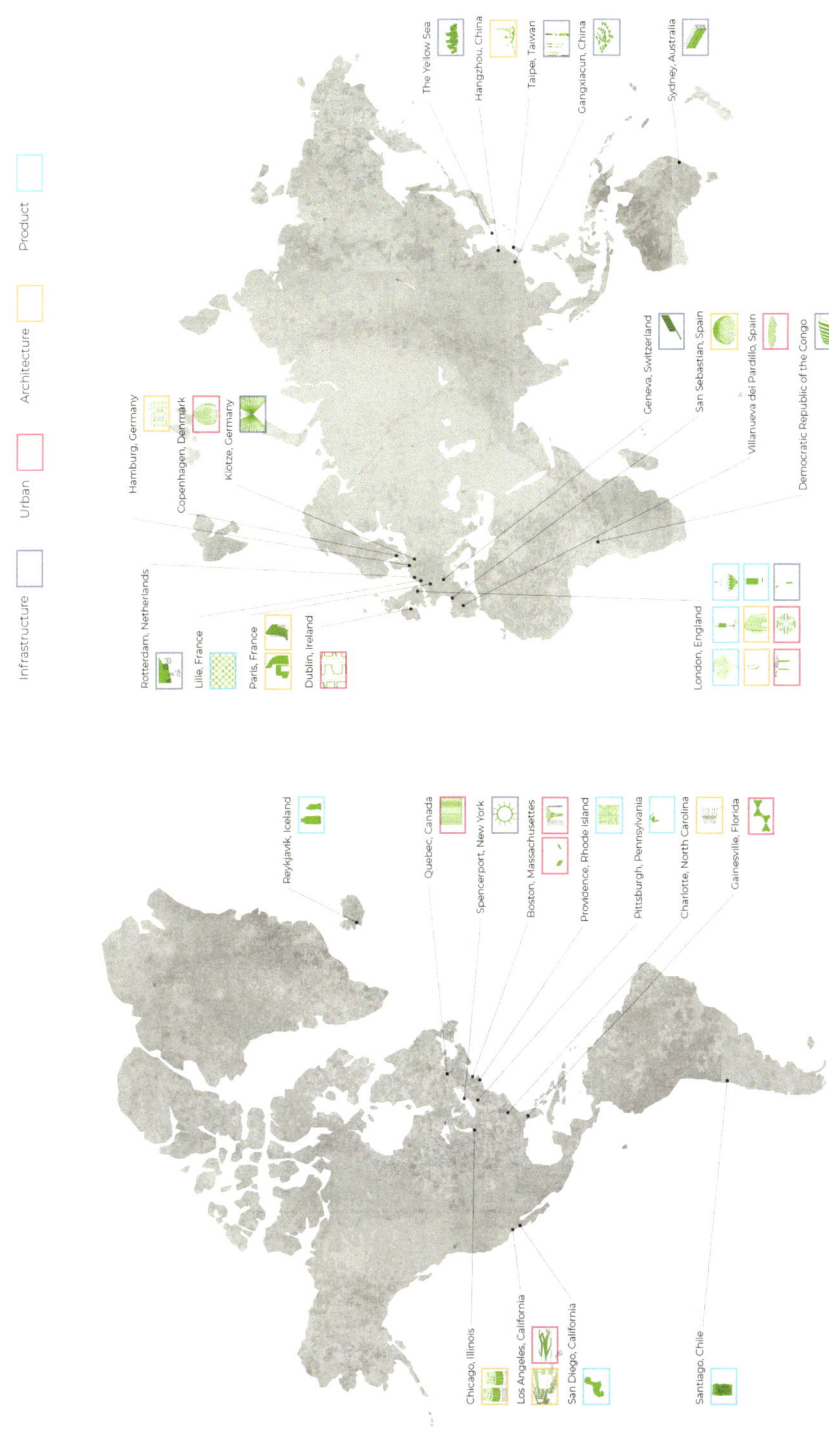

▲ Figure 4.1

Microalgae applications world map.

Precedent study list

	Title	Designer(s)	Location	Year	Key Performance
Chapter 4 Microalgae Infrastructure Intervention	I-Algomed	Roquette Klotze	Klötze, Germany	2000	World's largest algae farm composed of tubular bioreactors and greenhouse for chlorella production
	I-X SEA TY	XTU Architects	The Yellow Sea	2009	A floating city of structures integrated with microalgae to filter/improve air quality and water quality and enhance biodiversity
	I-Biolamps	Peter Horvath	London, England	2010	A network of microalgae-based streetlights to generate light for cities as well as to filter city air and produce biofuel
	I-Algal Scape	Federico Curiél	Rotterdam, Netherlands	2011	Large-scale microalgae harvesting and public park using the polders and reclaimed land enclosed by dykes
	I-(Infra)Structural Algae Ecology	Aleksandrina Rizova & Richard Beckett	Taipei, Taiwan	2011	A network between tall narrow structures to harvest microalgae within densely populated cities along with recreational spaces
	I-Algae-Powered Mushroom Farm	Frederick Givins	Democratic Republic of the Congo	2011	Mobile mushroom farming structures that are powered/cultivated by microalgae biofuel and fertilizer while supplying food and clean air
	I-Urban Algae Culture	Kady et al.	Shenzhen, China	2012	Modular algae system along the roofscape of tall buildings for wastewater treatment, biofuel production, and urban farming
	I-The Algae Tec. Facility	Algae Tec.	Sydney, Australia	2013	Grows microalgae to be used as biofuels in a large-scale production through the use of closed-system shipping containers
	I-Culture Urbaine	Cloud Collective	Geneva, Switzerland	2014	Microalgae bioreactor installed along a pedestrian overpass to grow biomass and filter emissions from vehicles on the road
	I-Algaewheel	Onewater Inc.	Spencerport, New York, USA	2015	A new method for treating wastewater in open air system with the combined use of microalgae and bacteria

	Name	Designer	Location	Year	Description
Chapter 5 Microalgae Urban Intervention	U-Perth Photobioreactor	Tom Wiscombe	Perth, Australia	2009	A set of public installations of seven bioreactors showcased to the public to produce biofuel and filter city air
	U-Flower Street Bioreactor	Tom Wiscombe	Los Angeles, California, USA	2009	Microalgae storefront system to raise awareness about biofuels produced from microalgae and animate the nightlife of the city
	U-Eco Pods	Howeler + Yoon	Boston, Massachusetts, USA	2009	A microalgae modular pod to enclose unfished city buildings and have the ability to link together to form larger living machine
	U-Algaegarden	Heather Ring and Thomas Kendall	Quebec, Canada	2011	A pubic installation made of flexible tubular curtain suspended from a structure for providing education through an interactive installation
	U-ALGA(e)zebo	Marcos Cruz and Marjan Colletti	London, UK	2012	Public pavilion that integrates bioreactor for social gathering and a viewing point toward integrated with local vegetation the surrounding environment
	U-Urban Algae Canopy	Marco Polleto and Claudia Pasquero	London, England	2014	A dynamic urban canopy for proving shades depending on weather characteristics and visitor's movement
	U-Photobioreactor Parking Canopy	Cervera and Pioz Architects	Villanueva del Pardillo, Spain	2014	Spiral bioreactor parking canopies for providing shading for cars/visitors as well as air filtration and biofuel production
	U-Algaevator	Tyler Stevermer and Jie Zhang	Boston, Massachusetts, USA	2016	Funnel-shaped bioreactor roof for maximum sun exposure, rainwater collection, and biofuel production
	U-The Algae Dome	Wadas et al.	Copenhagen, Denmark	2017	Bioreactor tubing attached to a dome structure for promoting and displaying the growth of microalgae for consumption as food and vitamins.
	U-AlgaeClad	Marco Polleto and Claudia Pasquero	Dublin, Ireland	2018	Lightweight digitally designed bioplastic bioreactors added to an existing office for carbon sequestration

(Continued)

	Title	Designer(s)	Location	Year	Key Performance
Chapter 6 Microalgae Architecture Intervention	A-Green Loop Towers	Influx Studio	Chicago, Illinois, USA	2011	Microalgae retrofitting residential towers to incorporate sustainable practices, including wastewater treatment and air quality improvement
	A-Process Zero	Sean E. Williams et al.	Los Angeles, California, USA	2011	Microalgae retrofitting office with a facade of various tubular bioreactors for building energy efficiency and air quality improvement
	A-algaeBRA	Marco Polleto and Claudia Pasquero	London, England	2011	Vertical tubular bioreactor cladding for passive cooling, thermal regulation, biomass production, and decarbonation
	A-FSMA Tower	Dave Edwards	London, England	2011	Mixed-use tower enclosed with bioreactors for CO_2 sequestration, energy cost reduction, and user comfort
	A-Alga Therapeia Center	Judit Aragones Balboa	Donostia-San Sebastián, Spain	2011	Research complex enclosed by tubular bioreactors to improve building performance and carry out research in medical and industrial application
	A-Urbanlab	Axel Schonert	La Defense, France	2012	Microalgae HQ enclosed with bioreactors to display and utilize microalgae for one of the country's largest providers of biofuel
	A-BIQ Building	Splitterwerk Architects	Hamburg, Germany	2013	Residential building with bioreactor cladding to grow biomass for heat and energy, sequester carbon, and improve building energy savings
	A-CSTB Building	XTU Architects	Champs-sur-Marne, France	2014	Microalgae curtain wall system that is integrated with photobioreactors to raise awareness, grow biomass, and reduce carbon
	A-In Vivo	XTU Architects	Paris, France	2016	Mixed-use complex to feature a biofaçade with photo-bioreactors for medical research and space and domestic water heating
	A-French Dream Towers	XTU Architects	Hangzhou, China	2018	Mixed-use towers with microalgae facades to provide natural insulation, as well as absorb carbon dioxide and produce oxygen
	A-Microalgae IVY	Ecoclosure + UNC Charlotte	Charlotte, NC, USA	2021	A full-scale prototype for retrofitting low-performing window for biological performance and environmental benefits

	Product	Designer	Location	Year	Description
Chapter 7 Microalgae Product Intervention	P-Algaerium Bioprinter	Marin Sawa	London, England	2010	Microalgae 3D printing for food nutrition and the aesthetic composition of the food using different microalgae strains
	P-Algae Curtain	Mathia Gmachl and Rachel Wingfield	Lille, France	2012	Transparent bioreactor tubing knotted into window drapes for the growth of microalgae in an aesthetic and educational way
	P-AlgaeBulb	Gyula Bodonyi	Budapest, Hungary	2013	Microalgae lightbulbs powered by the microalgae biofuel they grow while filtering room air and contributing to health and well-being
	P-Dino Pet: Algae Night Light	Yonder Biology	San Diego, California, USA	2015	Bioluminescent microalgae to create a night light as well as an interactive educational pet for children
	P-Algae Water Bottle	Ari Jonsson	Reykjavik, Iceland	2016	A water bottle created from microalgae byproducts that biodegrades immediately after use with a need to improve mechanical properties
	P-Living Things—Furniture	Jacob Doenius and Ethan Frier	Pittsburgh, Pennsylvania, USA	2016	Microalgae-based household lighting furniture as a way to grow microalgae as well as promote well-being within a household
	P-Bionic Chandelier	Julian Melchiorri	England, UK	2017	Aggregation of leaf-like bioreactor chandelier to provide illumination, intake CO_2, and grow microalgae/biofuel
	P-The Coral: Home Algae Farming	Hyunseok An	Providence, Rhode Island, USA	2019	Micro farming in home environments to provide environmental benefits and occupant health through carbon reduction and food production
	P-Indus-Algae Tiles	Jennifer Hahn	London, England	2019	Architectural cladding inlaid with microalgae for bioremediating heavy metals in the water and detoxifying surrounding environments
	P-Algae Packaging—Bioplastic	Margarita Talep	Santiago, Chile	2019	A biodegradable alternative to single-use plastics created from organic materials and microalgae with a need to improve durability and rigidity

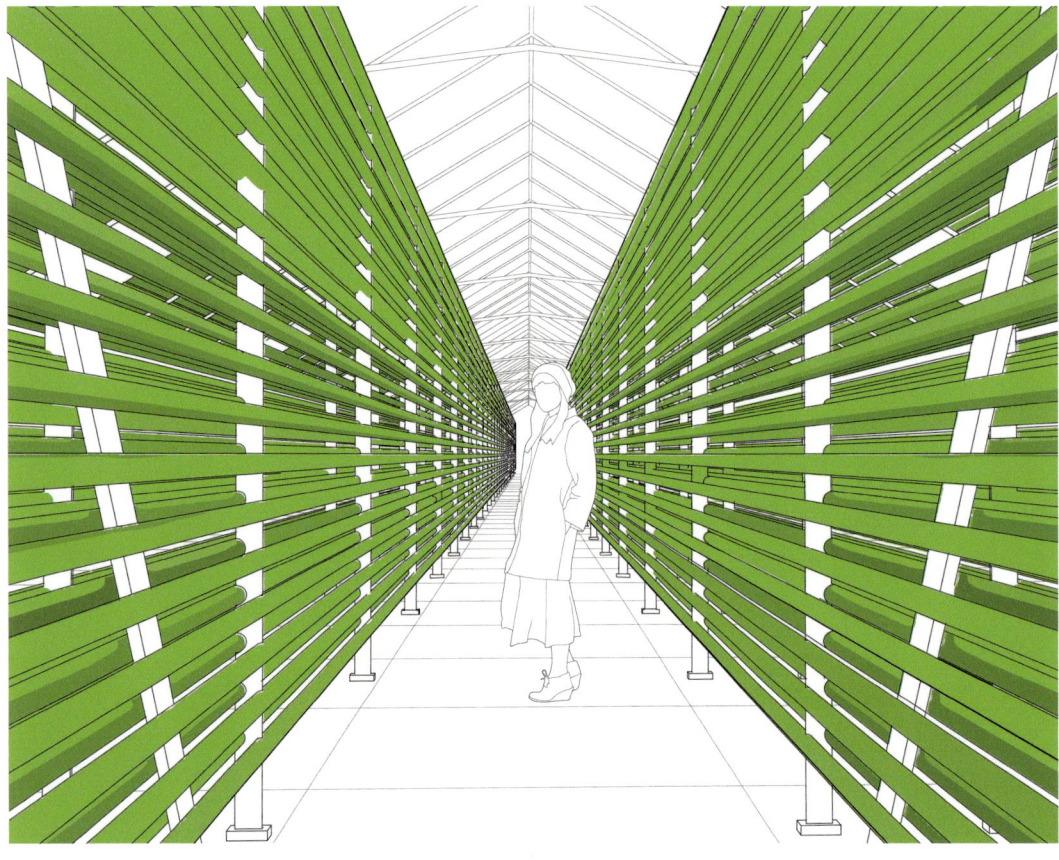

▲ Figure 4.2

Industry food production of Chlorella from 500 km long bioreactor housed in protected greenhouse with optimized cultivation and production process for maximum culture productivity and energy reduction (image reproduced from Roquette Klotze).

the floating island. The proposal can complement city needs from island cities like Hong Kong and Manhattan.

Biolamp (2010)[9] uses microalgae as a streetlamp to improve outdoor air quality and biofuel production. Street lighting is an essential element of the urban-scape and illuminates sidewalks, streets, and commercial buildings for pedestrians and cars. With a photovoltaic process, the localized air pump feeds outdoor air with street pollutants into the system and generate O_2 and biomass. The microalgae growth is increased by carbon dioxide from cars and factories. At the subterranean level, biolamps are connected to each other and tied back to a centralized opera-tion system where microalgae are harvested and turned into biofuel. Nighttime light from the biolamp will be dim, but it could employ a high-illumination light-bulb powered by biofuel or photovoltaic panels. The aesthetically pleasing design further inspires sustainable urban landscapes. The daytime cultivation, especially during summer time, will increase medium temperature that may need to incorpo-rate a heat exchanger or ways to mitigate heat stress. In line with the outdoor use, system durability under UV and other environmental conditions is an important consideration when selecting bioreactor materials (Figure 4.4).

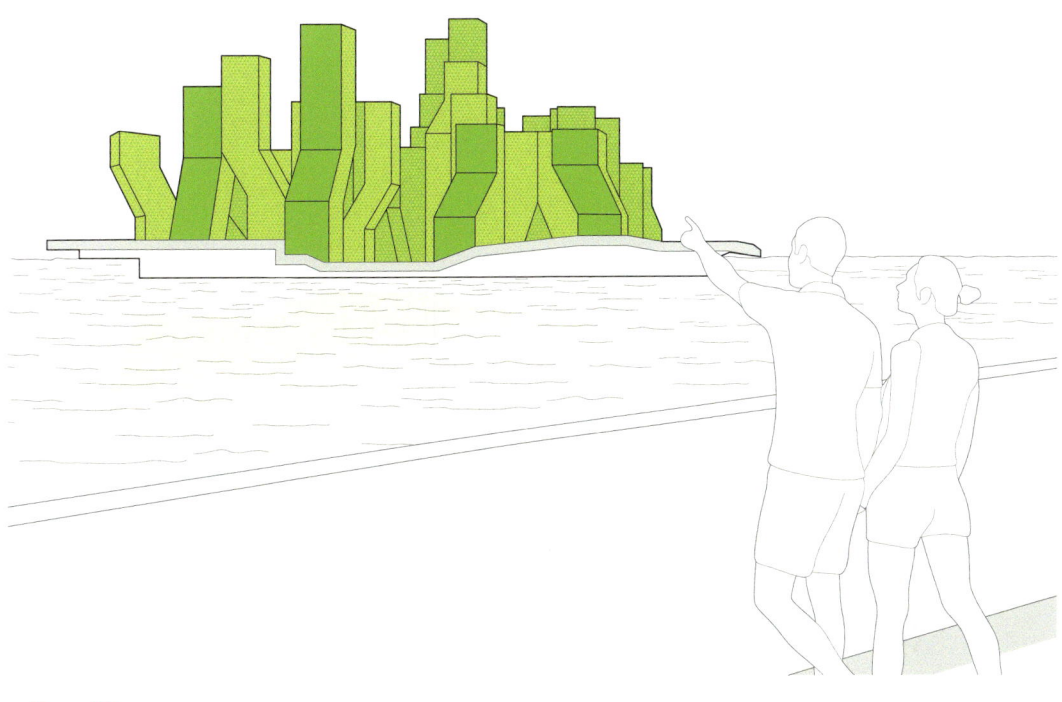

Speculative floating city buildings enclosed with microalgae systems for multiple benefits, including sustainable energy system, a food source, carbon sequestration, and wastewater treatment (image reproduced from XTU Architects).

Algal and Scape (2011)[10] serves as a sustainable strategy for transforming polders, reclaimed land, into a social and environmental infrastructure for Rotterdam, Netherlands. The proposal exemplifies reclaimed land development where microalgae-harvesting public parks are realized for large-scale biofuel production. The industrial-scale biofuel production is still cost-prohibitive from a microalgae cultivation and processing perspective, but having a vast open area like a polder can offset land requirements and construction costs. In addition, the reclaimed land is detached from the city's infrastructure and not well interconnected. Therefore, the microalgae farm on the reclaimed land can complement the city's land use planning by granting open pond production and encouraging public engagement. The transformation of polders into microalgae farms supplies biofuel production and environmental benefits such as carbon sequestration and wastewater treatment. As a green power plant and public park, it can accommodate the need for city expansion and reduce environmental stress. The proposal is open pond operation, and it needs close monitoring of cultivation status for reducing biological contaminants (Figure 4.5).

(Infra)Structural Algae[11] (2011) is a design proposal for microalgae towers and horticulture networks as sustainable infrastructure for dense cities like Taipei, Taiwan. The notion of porosity in the urbanscape encourages different layers of urban context to interact and integrate among transportation, commercial, residential, and educational networks. In addition to the permeable urban fabric, vertical microalgae towers and horizontal layers of hydroponics provide new

▲ Figure 4.4

Speculative microalgae street lamp to reduce outdoor pollutions and produce biofuels while illuminating nightscape (image reproduced from Peter Horvath).

urban ecology and agricultural infrastructure that further enrich the permeable urban fabrics. These ecological layers utilize recycled water, and the movement of fluid systems reflects the simultaneous movement of cars, cyclists, and pedestrians. When it comes to actualization of closed-loop bioreactors at the infrastructure scale, both technical and biotechnical challenges remain in an effort to develop more controllable, efficient, and economic culturing system. Bioreactors are typically constructed with transparent tubes made of plastic or glass. Scalization for infrastructural use can be achieved by increasing the diameter or length of reactor tube. However, bigger diameters limit light penetration and lengthy reactors exceed air saturation, leading to photoinhibition (i.e., limiting photosynthesis) (Figure 4.6).

Microalgae Power Mushroom Farm (2011) is a design proposal for microalgae-powered mobile mushroom farms in response to food shortages and environmental problems. In response to food shortages, Congo-Kinshasa in Africa was a proposed site for the first mobile farm location. The project consists of a mushroom farming pod enclosed with microalgae panels such that microalgae have access

▲ Figure 4.5

Speculative reclaimed land used for microalgae cultivation for encouraging social interaction, biofuel production, and carbon sequestration; upfront cost of biofuel production can be saved while complementing the city's land use planning (image reproduced from Federico Curiél).

to sunlight for photosynthesis while the mushroom farm is located inside of the pod where environmental factors such as light and moisture are easily controlled for the highest mushroom production. The project is a symbiotic relationship in that microalgae take nutrients from wastewater generated by the city and use the city's CO_2 for biomass production. The harvested microalgae provide renewable biofertilizer to grow mushrooms as a nutritious food source and supply biofuel to power the mushroom farm operation. The high photosynthetic performance of microalgae allows for greater carbon reduction and O_2 generation, improving the city's air quality. The novelty of this system is that it is a closed-loop system growing microalgae from anthropogenic wastes and powering the growth of mushrooms for human life and health in cities (Figure 4.7).[12]

Urban Algae Culture (2012) is a speculative masterplan for an urban village located in Shenzhen, China. The proposal seeks to create an elevated river across the roofscape and incorporate the modular algae system along the roofscape. As a community space, the network of the system on the roof offers the multiple functions of wastewater treatment, biofuel production, and urban farming. This microalgae roofscape exemplifies a new form of contemporary urban river, rearticulating the historic landform of Shenzhen. Following solar intensity and public accessibility, microalgae installations are zoned for four applications: (1) intensive algae production (biofuel production focused), (2) moderate algae production (water treatment focused), (3) lower-intensity algae production (water treatment focused), and (4) communal space for leisure gardens. The microalgae installations serve as a

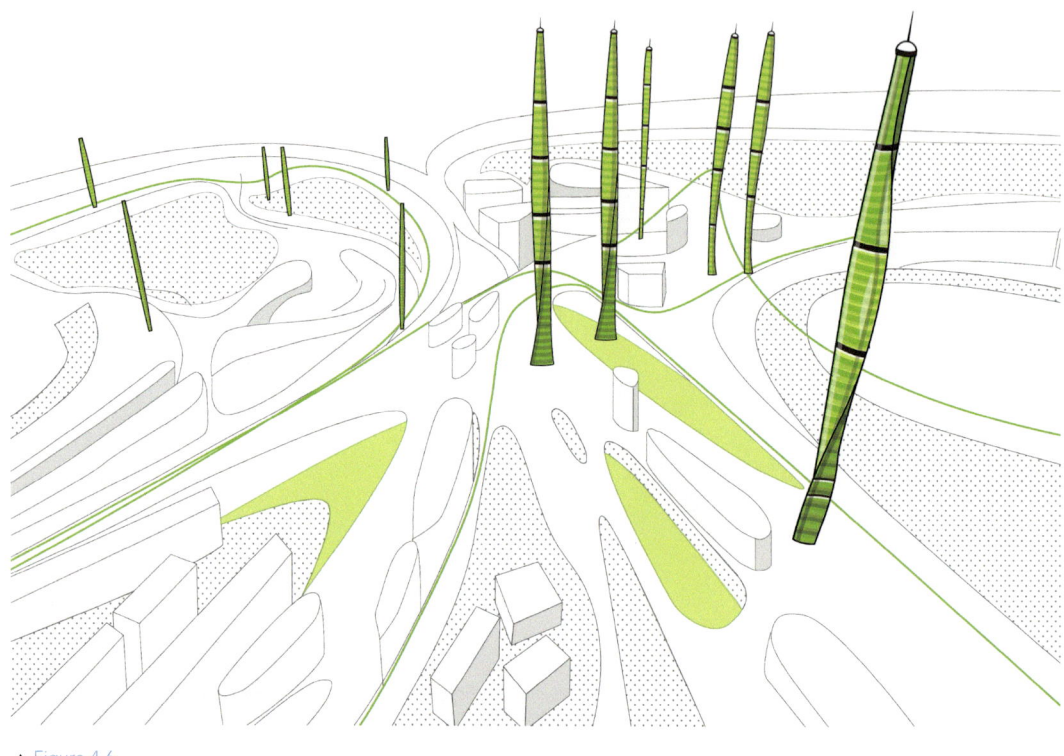

▲ Figure 4.6

A speculative urban buildings used for microalgae horticulture infrastructure integrated with different layers of transportation, commercial, residential, and educational networks (image reproduced from Rizova and Beckett).

decentralized wastewater infrastructure and networked with recycled water processed within the system. The microalgae are produced within a modular lightweight construction made of transparent plastic bottles. With nutrients supplied from wastewater, the grown algae will be transported to an algae processing facility to extract biofuel. Treated water is recirculated throughout the urban river integrating water pipes and pedestrian bridges creating an elevated urban hydro-landscape (Figure 4.8).[13]

Algae Tec (2013) is the first of its kind for an industry-scale carbon capture and utilization system by locating the bioreactor next to a power company. The carbon-fed microalgae determine the cost/revenue projections and the type of end-use production, such as biofuel and nutraceuticals. Unlike open pond operation, the patented 40 feet shipping containers offer protected environments supported by an enclosed modular engineering technology for high growth yield. It was estimated to produce 150,000 tons of algae and sequester 200,000–300,000 tons of CO_2 annually. Besides its high yield capability, the project is an outcome of successful collaboration between governments, power companies, shareholders, and stakeholders who aim to reduce environmental pollutants. Another application in China endeavored to capture 137,000 tons of CO_2 from manufacturing plants and produce 33 million liters of biofuel with 33,000 tons of biomass afterward. Another project located in Sri Lanka utilized CO_2 generated from cement

▲ Figure 4.7

A speculative mushroom farm powered by microalgae-based biofuel and fertilizers, supplying organic food, clean air, and clean power for the city (image reproduced from Frederick Givins).

manufacturing, sequestered 125,000 tons of CO_2, and produced 31 million liters of biodiesel with 31,000 tons of biomass. The project also exemplifies a global leader in carbon capture and renewable technology. With a long-term off-take agreement with airline company Lufthansa, it allows the airline to offset some of their kerosene aviation consumption (Figure 4.9).[14]

Culture Urbaine[15] (2014) is a photobioreactor installation along the viaduct over a heavily trafficked road. It exemplifies cohabitation of nature and urban settings where car highways play a role in producing biomass and bioproducts for microalgae. Reflecting the site characteristics and movements of pedestrians and cyclists, the bioreactor consists of a closed tubular system and lays horizontally attached to the viaduct. A steel structure supports the installation including pumps and growing apparatus which help the visual presence of the installation. With an abundance of sunlight and CO_2 on the site, the biomass cultivation can turn into a renewable fuel source or cosmetic and alimentary elements. Capitalizing on the existing urban infrastructure, the installation demonstrates future sustainable practices and symbiotic functions between natural and urban contexts in addressing food production, energy independence, and urban green space. The closed photobioreactor allows sterilization and limits contaminant risks. Because tube length and diameter affect photosynthesis, modular construction with optimum sizes is recommended (Figure 4.10).

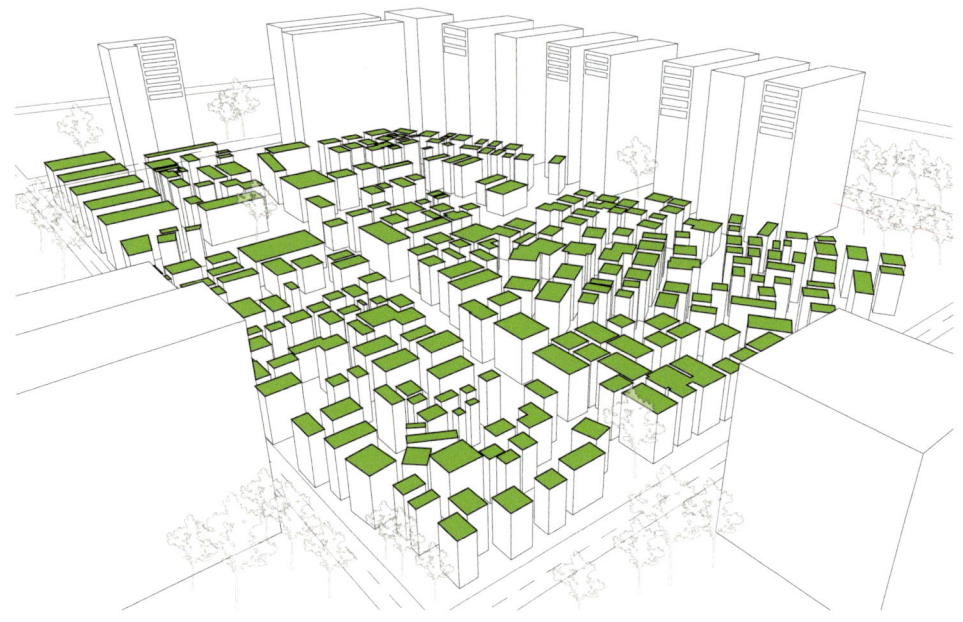

▲ Figure 4.8

A speculative urban intervention to use city roofscape to provide community space while producing microalgae-based biofuel and clean air quality for a densified city (image reproduced from Kady et al.).

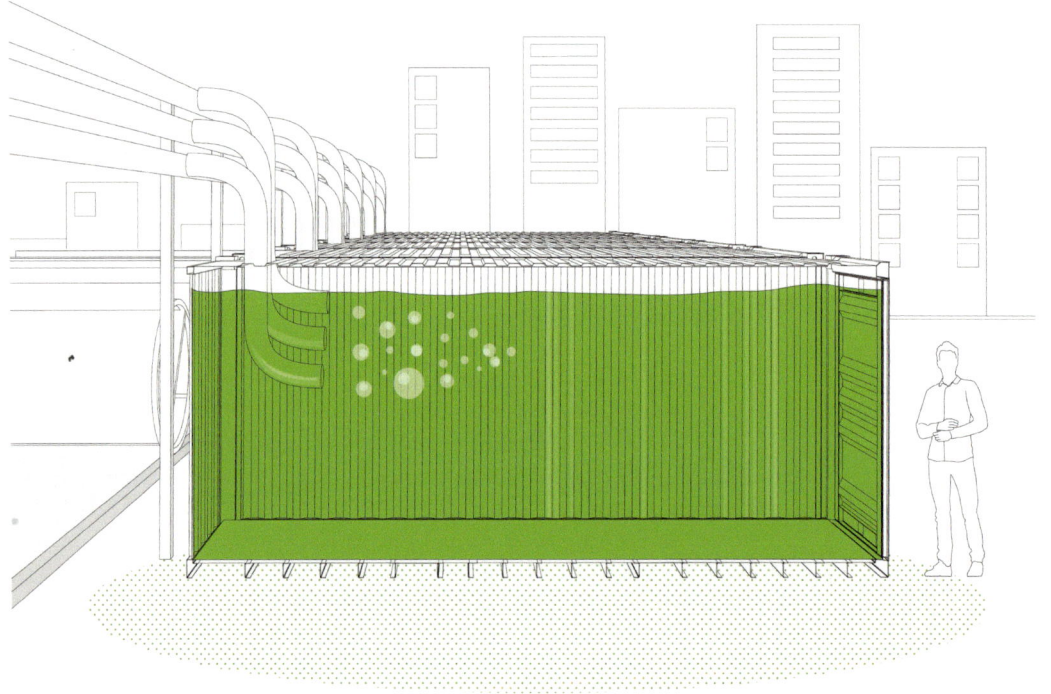

▲ Figure 4.9

Industry-scale carbon capture and renewable technology for microalgae biofuel production and uptaking nutrients from power plant flue gas (image reproduced from Algae Tec.).

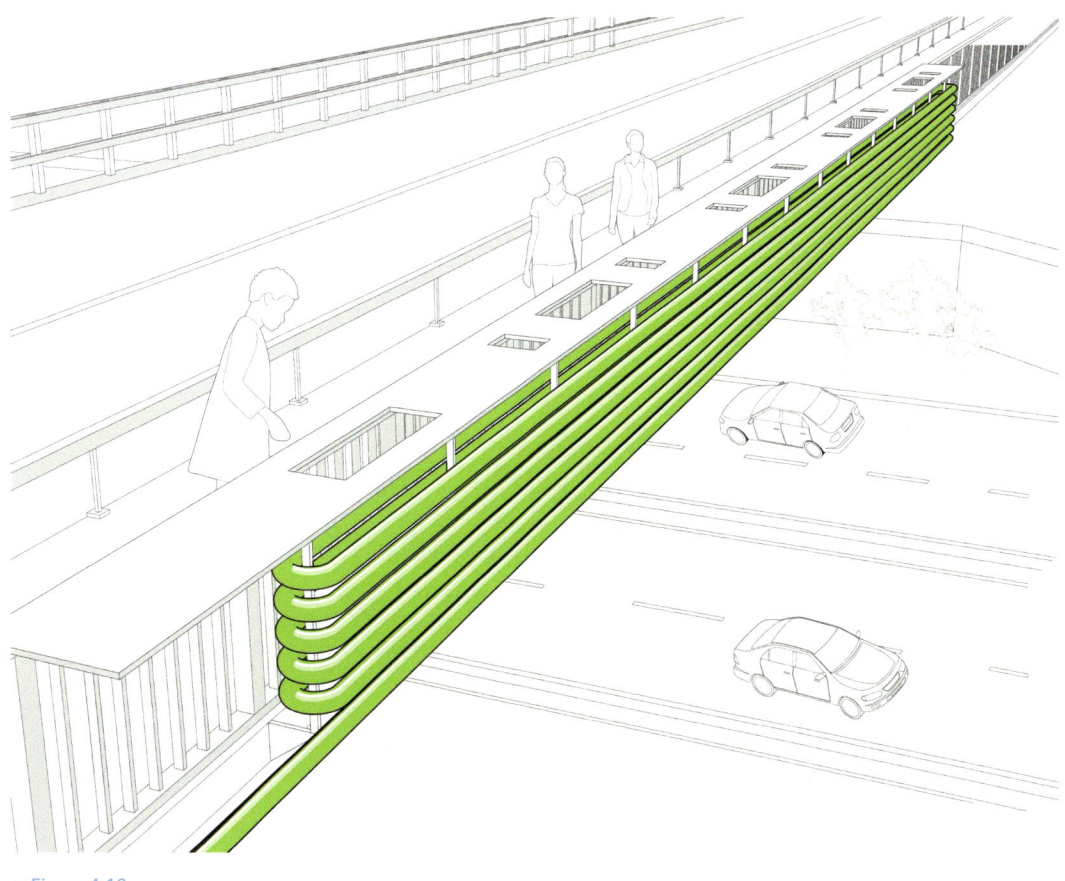

▲ Figure 4.10

A photobioreactor installed along a viaduct above a heavy traffic street to sequester CO_2 and renewable fuel source demonstrating symbiotic functions between natural and urban contexts (image reproduced from Cloud Collective).

Algaewheel[16] (2015) technology is an innovative sewage treatment plant in Naples. Microalgae use phosphorous and nitrate from wastewater and increase biomass. Algaewheels integrate a series of rotating algae contractor (RAC) rotated by a low-energy air system with 1–2 rpm rotation speed. Capitalizing on performance benefits from algae biofilms and moving bed biofilm reactors (MBBR), the external surface of the Algaewheel supports microalgae growth while the inside of the wheel provides MBBR encouraging bacteria growth to degrade contaminants in wastewater. The Algaewheels rely on the symbiotic relationship of dual microorganisms where the by-products of one group of organisms become the inputs for the other group. Microalgae film uptakes CO_2 (produced by bacteria), nitrate, ammonia, and phosphorus (from wastewater) and generate O_2 and carbohydrates (sugar). Bacteria in MBBR use the oxygen and sugars produced by microalgae and generate CO_2 for the microalgae. The rotational motion in the open pond cultivation offers good sunlight exposure with good regulation of temperature and air mixing (Figure 4.11).

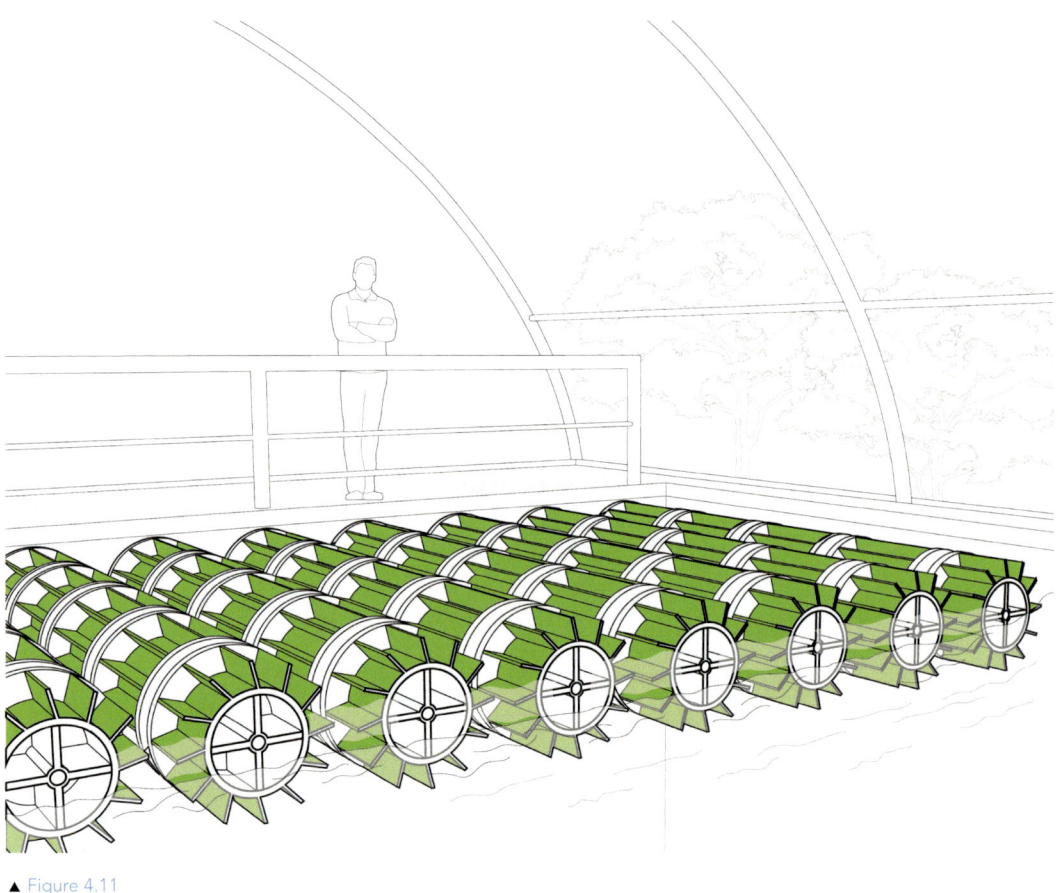

▲ Figure 4.11

An industry-scale rotating algae contractor (RAC) for wastewater treatment facility by utilizing bio microorganism of microalgae and bacteria cultivating in algae biofilms and moving bed biofilm reactors (MBBR) (image reproduced from Onewater Inc.).

Summary

As an essential part of decarbonation and oxygen generation for the ecosystem, microalgae could be integrated with urban infrastructure and help reverse environmental impacts, produce foods, supply a sustainable energy system, and encourage biodiversity. Global water resources are important for human health, ecosystems, and economic prosperity. Microalgae can play a role in protecting water resources by treating municipal, industrial, and agricultural wastewater. The biomass production using wastewater offers economic incentives for creating biofuel and bioproducts. In addition, with their fast growth and high lipid content, microalgae can serve as a future renewable energy system and mitigate anthropogenic climate change. Food production for growing population is another prospect of microalgae. Microalgae cultivation offers sustainable benefits such as effective land use and water utilization and environmental protection from no use of pesticides and fertilizers. The economic aspect of microalgae-based wastewater treatment coupled with biofuel or bioproduct creation, however, will largely depend on better understanding of the relationships between microalgae strains, system design (flow, temperature, light, pH, material, etc.), operation, microbiological processes, as well as cost-effective harvesting, dewatering, and commercial production.

Notes

1   Md Asraful Alam, Jing-Liang Xu, and Zhongming Wang, eds., *Microalgae Biotechnology for Food, Health and High Value Products* (Singapore: Springer, 2020), 34.

2   Michael A. Borowitzka, "Energy from Microalgae: A Short History," in *Algae for Biofuels and Energy* (Springer, Dordrecht, 2013), 1–15.

3   Pauline Spolaore, Claire Joannis-Cassan, Elie Duran, and Arsène Isambert. "Commercial Applications of Microalgae," *Journal of Bioscience and Bioengineering* 101, no. 2 (2006): 87–96.

4   Ibid., 87.

5   Ibid., 124.

6   Algomed, "Chlorella: Profile, Info and Interesting Facts, | ALGOMED®," accessed July 14, 2021, https://www.algomed.de/en/chlorella-4/.

7   XTU, "SEA TY," accessed July 14, 2021, https://www.xtuarchitects.com/seaty.

8   "X SEA TY Is a Carbon-Absorbing, Algae-Producing Floating City," accessed July 14, 2021, https://inhabitat.com/x-sea-ty-is-a-carbon-absorbing-algae-producing-floating-city/.

9   designboom | architecture & design magazine. "Biolamp | Designboom.Com," accessed July 14, 2021, https://www.designboom.com/project/biolamp/.

10   Robert Henrikson and Mark Edwards, "Imagine Our Algae Future," *Richmond, CA: Ronroe Enterprises* (2012): 86–87.

11   Ibid., 124–125.

12   Robert Henrikson and Mark Edwards, "Imagine Our Algae Future," *Richmond, CA: Ronroe Enterprises* (2012).

13   Ibid., 126–127.

14   Muriel Cozier, "CO_2 from One of the Largest Power Generation Companies to Provide Feedstock for Algal Biofuel," *Greenhouse Gases: Science and Technology* 3, no. 5 (2013): 315–317.

15   "Culture Urbaine: Co-habitation of the Urban and the Natural," August 14, 2019, https://urbannext.net/culture-urbaine/.

16   "Applications & Solutions," accessed July 14, 2021, https://algaewheel.com/index.php/applications-solutions/.

Chapter 5

Urban Intervention

Microalgae can have versatile applications in the urban landscape. Urban parks are extremely important spaces. They serve as connections to the natural world and provide places where people can grow food, engage with the community, and escape the busy traffic of city life. While many of the plants in a park are intentionally grown, water features are commonly populated with microalgae whether it is intentional or not. Microalgae play an important role in establishing a more positive relationship with people. Microalgae absorb carbon dioxide in the air, allowing them to grow and release oxygen. Through an interactive process such as a manual air pump, the public in turn learns more about microalgae and its benefits to the community and the environment. Other unique urban landscapes that can incorporate microalgae are urban farming and urban agriculture. Microalgae can be cultivated for consumption. By allowing elements such as climate, public interaction, and digital systems to influence the conditions of the canopy, a dynamic intervention, shifting with the ever-changing environment of the city, emerges.

Cities are pushing to be more compact and vertical to accommodate population increases, isolating urban dwellers from open green space. Urban green spaces provide multiple uses such as ecological (e.g., improving biodiversity), technical (e.g., providing leisure and mitigating urban heat island effect), aesthetic (e.g., enhancing historical gardens), and symbolic (e.g., attaining a symbol of the city) functions.[1] Active interaction with nature provides positive outcomes of well-being performance gain. There is an inborn human affiliation with nature and that contact with nature contributes to psychological and physical well-being. Microalgae technologies can serve as potential natural systems or biophilic elements in urban environments. Integrating microalgae systems with urban spaces can increase opportunities to interact and learn with nature. They can contribute to urban life and cultural tradition while enhancing environmental quality such as air quality, wastewater treatment, and detoxifying contaminants.

With their potential ecological, social, and economic roles, microalgae have sparked scientific and commercial interest and are an incrementally active and fast-growing area in the research and global market. Besides high-value bioproducts from biomass production, cultivating microalgae as part of bio-integrated building systems has gained popularity in niche experiments in the 2010s and increased various applications in built environments. Adding to the multifunctionality of microalgae, when it is integrated with densified urban environments, they could serve as a primary sink and phytoremediator for anthropogenic pollutants in air, water, and soil. Anthropogenic pollutants arise from a number of different

DOI: 10.4324/9780367814410-7

agricultural activities and manufacturing processes, including in the production of herbicides, pesticides, chemical by-products, pharmaceuticals, cosmeceuticals, textiles, plastics, and pigments. Due to high photosynthetic efficiency from the simple structures of unicellular microorganisms, microalgae effectively use light energy, fixate CO_2, and release O_2.

Helix BioReactor Perth[2](2009), a public art installation showcased seven photobioreactors based on Origin Oil's Helix BioReactor featuring microalgae colonies and LED helix. The bioreactor incorporates spinning microalgae transparent coils housed inside a polycarbonate shell, generating a dynamic visual effect. The aesthetics are further enhanced by color, luminosity, and space through an LED-integrated photosynthetic system. By being placed in a public space surrounded by buildings, the polycarbonate shell also has the functioning benefit of providing natural insulation and shading for the bioreactor, offering good microclimates for growing microalgae. The system utilizes the available light, CO_2, and wastewater to support the microalgae's growth process. Once the microalgae are fully grown, they can be harvested using OriginOil's Algae Appliance which can then collect the biofuel to be used within the building or elsewhere, making it a renewable resource. Outdoor installation is subject to environmental loadings. The longevity of the materials and reproducibility of the system are important design parameters. Practical issues such as good cultivating environments and cleaning should also be considered.

Flower Street Bioreactor[3] (2009) is a storefront project cladded with an "aquarium-like" bioreactor for producing biofuel. The bioreactor measures 9 by 19 feet in size and consists of a vacuum-formed transparent acrylic, uniquely shaped container imprinted with dynamic patterns and covered with a variety of LED lights. With the use of a biofeedback algae controller invented by the company OriginOil, LED lights regulate colors and intensity according to the need of growing microalgae for maximum productivity. In return, the storefront provokes visual dynamism and curiosity about the biology-integrated technology. The LED lights powered by solar cells complement daylight unavailability and increase the light intensity, making the operation off-grid and energy independent. The solar energy collected during the daytime can power the LED light usage at night, animating the nightlife of the city while the microalgae continue to grow. Through the storefront intervention, the biological system becomes part of the urban elements in a way that adds to the visual spectacle of cityscapes. The hydrostatic pressure for the flat bioreactor is one of primary design parameters. The bonds and the compatibility of adhesives and bioreactor materials will affect the longevity of the system (Figure 5.1).

Ecopods[4] (2009) is a concept proposal that suggests that unfinished buildings in cities could be enclosed by modular pods where microalgae can be cultivated for biofuel production. Urban densification with the rise of tall buildings may have more benefits from decentralized local regeneration systems. This can minimize infrastructural network and centralized plants powered by non-renewable resources and. The Ecopod would be installed by on-site robotic arms powered by microalgae biofuel while providing optimum growing conditions. The Ecopod is operated with a closed-loop system of material and energy flows between human, ecosystem, and the built environment. As a living machine, the pods inform the public about biofuel production while hosting research projects inside the pod. The plug-and-play nature of the pod can be applied to adjacent buildings through suspended construction, fostering widespread application. While the modular

A speculative public art installation utilizing OriginOil's Algae Appliance filled with microalgae bioreactor and LED helix for environmental benefits and biofuel production (image reproduced from Tom Wiscombe).

construction of the ecopods help industrial scalability, the practical issues of automating the cultivation process through close monitoring and control of growing conditions are important in attaining maximum cultivation and environment benefits of carbon sequestration and biofuel production (Figure 5.2).

AlgaeGarden[5] **(2011)** located in Quebec, Canada celebrates the synthesis of art and science and the aesthetics and productivity of microalgae. The installation consists of a bespoke flexible tubular curtain suspended from a primary structure. This bioreactor can grow different microalgae from red to green to bioluminescent microalgae for nighttime glowing. Often only negatively perceived as the nuisance of algae blooms, this project delivers beauty, scientific curiosity, and social interaction. The visitor learns the potential use of microalgae as renewable biofuel energy and natural sources for nutrition and food. Connecting with a nearby pond, the AlgaeGarden grows microalgae obtained from the adjacent pond and accommodates the extension of pod grasses. The installation explores possibilities and wonder from nature often overlooked in our daily lives. Maintaining stable algal cultivation is challenging in outdoors due to changing environmental conditions. Automated monitoring and control systems will be required to provide optimum operation and maintain a healthy culture. Practical issues of cleaning biofouling and durability are important for large-scale deployment (Figure 5.3).

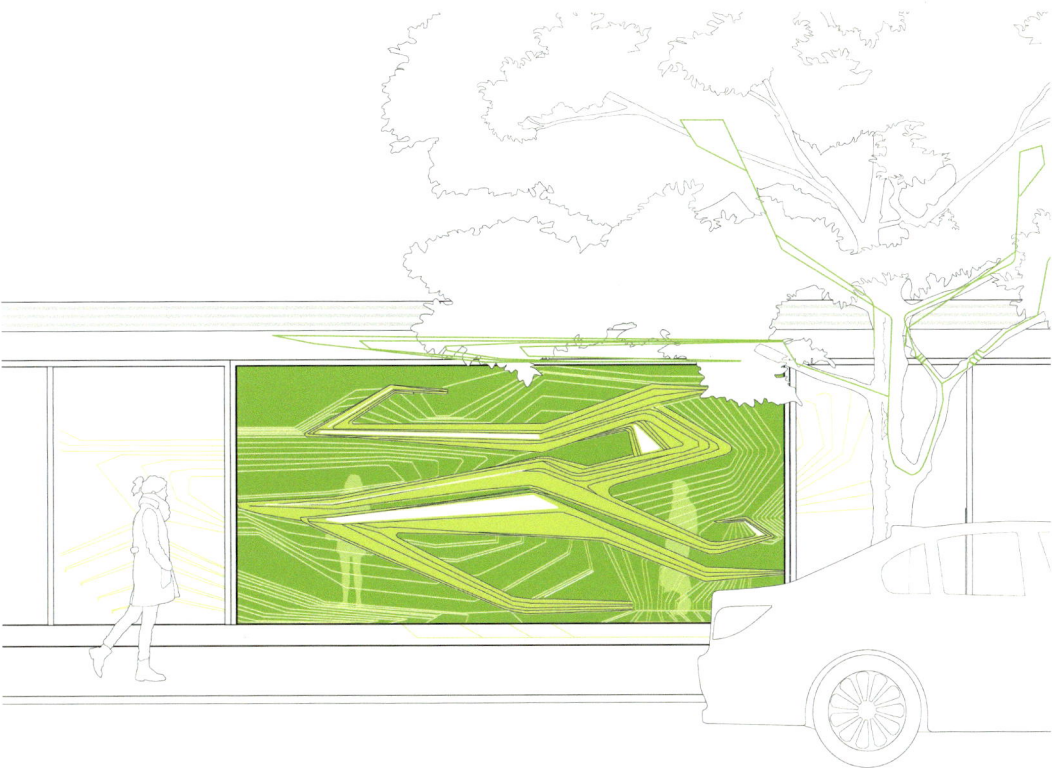

▲ Figure 5.2

A speculative storefront filled with microalgae cultivated with solar cell-powered LED for enhancing biofuel production and visual spectacle of sustainable cityscapes (image reproduced from Tom Wiscombe).

ALGA(e)zebo[6] (2012) is a public pavilion located in Euston Square Gardens in London. Similar to the English tradition of a jewel surrounded by filigree, the Alga(e)zebo offers a space for social gathering and a viewing point toward natural settings. The leaf-like double-curved metalwork evokes delicacy and relevance of traditional metalwork in gates, fences, and fountains in the U.K. cities. The bioreactor hugged by the leaf-like structure augments social and environmental potentials. The organic metal structure and organic lifeforms are interrelated and influence each other. The microalgae strains demonstrated in the system include *Chlorella vulgaris*, *Chlorella sorokiniana*, and *Pyrocystis lunula*, showcasing the promise of microalgae for all kinds of environmental and social value. People have affinity with nature and get inspired by nature. Interaction with nature raises awareness about a multitude of environmental and social values. The reproducibility of the project can facilitate widespread awareness in other public spaces and contribute to social interaction and integration. It serves a social gathering and art installation with ecological systems, attracting people to interact with each other (Figure 5.4).

Urban Algae Canopy[7] (2014) is a 1:1 scale prototype presented at 2014 Milan Design Week that integrates microalgae culture with real-time climate response digital control. The bioreactor consists of multilayered ETFE (ethylene tetrafluoroethylene) tubing with CNC (computer numerical control) in which aeration and water flow are controlled and regulated according to weather characteristics and

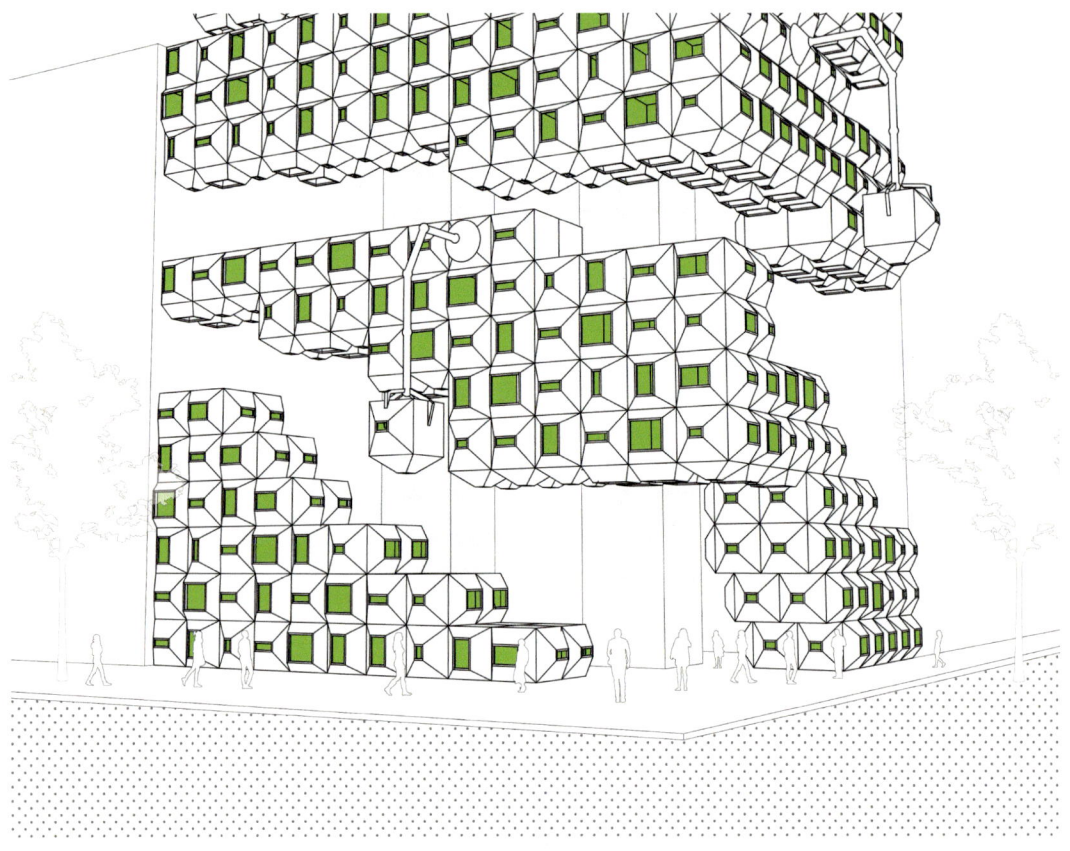

▲ Figure 5.3

A speculative R&D Ecopod enclosed with microalgae to be used for unfinished urban buildings for biofuel production and public engagement (image reproduced from Howeler and Yoon).

visitors' movements. Intense sun increases the growth of microalgae with a sub-sequent increase in opacity of the system and a shading effect. Its adaptation to the sunlight showcases real-time changes in a natural shading property. The electrovalve of the pumping system responds to visitors' movements and changes the algae flow rate, "provoking emergent differentiation across the space." The density, color, and shading efficacy is a product of multiple factors of climates, microalgae, visitors, and automatic controllers. The installation demonstrates urban agriculture and high yield of biomass production enabled by a custom bio-reactor, the morphology of which is controlled by the pressure of the cushion and fluid dynamic behaviors of the medium (Figure 5.5).

Algae Photobioreactor Parking Canopy[8] (2014) is an example of a parking canopy integrating a closed spiral photobioreactor to minimize the negative impacts from ever-increasing automobiles and parking lots. Parking is an essential element of urban cities, and the increasing influx of traffic shapes urban environ-ments. The increased construction of impervious surfaces becomes a by-product of unsustainable urban development. The tall tree-like microalgae parking canopy filters contaminated air generated by vehicles while providing shade and protec-tion from harsh weather for cars and users. The canopy surface funneling toward

▲ Figure 5.4

An algae garden installation at a public park for promoting beauty, scientific curiosity, and social interaction (image reproduced from Ring and Kendall).

the center is covered by 21 circular bioreactor tubes that are connected back to the center of the structure. The funneling geometry also helps rainwater collection that can mitigate surface run-off and be used for microalgae cultivation. The length, diameter, and surface area-to-volume ratio of the spiral bioreactor affect culture productivity and carbon sequestration rate. Smaller diameter of the tubular bioreactor for light penetration may reduce the residence time of CO_2 due to limited gas–liquid contact time (Figure 5.6).[9]

Algaevator[10] (2016) is a roof system for growing microalgae while providing sheltered space underneath. The spiral bioreactor is a gravity-based system allowing for the fluid's slow movement, CO_2 aeration, and increased exposure to sunlight. Carbon dioxide is introduced from the bottom of the spiraling coil, and oxygen is released from the top of the coil. The final yield of microalgae supplies consumer bioproducts and biofuel production. The microalgae roofing system is made of transparent flexible pipes secured to a clear membrane, exhibiting agricultural productivity in the built environment. The spiral bioreactor is shaped into a funnel, which optimizes access to solar intensity while harvesting rainwater for algae growth. The tube length and diameter of the bioreactor are important for mixing the culture for even distribution of nutrients and light penetration. The use of collected rainwater for cultivation needs sterilization before entering into the system to eliminate potential biological contaminants. The system successfully operated for three months, yet the type of the production modes used for the project (e.g., continuous, semi-continuous, and batch production) was not discussed (Figure 5.7).

▲ Figure 5.5

A microalgae pavilion installed at a public park for encouraging social gathering and viewing natural settings surrounding it (image reproduced from Cruz and Colletti).

▲ Figure 5.6

A microalgae canopy prototype installed at an exhibition fair with automatic controlling cell concentration based on climate conditions and visitor interactions (image reproduced from Polleto and Pasquero).

Algae Dome[11] (2017) is a food-producing pavilion in Copenhagen, Denmark. It consists of continuous flexible bioreactor tubing attached to a plywood dome structure. As a future nutrient-rich food source, the pavilion grew 450 liters of *Spirulina* during the three days of the exhibition fair. The installation raises the awareness of the role of biology in alleviating pressing global warming and anthropogenic pollution. The coiled bioreactor shields direct sunlight and filters the ambient air, creating comfortable microclimatic conditions. The tapered shape of the dome maximizes the solar exposure to the bioreactor. The dynamic appeal of the green dome encourages visual curiosity and interaction with visitors. In 1974, the UN declared *Spirulina* a super food for humans with a higher nutritional content than carrots, wheatgrass, and spinach combined.[12] It also contains more protein than soybeans and meat per unit weight. As a future for bioresources and high-value bioproducts, the project also sparked a sustainable development plan for using protein-rich microalgae for animal feed because soybean meal, the world's largest animal feed, causes impacts on agriculture land, water, and soil contamination from fertilizer (Figure 5.8).

▲ Figure 5.7

A speculative parking canopy filled with tubular microalgae bioreactor to alleviate surface run-off, provide shades for vehicles and users, and filter air pollutions on site (image reproduced from Cervera and Pioz Architects).

AlgaeClad[13] (2018) located in Dublin, Ireland retrofits a building skin with a lightweight translucent microalgae system as an urban skin. The installation encloses an office building to achieve a CO_2 sequestration rate of 1 kg CO_2 per day, equivalent to absorption by 20 large trees. The enclosure contains 16 modules, each a 6 feet wide by 21 feet tall photobioreactor, a digitally designed bioplastic container. Unfiltered outdoor air is aerated from the bottom of each module and filtered through the microalgae culture to the top and is released back to the urban air. The lengthy bioreactor design increases CO_2 dissolving time, which increases the efficiency in decarbonation and biomass production. Material selection of the lightweight membrane bioreactor and simple fabrication techniques contribute to reproducibility of the design and large-scale deployment. The custom shape and the bioreactor length may need attention on mixing of the culture and venting of photosynthetically generated dissolved oxygen. The placement of field sensors and control system is necessary to support autonomous cultivation of the system (Figure 5.9).

▲ Figure 5.8

A microalgae-filled open roof installed on a natural setting for uptaking carbon and producing bioproducts and biofuel (image reproduced from Stevermer and Zhang).

▲ Figure 5.9

A microalgae food pavilion installed at an exhibition fair, raising awareness of the important environmental role of biology and showcasing sustainable nutrient-rich food made of microalgae (image reproduced from Wadas et al.).

▲ Figure 5.10

An urban algae curtain installed on an existing office building that filters outdoor air and produces biomass (image reproduced from Polleto and Pasquero).

Summary

Microalgae as an urban intervention could address environmental, economic, and social challenges we face today. Over the past century, cities have evolved to cater to construction and automobile use with impervious surfaces. With their excessive use of concrete and nonpermeable urban surfaces, these spaces are the product of unsustainable practices, contributing to environmental problems such as rising temperatures caused by the heat island effect and irresponsible surface run-off contaminating soil and water systems. Designers have begun to explore the use of microalgae photobioreactors as rainwater collectors or solar regulators, employing strategies to minimize these negative effects. Water is a necessity for everyone on the planet, and there has been an escalation of water shortages and the continued spread of pollution into natural water sources. While many of the processes to clean contaminated water require the use of expensive technology, microalgae can serve as a cost-effective phytoremediator. Urban intervention showcasing micro-algae can also help raise awareness of alternative fuel sources. Microalgae-based biofuel as part of the visible landscape with features such as canopies, bus stops, and storefronts has environmental merit in that it does not require arable land, agricultural water, or fertilizer. Through the use of this technology, the growth of microalgae is inserted into the urban landscape in a way that adds to the visual spectacle of sustainable cities.

Notes

1   Ulf G. Sandström, "Green Infrastructure Planning in Urban Sweden," *Planning Practice and Research* 17, no. 4 (2002): 373–385.
2   "PERTH PHOTOBIOREACTOR—Tom Wiscombe Architecture," accessed July 14, 2021, https://tomwiscombe.com/PERTH-PHOTOBIOREACTOR.
3   "FLOWER STREET BIOREACTOR—Tom Wiscombe Architecture," accessed July 14, 2021, https://tomwiscombe.com/FLOWER-STREET-BIOREACTOR.
4   "International Algae Competition," accessed July 14, 2021, http://www.algaecompetition.com/x1172/.
5   Wayward, "Algaegarden," accessed July 14, 2021, https://www.wayward.co.uk/project/algaegarden.
6   "MAM—AlgaeZebo," accessed July 14, 2021, http://mam-arch.com/algaezebo.html.
7   "EcoLogicStudio," accessed July 14, 2021, http://www.ecologicstudio.com/v2/index.php.
8   Rosa Cervera Sardá and Javier Gómez Pioz, "Architectural Bio-photo Reactors: Harvesting Microalgae on the Surface of Architecture," in *Biotechnologies and Biomimetics for Civil Engineering* (Springer, Cham, 2015), 163–179.
9   Ying Shen, W. Yuan, Z.J. Pei, Q. Wu, and E. Mao, "Microalgae Mass Production Methods," *Transactions of the ASABE* 52, no. 4 (2009): 1275–1287.
10 "Algaevator," accessed July 14, 2021, https://www.domusweb.it/en/news/2016/07/01/algaevator_zhang_stevermer.html.
11 SPACE10, "The Algae Dome: A Food-producing Pavilion," September 4, 2017, https://space10.com/project/algae-dome/
12 Ernest Small, "37. Spirulina—Food for the Universe," *Biodiversity* 12, no. 4 (2011): 255–265, 261.
13 photoSynthetica, "CLADDING," accessed July 14, 2021, https://www.photosynthetica.co.uk/cladding.

Chapter 6

Architecture Intervention

With the increase in technological advancement, achieving net zero energy buildings is important for conserving resources and preventing climate changes. Despite improvement in building energy efficiency, buildings' overall energy use and pollutant emissions are growing because efficiency couldn't outpace growing construction activities. To cope with this imbalance, technology needs both efficiency and regeneration. To that end, building enclosures serve as a prime location to harness solar energy. Therefore, it is optimum to integrate microalgae cultivation due to the direct access to sunlight and air.

In 2013, a microalgae bioreactor cladding system was installed and began generating power for a four-story apartment building in Hamburg, Germany. Since then, additional real-world applications have not been implemented other than small-scale prototyping and experiments to test feasibility. When microalgae are integrated with building envelopes, their biomass growth response to solar intensity provides good summer shading efficacy and winter solar heating, reducing heating and cooling loads. They provide renewable fuel stocks, such as biomass or biofuel, to offset building energy consumption. Indoor air quality typically has higher CO_2 concentrations than outdoors, and using indoor air for photosynthesis reduces carbon concentrations and improves indoor air quality. They are a key part of the ecosystem and provide a great potential for promoting physical and psychological health. In return, the microalgae-integrated building system provides optimum physiochemical factors, such as irradiance, temperature, pH, nutrients for maximum biomass production, and by-product utilization from building operations as their living resources (e.g., daylight, waste heat, wastewater, flue gas).

Microalgae can be cultivated in open ponds or closed photobioreactors (PBR). Open ponds grow microalgae under natural light and temperature conditions with an aeration system as needed. The majority of microalgae production worldwide utilizes an open pond due to its economics and ease of scale-up despite disadvantages such as low growth rate, possibility of contamination, and large water evaporation from the open water surface. Closed PBRs yield greater productivity, controllability, and cultivation of various microalgae strains, but there is a high premium upfront cost. Symbiosis between microalgae applications and the built environment could bring benefits to both natural and built environments in the triple bottom line of sustainability. Environmental sustainability is achieved through reduced energy consumption, reduced pollution, and biofixation of pollutants. Biodiversity and green energy production are encouraged because of their ability to enhance occupants' physical and psychological well-being. There are three

DOI: 10.4324/9780367814410-8

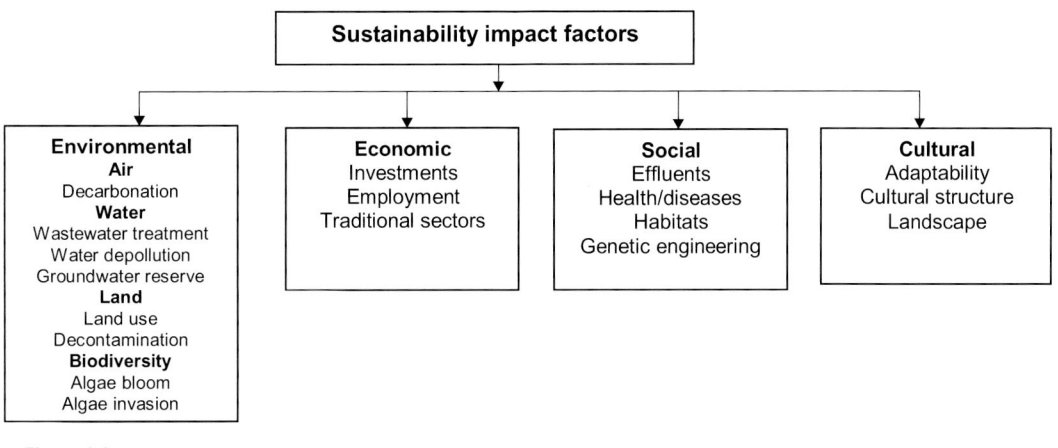

▲ Figure 6.1

Microalgae's sustainability framework (reproduced from Zhu and Ketola, 2012).[1]

operational modes for commercial microalgae production: (1) batch cultivation, (2) continuous cultivation, and (3) semi-continuous cultivation. Batch cultivation refers to microalgae produced in a confined batch that is relatively cost-effective with low maintenance and control of contaminants for PBR. To improve low biomass productivity, fed batch modes intake nutrient supplements which results in a growth boost. In continuous production mode, grown algae is discharged and new nutrient-rich media is added. Due to its continuous operation, it is costlier and needs higher maintenance, and there is an increased chance of introducing contaminants. The semi-continuous operation outperforms batch and continuous modes where the grown microalgae is discharged periodically and a new medium is added. Post-cultivation phase requires dewatering, drying, and downstream processing, which involves high energy and resource input (Figure 6.1).

Green Loop Tower[2] (2011) is a sustainable development strategy to retrofit old building enclosures with green technologies. Committed to the Chicago climate action plan to reach an 80% greenhouse gas (GHG) emission reduction from 1990 levels by 2050, the proposal integrates sustainable technologies to achieve net zero carbon practice. The balcony of the Marina City Tower is retrofitted with a photovoltaic system, and the parking levels and roof top are enclosed by microalgae systems, contributing to CO_2 reduction. A wet garden installed along the spiral ramp treats wastewater generated from apartment units that can be reused. The wind turbine on the rooftop harnesses kinetic energy while assisting ventilation and air circulation. Primary sources of CO_2 in cities like Chicago come from buildings, and the CO_2 is absorbed by microalgae and in return oxygen is released to the city. It is a closed carbon cycle where anthropogenic carbon is stored in microalgae biomass that can be used in bio-industry applications. Microalgae biofuel or electricity from solar cells will power vehicles. The project exemplifies the closed resource use of water, energy, and air, leading to carbon neutrality (Figure 6.2).

Process Zero[3] (2011) is a General Services Administration (GSA) retrofit concept proposal located in Los Angeles (Figure 6.3). The eight-story building's enclosures are covered by a series of transparent tubular bioreactors growing

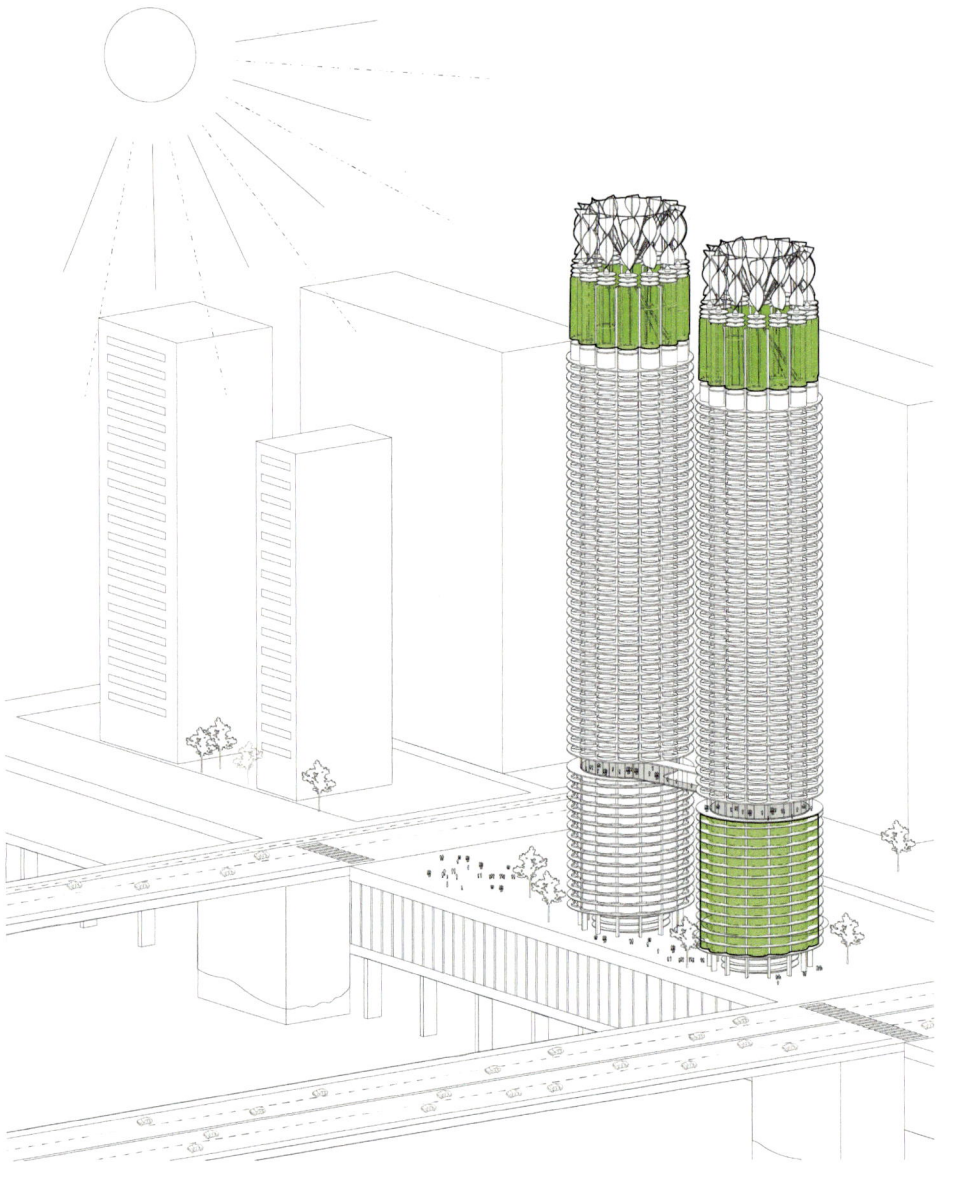

▲ Figure 6.2

A speculative building intervention of the Marina City Tower where the parking deck and rooftop are enclosed with microalgae systems, contributing to wastewater treatment, CO_2 reduction, biofuel production, and net zero energy building (image reproduced from Influx Studio).

microalgae. The density of the microalgae tubes varies at 25%, 60%, and 80%, depending on open view provision and solar availability according to building orientations. It filters wastewater and absorbs CO_2 generated by the nearby freeway while serving as a micro climate controller by shading the building. A fraction of the solar energy is absorbed by microalgae for photosynthesis. The majority of the solar energy is converted to heat, which can be reclaimed for space and domestic water heating. It grows microalgae with high lipid content that can be processed

A speculative office retrofitting with microalgae reactors with varying density depending on open view provision and solar availability (image reproduced from Williams et al.).

into biofuel. Microalgae as third-generation biofuel show a promise because they do not rely on arable land and fresh water. A centralized pump system supplies CO_2, nutrients and reclaimed water to maximize biomass production. Maintenance and cleaning of the bioreactor can be carried out from the outside. The oxygen generation benefits outdoor pedestrians and building occupants.

AlgaeBRA[4] (2011) is a concept proposal for a fashion company housing offices and commercial spaces. The project revolutionizes the traditional architecture model with material reinvention and regenerative prototypes that are rooted in the culture of the location and fitted to local climates. The transparent tubular PBRs are connected from the exterior surface to the interior partitions where they provide innovative passive cooling strategies, thermal regulation, biomass production, and decarbonation, while the interior PBRs allow privacy and flexible spatial organization. Not only does this microalgae system provide a thermal regulatory system from thermal storage of water; it is also closely connected with the architectural and spatiality it produces. The microalgae system also offers unique spatial effects from gradients of daylight, color, and coolness, psychologically or physiologically. Designers and scientists incorporate microalgae as part of their design

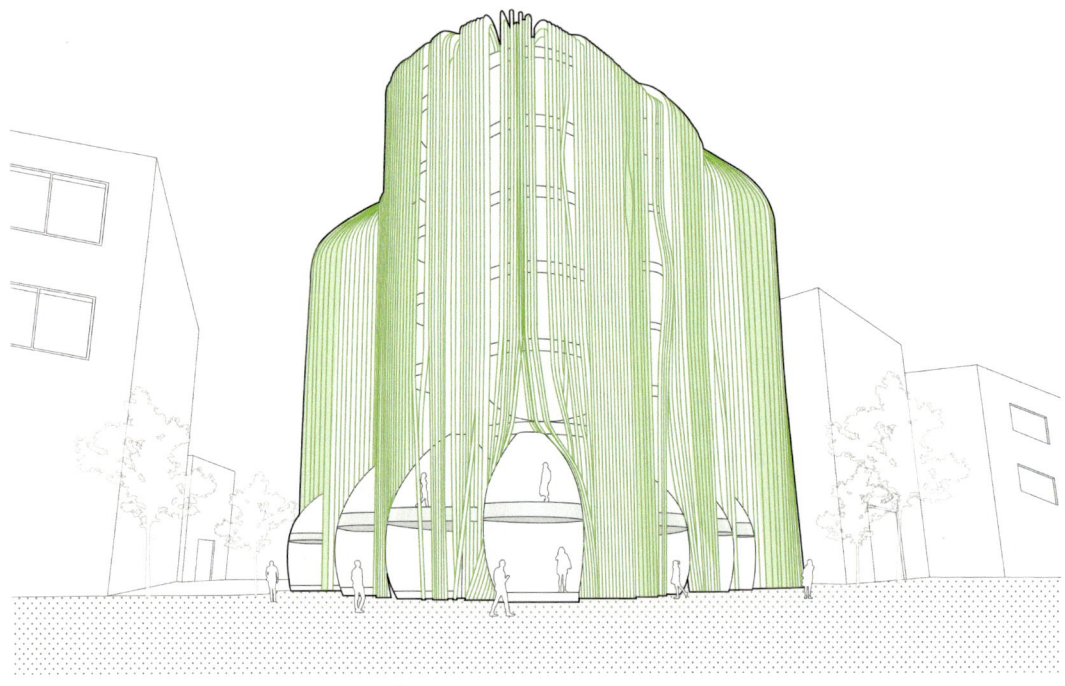

▲ Figure 6.4

A speculative office building installed with exterior and interior tubular bioreactors, offering passive cooling, thermal regulation, biomass production, decarbonation, and flexible spatial organization (image reproduced from Polleto and Pasquero).

constituents not only for taking inspirations of various shapes but also delivering benefits, functions, and process. It also has to augment the symbiotic relationship between people, nature, and the built environment (Figure 6.4).

The FSMA Tower[5] (2011) is a speculative project researching vertical greening and the integration of biological systems within skyscrapers (Figure 6.5). Instead of being a singular edifice, the project promotes diversity and interdependency of programs and users throughout the vertical surfaces of the tower. The tower located in the center of the financial district provides public spaces that are vertically oriented and filled with the street life of new living and working life in cities. Public space, housing, retail, community, and school programs are dispersed across the floors where the public can gather surrounded by nature. The tower is enclosed by PBRs, extending vertical landscapes while sequestering CO_2 and providing energy and bioproducts to the building and users. It is estimated that biofuel generated by the PBR can power 120 homes. The nutrition for algae growth is supplied by wastewater gathered from London's subterranean transit system, and, in return, it helps wastewater treatment. The solar energy stored in the façade and the waste heat from biofuel processes can be used for space heating or domestic hot water.

Algae Therapeia[6] (2011) is a research complex of seawater and algae for use in medical, nutritional, and industrial applications. The research complex near the coastline is a network of multiple buildings with different building programs. The buildings have PBR systems as external environmental skins where light, heat, sound, and air are filtered into the space while color and tactility of the buildings are brought to the city. The external skin consists of a transparent bioreactor

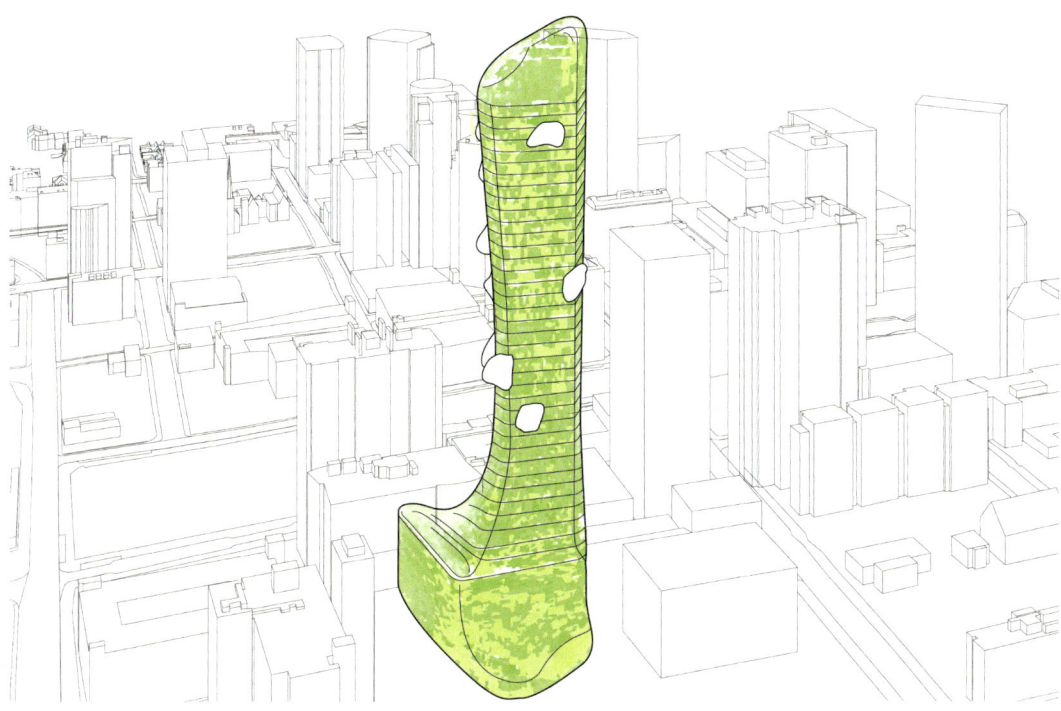

A speculative skyscraper enclosed with microalgae systems dispersed across the vertical surface supporting social interaction and environmental benefits (image reproduced from Dave Edwards).

that utilizes available sunlight and carbon dioxide for photosynthesis and biomass growth. The tubular bioreactors are curved in plane, stacked horizontally, and slightly offset from one another. The dome-shaped building geometry enclosed by a curved tubular bioreactor offers fluidity in nature. The research center focuses on seawater and algae, and the biomass is used as an energy resource for innovation and experimentation toward sustainable development. The strategic insertion of hybrid facilities serving both public access and research activities could offer potential environmental, social, economic, and educational benefits. This allows to increase the public awareness about ocean sustainability research and contributions to social well-being (Figure 6.6).

UrbanLab[7] (2012) is a speculative building proposal for enclosure by a transparent plastic PBR growing microalgae. Ennesys, a French-based startup, in collaboration with OriginOil, a company dedicated to transforming algae into biofuels, partnered with a forward-looking architect to develop a microalgae technology building in La Defense, France. Using approximately 100,000 ft^2 (10,000 m^2) of bioreactor panels, the building is expected to reduce energy usage by 80% with a water-saving capability of 80%. It is also estimated to yield 150 tons of microalgae per year which can produce 70 tons of biofuel for use in a generator in the building. The dry residue after extracting lipid for biofuel can be used for heat and electricity generation (although this process generates carbon from burning) or high-value bioproducts. The wastewater generated from the building supplies nutrients for the microalgae, and, in return, treated water can be used for domestic use such

A speculative research complex with the external building skin being enclosed with tubular bioreactors to filter light, heat, sound, and air (image reproduced from Judit Aragones Balboa).

as toilet flushing. The integration of microalgae system within urban fabrics can alleviate environmental stress due to demographic growth and urban densification. Post-cultivation phase is known to be energy-intensive and has challenges in economic production (Figure 6.7).

BIQ house[8] **(2013)** is the world's first building enclosed with a PBR growing microalgae. As many as 129 microalgae glass panels are installed on two sides of the building enclosures (southeast and southwest) in front of an opaque skin providing solar shading, thermal regulation, and acoustic performance (Figure 6.8). The solar energy stored in the panel and biomass-grown microalgae supply renewable energy for building operation such as space heating and hot water. The heat energy can be used for a geothermal heat exchanger with an 80-meter-deep borehole underground. The apartment residents have a visual connection to the microalgae enclosures where visible density change through microalgae growth provides the natural process of carbon sequestration and biomass production. The centralized air compressor delivers carbon dioxide and separate water pipes aid harvesting grown algae and supplying media and nutrients. The installation (approximately 2,500 ft^2 algae façade area) sequesters approximately 16 kg of CO_2 per day or 6.3 g/ft^2 with a biomass production of 9.5 kBtu/h-ft^2 year (30 kWh/m^2 year) and heat production of 47.5 kBtu/h-ft^2 year (150 kWh/ m^2 year).[9]

The CSTB prototype (2014) is a 200 m^2 PBR curtain wall located at a CSTB (Scientific and Technical Centre for Building) site in Champs-sur-Marne, a town

▲ Figure 6.7

A speculative R&D office building enclosed with microalgae photobioreactors focusing on the development of microalgae technology for biofuel production coupled with wastewater treatment (image reproduced from Axel Schonert).

▲ Figure 6.8

A real-world application of flat bioreactors installed on a residential building for saving building energy use, carbon sequestration, and biomass production (image reproduced from Splitterwerk Architects).

▲ Figure 6.9

A technology demonstration project of bioreactor curtainwalls installed on an office building in France for carbon sequestration and air quality improvement (image reproduced from XTU Architects).

slightly east of Paris, France (Figure 6.9). Built on microalgae façade experience since 2009, this project became the first technology demonstration installed in a real-world application, experimenting with different configurations and density effects on daylight penetration. The project capitalizes on a high growth rate and superior carbon sequestration in which 1 m^3 of microalgae absorbs the same amount of carbon dioxide as 80–100 trees. The operation and monitoring system assists the year-around algae growth, and such technological demonstrations help raise awareness of its possibility for benefiting human and built environments.[10]

In Vivo[11] (2016) is a design competition winning project consisting of three buildings housing dormitories (13,000 m^2), public space (1,200 m^2), cafes (255 m^2), research lab (1,000 m^2), and vegetable gardens (2,000 m^2). The project aims to promote social interaction and openness toward a more sustainable resilient city. Each building has a unique façade integrated with different biological systems and functions. One raises earthworms to compost organic waste generated from the building, and another grows microalgae for medical research. The solar energy collected by the bioreactor will be used for space heating and domestic hot water, and it is estimated to reduce energy consumption below 15 kBtu/ft^2 (48 kWh/m^2-yr). Tall building surfaces are the prime location to integrate renewable energy

▲ Figure 6.10

A speculative commercial complex growing microalgae for medical research with potential solar energy reclamation for space hating and domestic hot water (image reproduced from XTU Architects).

technologies that add aesthetic creativity and practical functions such as solar regulation, daylight transmission, and clean power production. The integration of microalgae system should enhance the visual appeal of a building for advertising their environmental, economic, and social benefits, resulting in increasing adoption for public use (Figure 6.10).

French Dream Tower[12] (2018) is a mixed-use development concept with hospitality tower, art tower, and office tower embracing user well-being, ecosystem, culture, and economic prosperity. Combating environmental problems associated with glass towers, microalgae bioreactors enclose the towers and regulate solar energy while providing good thermal insulation. Glass tall towers have gained popularity due to their lightweight and contemporary appeal. However, the glass skin transmits energy quickly depending on climate, resulting in high energy consumption. The high specific heat and dynamic biomass of the microalgae façade can offer superior insulative values. The building also encourages natural ventilation through a layer of microalgae which in the meantime offsets CO_2 and generates oxygen. The microalgae façade is a patent-pending system in development by XTU Architects for several years. The building enclosure is also covered by vegetation and greenhouses at the top of the building, providing additional natural air filtration for occupants. The rippling podium connects the towers at the lower level and collects rainwater in cisterns below. This reflects the hydro landscape of Hangzhou, China where water features are always present in the cityscapes (Figure 6.11).

▲ Figure 6.11

A speculative mixed-use towers enclosed with flat bioreactors for regulating solar energy and thermal insulation, collecting rainwater, and filtering outdoor air (image reproduced from XTU Architects).

Microalgae IVY (2021) is a patent-pending technology for retrofitting low-performing windows. A full-scale prototype (8 feet tall by 12 feet wide) was developed and installed at the School of Architecture at the University of North Carolina at Charlotte (Figure 6.12). The system consists of a network of interlocking bioreactors, enabling the cultivation of different strains for multi-functional use and aesthetics. Commercial windows are responsible for more than half of the building energy use and pollutant emissions. The prototype was developed as an alternative to energy-efficient retrofitting for low-performing windows. Five strains (*Chlorella, Chlorococcum, Haematococcus, Scenedesmus, and Spirulina*) were cultivated in the system using a semi-continuous production mode. Their biological performance and environmental benefits (e.g., biomass production, CO_2 reduction potentials) were monitored and measured using various environmental sensors and biological measuring tools. By utilizing 26 valves, the microalgae can easily be transported and extracted once the micro-algae are ready for harvesting. The system was able to sequester CO_2 produced by three occupants and output 500 g of biomass per day (~200 kg of biomass per year).

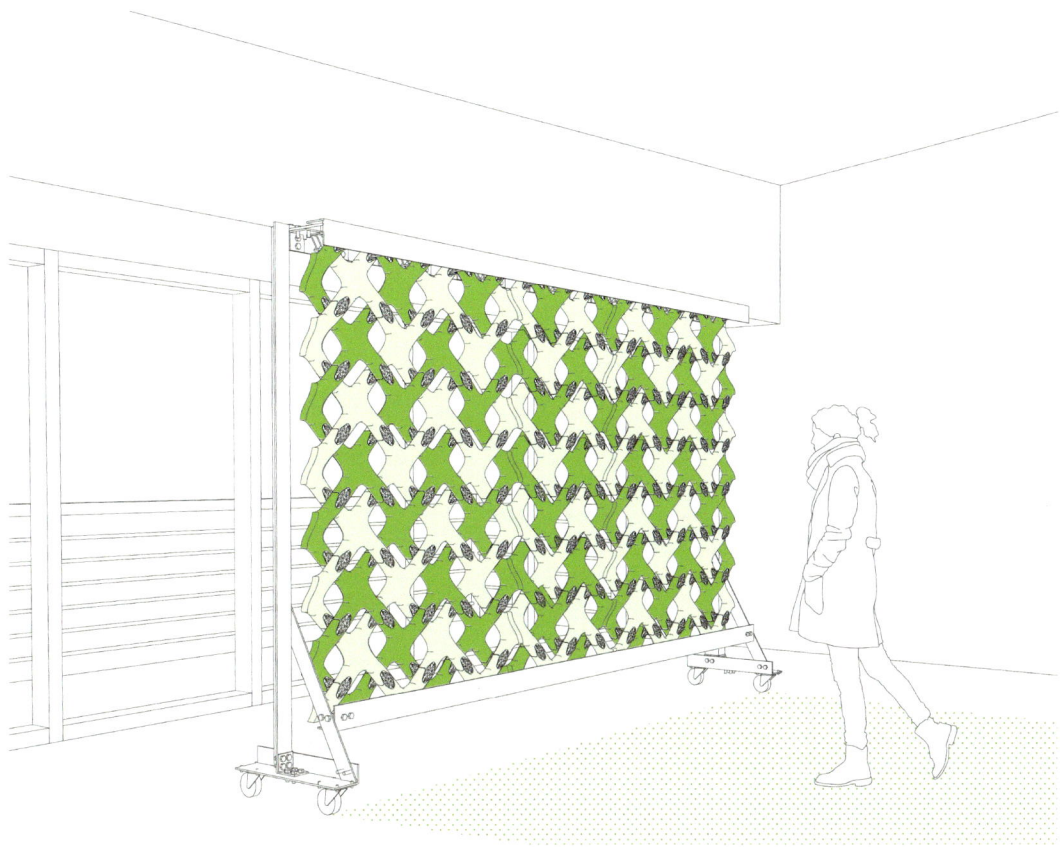

▲ Figure 6.12

A full-scale prototype installed at a school of architecture building at UNC Charlotte for technology demonstration filled with five strains for energy efficiency, biofuel production, and indoor air quality enhancement (image reproduced from Ecoclosure + UNC Charlotte).

Summary

Microalgae offer multi-performance features and dynamic aesthetics. With performance-based design and parametric design approaches, microalgae buildings further enhance sustainability and remediate environmental impacts. Microalgae integrated with building enclosure systems offer energy savings and good indoor air quality through thermal insulation, solar control, shading, carbon sequestration, and wastewater treatment. By allowing the microalgae to serve as the primary building envelope, the exterior skin is ever-changing, displaying changes in natural light, air flow, density, and color as the microalgae grow. The microalgae system could also be part of building service systems such as HVAC (heating, ventilation, and air conditioning) where solar energy stored in the system could be reclaimed for space heating and domestic hot water heating. For indoor applications, a microalgae interior system can divide space within the building, aiding in good microclimates of the interior. The presence of the microorganism in our living and working environment can also provide a biophilic quality that can make occupants more productive, healthy, and enhance their well-being. While CO_2 from the city or interior room air is fed into microalgae systems, rainwater can also be recycled for cultivation, and collected through a series of building surfaces, providing a closed-loop cultivation system.

Notes

1   Liandong Zhu and Tarja Ketola, "Microalgae Production as a Biofuel Feedstock: Risks and Challenges," *International Journal of Sustainable Development & World Ecology* 19, no. 3 (2012): 268–274, 269.
2   ArchDaily, "Algae Green Loop / Influx Studio," December 12, 2011, https://www.archdaily.com/191229/algae-green-loop-influx-studio.
3   Archinect, "Process Zero: Ideas Competition for Metropolis Magazine | Sean E Williams," accessed July 14, 2021, https://archinect.com/sewilliams/project/process-zero-ideas-competition-for-metropolis-magazine.
4   "AlgaeBRA—EcoLogicStudio," accessed July 14, 2021, http://www.ecologicstudio.com/v2/project.php?idcat=3&idsubcat=59&idproj=53
5   Dezeen, "FSMA Tower Algae Skyscraper by Dave Edwards," August 1, 2012, https://www.dezeen.com/2012/08/01/fsma-tower-by-dave-edwards/.
6   dinamik, "Alga Therapeia. Centro Termal. Donosti. País Vasco," Estudio Dinamik (blog), accessed July 14, 2021, https://www.estudiodinamik.com/en/portfolio/alga-therapeia-centro-termal-donosti-pais-vasco/.
7   architectes, Axel Schoenert, "Axel Schoenert—Urbanlab," accessed July 14, 2021, http://www.as-architecture.com/en/projects/urbanlab-1314.html.
8   "BIQ," accessed July 14, 2021, https://www.internationale-bauausstellung-hamburg.de/en/projects/the-building-exhibition-within-the-building-exhibition/smart-material-houses/biq/projekt/biq.html.
9   The European Portal for Energy Efficiency in Buildings. The BIQ House: First Algae-powered Building in the World | Build Up. Eur Portal Energy Effic Build 2015, accessed March 27, 2017, http://www.buildup.eu/en/practices/cases/biq-house-first-algae-powered-building-world
10 S.Ş. Öncel, A. Köse, and D.Ş. Öncel, "Façade Integrated Photobioreactors for Building Energy Efficiency." In *Start-Up Creation* (Woodhead Publishing, 2016), 237–299.
11 XTU, "XTU," accessed July 14, 2021, https://www.xtuarchitects.com.
12 XTU, "#FrenchDreamTowers | Hangzhou | China," accessed July 14, 2021, https://www.xtuarchitects.com/french-dream-towers-hangzhou-china-1.

Chapter 7

Product Intervention

We spend the majority of our time indoors. Designers and researchers are finding ways to extend microalgae systems as part of our daily working and living environments. Drawing inspiration from existing commodities, some interventions are additional products to our environments or are nested within the existing commodity. Plastics are an environmental crisis that has a detrimental impact on the Earth's ecosystems. Every year, millions of tons of plastic enter the ocean, and it takes a long time to decompose. Microalgae-based biodegradable plastic adopts a circular economy principle, unlike the petroleum-based plastic's cyclical system which is primarily landfilled or incinerated. Furthermore, designers use microalgae as a base material to turn them into different products like 3D printed foods, textile yarns for clothes, or color ink for writing pens. In addition, macroalgae seaweed can become products like lampshades and architectural cladding. Overall, product intervention hopes to create a fun interactive experience which, in turn, can educate users about microorganisms and their benefits. Due to a prolonged period of staying indoors, people can expand the incorporation of biological systems into their lifestyles. Micro farming in a home environment can make a positive impact equivalent to urban farming.

About 300 million tons of plastic waste is produced every year.[1] Over 8.3 billion tons of plastic have been produced since 1950s, and it is estimated that by 2050, around 12 billion tons of plastic litter will be landfilled or disposed to the natural environment.[2] Plastics can be categorized as fossil-based or partly bio-based or bio-based. They are also divided into nonbiodegradable and biodegradable plastics. Using the possible combinations, for example, bio-based plastics can be biodegradable or nonbiodegradable. Bioplastics takes only around 1% of the annual plastic production, but its market share is projected to increase.[3] Fishing nets and lines are a primary contributor for ocean plastic pollution. Marine ecosystems would greatly be protected from such biodegradability. Research conducted that cyanobacteria (blue-green algae) can be a renewable source to produce potential biodegradable plastic (PHA).[4] However, additional research is required to make more economic scalization through biotechnical engineering and optimum manufacturing.[5]

The integration of biological system as a building material has the potential to create new appearances of buildings while implementing a functional innovation through the intricate relationship between microalgae and the built environment. Microalgae can be integrated as a building enclosure (e.g., roof, façade, window, and canopy), interior partitions, and furniture where microalgae grow

DOI: 10.4324/9780367814410-9

	Petrochemical	Partly bio-based	Bio-based
Non-biodegradable	PE, PP, PET, PS, PVC	Bio-PET, PTT	Bio-PE
Biodegradable	PBAT, PBS(A), PLc	Startch blends	PLA, PHA, Cellophane

▲ Figure 7.1

Comparison of fossil-based versus bio-based plastics and their biodegradability (reproduced from Van den Oever et al.[6]) and potential biodegradable plastic (PHA) from microalgae.

under diffused light and artificial lighting. The building and occupants cohabitate with living organisms; while human respiration supplies CO_2 for microalgae growth, microalgae generate oxygen and absorb CO_2. When they are integrated within the building enclosure, microalgae provide dynamic solar shading. A dynamic appearance in response to solar intensity could make occupants curious and wonder about the multiple functionality of microorganisms (Figure 7.1).

Algaerium Bioprinter (2010) is a revolutionary 3D printing technology that prints microalgae food in a layer-by-layer manner (Figure 7.2). The system consists of two parts: the algaerium and the bioprinter. The algaerium is a bioreactor containing algae sources as 3D printing ink used by the bioprinter. For food aesthetics and nutrition, *Chlorella*, *Spirulina*, and *Hamaetococcus* were used for 3D printing. *Chlorella* has high chlorophyll that produces a bright green, while *Hamaetococcus* is red from the antioxidant astaxanthin. *Spirulina* is filamentous cyanobacteria which contain high protein for novel functional foods. These physical appearances contribute to the creative and aesthetic composition of the food. The chemical compounds from microalgae are a good source of nutrients. Furthering the 3D printing application, the designer aims to 3D-print a microalgae-based energy system and water purification technology.[7] The project demonstrates microalgae as a promising ingredient for innovative and healthier foods with a novel production technique and better visual appeal. The printability of food has gained in popularity as people tend to spend less time in preparing food.[8]

Algae Curtain (2012) is the living laboratory project where biology and technology are combined to produce future sustainable energy systems. The installation consists of transparent bioreactor tubing knotted into window drapes. The bioreactor drape is then networked with hanging silicone pouches along the ceiling, cascading down toward the window. The pouches, along with the woven bioreactor, cultivate *Nannochloropsis* with rich lipid content for biofuel production. The grown algae are collected in containers at the bottom of the woven bioreactor along the window sill. Microalgae's growth rate can be ten times faster than trees, and some strains double their biomass every six hours. Capitalizing on superior photosynthesis and high lipid content, microalgae can generate biofuel of 75,000 liters per acre compared to corn-based biofuel of 68 liters. The project envisions microalgae biofuel as a renewable resource to power the city grid.[9] Scale-up and commercialization for biofuel production require continuous research tackling practical issues and biotechnical challenges (Figure 7.3).

Algae Bulb[10] (2012) is to integrate microalgae within a lighting system that illuminates ambiently and grows microalgae, and in return, filter air and produce biomass (Figure 7.4). The algae bulb consists of LED bulb, air pump above the LED, hydroponic container, plastic shells, air inlet at the top, and air outlet at the bottom. *Chlorella pyrenoidosa* and *Spirulina* are cultured in the container by intaking room air from the top and giving off oxygen into the room. The algae can turn

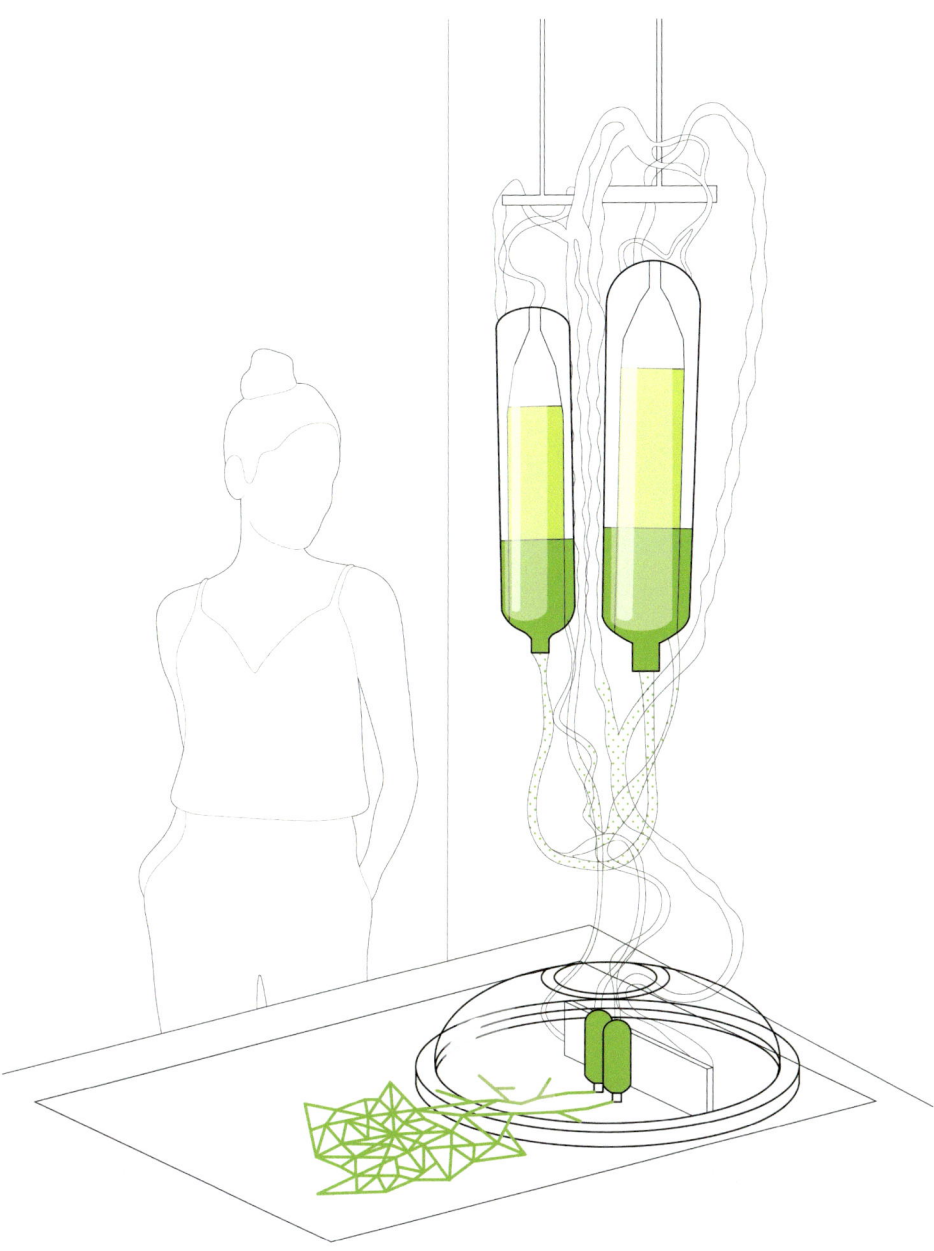

▲ Figure 7.2

3D bioprinter prototype's layer-by-layer printing of healthy and creative microalgae foods (image reproduced from Marin Sawa).

into biofuel that can power the LED lightbulb. The lightbulb creates an interesting spatiality when the light is off and dark green appears in the dark, whereas when the light is on, bright green literally illuminates the interior. People spent more than 90% of their time inside buildings, and indoor air is often more contaminated. The algae bulb is an innovative noninvasive way to sequester CO_2 and improve user health and well-being. The biomass growth rate varies depending on light intensity and can turn into a renewable energy supply for the home.

▲ Figure 7.3

Bioreactor drape prototype installed in a laboratory setting to produce microalgae-based biofuel to power the city grid (image reproduced from Gmachl and Wingfield).

Algae Night Light (2015) is a synthesis of art and science and an interactive toy for children's education. The dinoflagellates, a bioluminescent algae, need interaction with users to subtly illuminate when shaken, not something you simply plug in when needed. Children also need to care for the photosynthesis requirements during the daytime using appropriate light exposure and keep them away from light at night. Their optimum growing temperature is around 65–75°F within ocean temperature ranges. The Dino Pet consists of three parts. The dino-shaped container is a polyethylene-based 3D print and is able to hold 500 ml (917 oz) of the mixture of dinoflagellates, nutrients, and salt water. Shear stress from fluid affects the intensity and decay rate of bioluminescence. An anti-leak port is placed at the bottom of the toy which allows for the addition of media and microalgae. The dino shape is based on the *Apatosaurus* as the dinosaur resembles the dino-flagellate cellular structure of the microalgae inside (Figure 7.5).

Algae Water Bottle[11] (2016) is a microalgae-based biodegradable plastic in response to the massive use of plastics in our daily lives. Fossil-based plastic takes a long time to degrade, and a large volume of wastes ends up getting incinerated, contaminating soil or endangering aquatic ecosystems floating in the oceans. Heated algae paste is added to a mold and set cold in a refrigerator. This cooling process hardens algae paste and helps retain its shape. The project was to prototype an algae plastic that can not only hold water but can also be eaten after

▲ Figure 7.4

Algae bulb prototype that illuminates the space with a LED light filtered through microalgae cells while growing microalgae for improved indoor air quality and biomass production (image reproduced from Gyula Bodonyi).

drinking the water. Although its mechanical properties still need improvement, the project sparked creativity in tackling plastic problems in society (Figure 7.6). Biopolymers are mostly produced by starch, and poly lactic acid (PLA)-based polymers take around 1% of the annual global plastic production.[12] Microalgae-based biopolymer shows a promising future solution. Cost-effective production process is critical in reducing biopolymer prices, which needs long-term strategies such as streamlined cultivation and harvesting techniques and biotechnical engineering.[13]

Living Things—Furniture (2016) is a set of household furniture that illustrates a symbiotic life with *Spirulina* in a home environment where indoor air supplies inorganic carbon for microalgae growth, and, in return, food and fuel are produced in a domestic household. Microalgae-based lighting and decorative products create biophilic home environments while creating a unique ambience (Figure 7.7). Living furniture sets are displayed for the kitchen/controller, dining room, and living room, staging different microalgae furniture applications. Nine glass bioreactors are made of blown glasses that provide good environments for *Spirulina*. *Spirulina* has high protein with the remaining lipid, carbohydrates, and

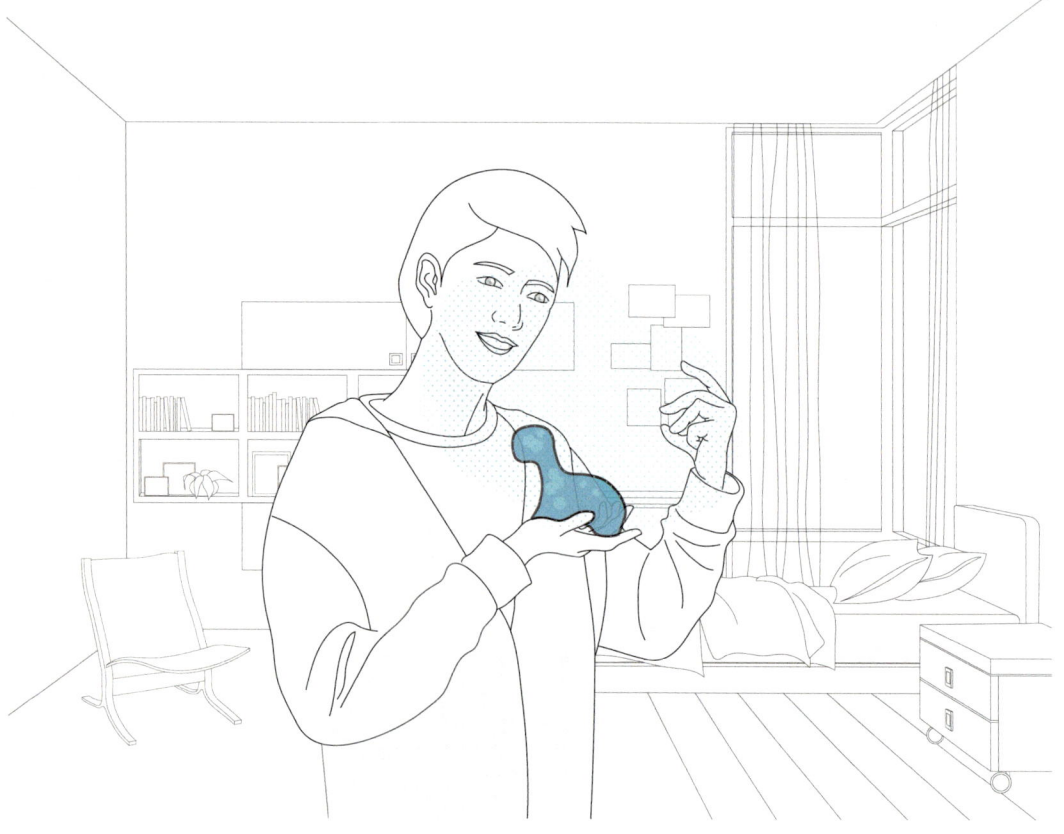

Interactive bioluminescent microalgae that photosynthesize during the daytime and illuminate at night when shaken (image reproduced from Yonder Biology).

other natural antioxidants. The light and heat generated by the bioreactor illuminate and heat the space. The glass bioreactors are connected with aeration and plumbing systems, and microalgae can be easily extracted as density increases. Major high-value products from microalgae can include proteins, lipids, carbohydrates, carotenoids, fatty acids, and polysaccharides which are used in different sectors such as pharmaceuticals, nutraceuticals, food and beverage, agriculture, and aquaculture.

Bionic Chandelier[14] (2017) is a daily commodity in combining biotechnology and engineering to improve quality of life and enhance indoor air quality (Figure 7.8). The chandelier provides illumination and intakes CO_2 generated by occupants. In return, microalgae generate oxygen and grow biomass. Indoor air quality typically has higher CO_2 concentrations than outdoors, and using indoor air for photosynthesis reduces carbon concentrations and improves indoor air quality. The relationship demonstrates mutual functionality in which by-products/wastes from human, object, and biological system become valuable life-giving resources for each other. The leaf-like bioreactor is arrayed radially around the metal structure. The microalgae leaf is connected to the microalgae-growing device that provides nutrients and maintains living organisms. The biomass growth responds to the lighting intensity

▲ Figure 7.6

Microalgae based biodegradable plastic bottle prototypes (image reproduced from Ari Jonsson).

as well as indoor CO_2 concentrations. The leaf-like bioreactor is designed in a way that can minimize biofouling, and the modular construction facilitates easy system assembly. Given the size and the number of bioreactors, the project may use batch cultivation which cultivates microalgae in a confined batch. This is relatively cost-effective as well as low maintenance and control of biological contaminants.

The Coral: Home Algae Farming (2019) exemplifies future personal algae farming that is welcoming, beautiful, and functional for home environments. The project consists of a wall mounted 4 X 4 mosaic grid constructed of 16 individual rectilinear bioreactors resembling agricultural land use. The imprint of the bioreactor took its design motif from coral in which organic coral patterns are engraved on the surface of the bioreactors. Literally this also illustrates a symbiotic connection between coral and microalgae and the environmental role of microalgae in protecting living coral environments. In the symbiotic relationship, coral provides nutrients and shelter for microalgae, and microalgae provide oxygen and uptake waste from the coral. The micro farming produces environmental benefits and occupant health through carbon reduction and nutraceutical products. The harvested microalgae can be used for nutritional ingredients and enhancements for health and well-being. The prototype replenishes biweekly and produces an average of 2 g of

▲ Figure 7.7

Microalgae-based lighting and decorative products for home environments, illustrating a symbiotic life with biological systems (image reproduced from Doenius Ethan Frier).

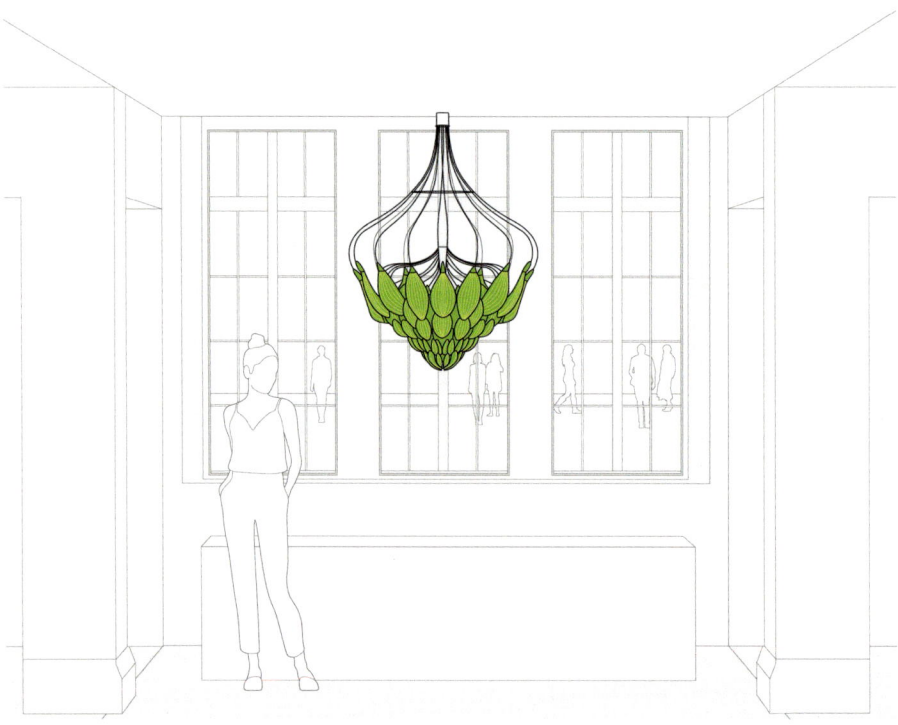

▲ Figure 7.8

Microalgae bionic chandelier installed in a museum to address decarbonation potentials and food production (image reproduced from Julian Melchiorri).

▲ Figure 7.9

Private microalgae farming installed in a home environment for occupants' health, environmental benefit, and food production (image reproduced from Hyunseok An).

Spirulina per day.[15] Growth conditions as well as nutrient starvation and limitation can alter the biochemical composition of *Spirulina* (Figure 7.9).

Indus Algae Tiles (2019) is an architectural cladding system made by pressing clay inlaid with microalgae and hydrogel (Figure 7.10). The tile geometry resembles tree leaves and their veins where the ravines are filled with wet microalgae. The project utilizes chemical compounds of phytochelatins to bioremediate heavy metals in the water and store them within cells. Anthropogenic activities contaminate water with heavy metals and cause inactivation and disturbance to plant growth. Microalgae can uptake heavy metals up to 10% of their biomass into cells for their metabolism.[16] It is an affordable and noninvasive process of uptaking contaminants with aesthetically pleasing ecosystems. With bio-assimilation and bio-adsorption techniques, detoxifying capability may vary depending on algae type, environmental conditions, and contaminant types. The vertically laid algae cladding integrates microalgae with hydrogel that provides water and

▲ Figure 7.10

Cladding titles infilled with microalgae and hydrogel for purifying water and detoxifying surrounding environments by uptaking heavy metals (image reproduced from Jennifer Hahn).

nutrients for microalgae growth. Hydrogel can be easily reproduced with a simple mix of water and nutrients. The modular units are highly adaptive in that they can be easily assembled into various sizes of arrays to enclose building skins while purifying contaminated water and detoxifying surrounding environments.[17]

Algae packaging[18] **(2019)** is a biodegradable algae-based plastic (Figure 7.11). Plastic waste is a global pollution, and recycling plastic alone cannot solve increasing waste problems. Biological feedstock can serve as a sustainable solution with similar performance characteristics as fossil-based plastic products.[19] Agar was first discovered in Japan in 1658 and it is released when boiling red algae. Agar—jelly-like polysaccharides—serves as a polymer substitute. The gel-like substance can be achieved after the heated agar is stored at a cold temperature. After being stored in a dry and ventilated environment, the substance becomes a thin plastic. By changing the proportions of plastic, water, and other additives, the final algae plastics have the potential to generate different types of bioplastics. The proportion of the mixture affects rigidity and durability. The major global issue of plastic packaging is it is non-biodegradable, but algae packaging is able to fully decompose within a few months. Bioplastic's biodegradability favors warm temperatures and so is less effective in winter.

▲ Figure 7.11

Biodegradable microalgae-based plastic (image reproduced from Margarita Talep).

Summary

Designers and researchers are finding ways to extend microalgae systems as part of our daily working and living environments. Drawing inspiration from existing commodities, some interventions are additional products to our environments or are nested within the existing commodity. Microalgae products set out to create furniture, commodities, or micro farming that, in turn, establishes a symbiotic relationship between people and microalgae in private environments. Having both adaptive and interactive capabilities, the microalgae product provides the heat, light, air, and nutrients needed to ensure a habitable environment for growing microalgae, while the light, heat, and air (oxygen) generated by microalgae can also be used to elevate the comfort and well-being of human inhabitants. Microalgae plastics could alleviate the plastic crisis because they decompose at a much faster rate. Private micro farming could be utilized in home environments and once extracted, microalgae can be prepared as food for consumption. This concept can be extended to urban agricultural systems where cities continue to develop. All microalgae interventions differ, but they all share a commonality through the way they visually celebrate the inclusion of microalgae. What arises is a unique aesthetic where the rich green biomass glows, drawing human inhabitants and microalgae closer together.

Notes

1   UNEP, Single-use Plastics: A Roadmap for Sustainability, 2018 (Rev. ed., pp. vi; 6).

2   Ibid., 8.

3   Martien Van den Oever, Karin Molenveld, Maarten van der Zee, and Harriëtte Bos, *Bio-based and Biodegradable Plastics: Facts and Figures: Focus on Food Packaging in the Netherlands*, No. 1722, Wageningen Food & Biobased Research, 2017.

4   Akhilesh Kumar Singh, Laxuman Sharma, Nirupama Mallick, and Jyoti Mala, "Progress and Challenges in Producing Polyhydroxyalkanoate Biopolymers from Cyanobacteria," *Journal of Applied Phycology* 29, no. 3 (2017): 1213–1232.

5   Ibid., 1226.

6   Ibid., 17.

7   Marin Sawa, "Algaerium Bioprinter," Alive: New Design Frontiers. Alive, 2013, http://thisisalive.com/algaerium-bioprinter/.

8   Zaida Natalia Uribe-Wandurraga, Marta Igual, Javier Reino-Moyón, Purificación García-Segovia, and Javier Martínez-Monzó, "Effect of Microalgae (Arthrospira platensis and Chlorella vulgaris) Addition on 3D Printed Cookies," *Food Biophysics* 16, no. 1 (2021): 27–39.

9   Mathia Gmachl and Rachel Wingfield, "Algae Curtain," Loop.pH, October 2012, http://loop.ph/portfolio/algae-curtain/.

10 Lori Zimmer, "Gyula Bodonyi's Algae Powered LED Is Truly a 'Green' Light Bulb," Inhabitat Green Design Innovation Architecture Green Building, Inhabitat, May 31, 2013, https://inhabitat.com/gyula-bodonyis-algae-powered-led-is-truly-a-green-light-bulb/.

11 Alice Morby, "Ari Jónsson Uses Algae to Create Biodegradable Water Bottles," Dezeen, July 19, 2019, https://www.dezeen.com/2016/03/20/ari-jonsson-algae-biodegradable-water-bottles-iceland/.

12 Senem Onen Cinar, Zhi Kai Chong, Mehmet Ali Kucuker, Nils Wieczorek, Ugur Cengiz, and Kerstin Kuchta, "Bioplastic Production from Microalgae: A Review," *International Journal of Environmental Research and Public Health* 17, no. 11 (2020): 3842.

13 Akhilesh Kumar Singh, Laxuman Sharma, Nirupama Mallick, and Jyoti Mala, "Progress and Challenges in Producing Polyhydroxyalkanoate Biopolymers from Cyanobacteria," *Journal of Applied Phycology* 29, no. 3 (2017): 1213–1232.

14 Julian Melchiorri, "Bionic Chandelier," accessed July 14, 2021, https://www.julianmelchiorri.com/Bionic-Chandelier.

15 "Nature Lab—Student Work: Hyunseok An's Algae Farming Design," accessed July 14, 2021, https://naturelab.risd.edu/discover/student-work-hyunseok-ans-algae-farming-design/.

16 Sathish Rajamani, Surasak Siripornadulsil, Vanessa Falcao, Moacir Torres, Pio Colepicolo, and Richard Sayre, "Phycoremediation of Heavy Metals Using Transgenic Microalgae," *Transgenic Microalgae as Green Cell Factories* (2007): 99–109.

17 Jennifer Hahn, "Bio-ID Lab Designs DIY Algae-Infused Tiles That Can Extract Toxic Dyes from Water," Dezeen, September 23, 2019, https://www.dezeen.com/2019/09/21/bio-id-lab-indus-algae-tiles-water/.

18 Dezeen, "Margarita Talep Develops Algae-based Alternative to Single-use Plastic Packaging," January 18, 2019, https://www.dezeen.com/2019/01/18/margarita-talep-algae-bioplastic-packaging-design/.

19 Senem Onen Cinar, Zhi Kai Chong, Mehmet Ali Kucuker, Nils Wieczorek, Ugur Cengiz, and Kerstin Kuchta, "Bioplastic Production from Microalgae: A Review," *International Journal of Environmental Research and Public Health* 17, no. 11 (2020): 3842.

Part III | Microalgae Building Enclosure Design

Chapter 8

Biotechnical Design Criteria

8.1 Environmental Conditions

Microalgae have drawn substantial attention in the scientific field and industry due to diverse opportunities in environmental restoration and biofuel production. Despite the various benefits from microalgae, it is still challenging when it comes to scaling up commercialization because the harvesting process requires energy-intensive, time-consuming post-processing. One way to make microalgae commercially viable is to offset the operational cost by supplying free nutrients from wastewater or flue gas. The other way is to increase biomass productivity and lipid content through bioengineering of microalgae. Together with optimum strains, the cultivation operation coupled with wastewater or flue gas processing needs to be optimized to provide good growing conditions (e.g., solar intensity, temperature, aeration, pH) as well as cultivation schedules and harvesting techniques.

U.S. Department of Energy's (DOE) Bioenergy Technologies Office, for example, aims to meet a cost target of producing biofuels at \$2.25/GGE (gasoline gallon equivalent) by 2030.[1] In order to meet the target, microalgae growth aims to reach 25 g/m^2/day (~2.5 g/ft^2/day) with a projection of 50 g/m^2/day (~5 g/ft^2/day) for industry scalization.[2] Although the high productivity rates in the laboratory scale have been reported, the current productivity rates in industrial scale are far below the theoretical potential. Bioengineering involves strain improvement and targets biochemical composition with stress-tolerant microalgae. Productivity and biochemical compositions are influenced by major factors such as daily and seasonal fluctuations in light intensity, aeration rate, solar intensity, temperature, pH, nutrients, salinity, and so on. Keeping strains safe from external contamination is another important requirement during cultivation to improve productivity.

Environmental indicator: Microalgae in nature generally indicates high concentrations of nutrients in water bodies, called algal blooms. Anthropogenic activities such as agriculture (e.g., fertilizer), surface run-off from impervious urban settings, storm water management, wastewater treatment, and so on play an important role in preventing algal bloom in a water system. The visual presence of algal blooms can cause drinking water problems, unattractive aesthetics, and economic loss through hindering the recreational use of the water body.[3] Algae blooms also produce toxins that are harmful to ecosystems and human health.[4]

Light: Light is one of the key environmental factors for photoautotrophic microalgae growth. Microalgae convert light, CO_2, and water into biomass and O_2 using photosynthesis. Depending on how microalgae supply energy for growth, they are divided into photoautotrophic, heterotrophic, and mixotrophic. Photoautotrophic microalgae

DOI: 10.4324/9780367814410-11

Environmental factors affecting maximum microalgae growth

Abiotic factors	Range	Optimum
Temperature	50–90°F (10–32°C)	100–200 µmol/m²s
Light intensity, PPFD	50–400 µmol/m²s	70–80°F (21–27°C)
Photoperiod	12h/12h, 14h/10h, 16h/8h, continuous light	16h light/8h night
Aeration	1–7 mm diameter bubble size	0.5 m/s velocity
pH	7–9	8.2–8.7
Salinity[5]	35–223 ppt	55–85 ppt

Source: Reproduced from Lavens et al. and various sources.[6]

have superior photosynthesis and use photorespiration under high concentrations of dissolved oxygen, low CO_2, and/or low light conditions.[7] Heterotrophic microalgae grow on organic carbon using respiratory and fermentative processes without relying on light. Mixotrophic algae metabolize both CO_2 and organic carbon during both respiration and photosynthesis. Photoautotrophs yield higher CO_2 sequestration, whereas heterophs and mixotrophs yield higher biomass.[8] Photoperiod and intensity affect biomass productivity and biochemical composition.[9] When light is unavailable, microalgae uptake O_2 during photorespiration. For the balance of photosynthesis, both light intensity and photoperiod is important. The maximum CO_2 assimilation would occur at the highest photosynthesis rate and lowest photorespiration with as little photoinhibition as possible.[10] Photoinhibition occurs when light intensity exceeds a certain threshold and microalgae stop uptaking CO_2. The light supply of the light/dark cycle has a vital role in growing microalgae. Different light/dark cycles such as 12h/12h, 14h/10h, 16h/8h, and continuous light exposure have been applied in algae culture. The 16h light/8h dark cycle is an optimum cycle for maximum algae growth.[11]

It is important that bioreactors receive uniform illumination with optimum sunlight intensity because the unused light across the bioreactor is released as heat inside the culture. It should take into consideration that self-shading caused by front layers of microalgae diminishes light intensity as it penetrates deep inside the culture. Cultivation systems up to 4 inches deep are recommended for good light penetration. In addition, it should be considered that different microalgae require different levels of light intensity. Plant growing light is called PPFD (photosynthetic photon flux density measured in µmol/m²/s) in the 400–700 nm visible light spectrum that plants use in the process of photosynthesis. High PPFD results in an increase of algae productivity until photoinhibition of light saturation is reached. Low PPFD causes photolimitation that reduces growth rate. Microalgae productivity was measured under different PPFD ranges under natural light or artificial light conditions, and microalgae growth reduction was observed at extreme high and low light intensities.[12] PPFD varies from the lower range of 10 µmol/m²/s to the upper range of 800 µmol/m²/s. Different algae species demand different levels of PPFD as an optimum condition. For example, 1 µmol/m²/s from daylighting is approximately 54 lux.[13] As shown in Table 8.1, *Chlorella*, for example, was tested under a light intensity range of 60–150 PPFD (~3,200–8,000 lux), *Haematococcus* under 0–170 PPFD (0–9,000 lux), *Scenedesmus* under 55–150 PPFD (3,000–8,000 lux), and *Spirulina* under 85–430 PPFD (4,500–23,000 lux).

Light color affects productivity. Red light color produces the highest cell numbers, while blue light produces the largest cell size.[14] Green microalgae do not use the green light spectrum at 550 nm.

When bioreactor design is considered, metrics to describe good light availability is a high surface area-to-volume ratio. Bioreactors with a high surface area-to-volume ratio have a greater amount of light availability and penetration on an enlarged surface. Light can come from both natural and artificial light sources. Sunlight is cost-effective but controlling light intensity in outdoor bioreactors is challenging. Artificial light sources include LEDs, fluorescence lamps, specialized light sources, and solar collectors in which additional input of energy is required. Fresnel lenses can be used to intensify light illumination, especially for vertical applications where light intensity diminishes from the tilted incident angle. Another potential application is for *Haematococcus* where a two-stage cultivation process is required. The initial growth was exposed to low light intensity, and the second stage was exposed to super high levels of 6,000 PPFD (324,000 lux), resulting in higher astaxanthin production and content compared to a counterpart exposed to ambient conditions.[15] The following figures illustrate PPFD ranges impinging on different building surface orientations and site locations.

Temperature: Temperature is another environmental factor affecting biomass productivity and photosynthesis. Different strains have their own optimum temperatures for growth. Because microalgae utilize a fraction of the solar energy for synthesis, the majority of solar energy is converted to heat inside the culture. It has been shown that the photosynthetic efficiency under low light levels (e.g., 100–300 $\mu mol/m^2/s$) is typically below 5% and efficiency under high light levels is under 2%, meaning that 95% of the solar energy is converted to heat.[16] Therefore, the culture inside the photobioreactor could experience high temperatures if no active heat exchanger is introduced. The optimum growing temperature range is 10–32°C (50–90°F) in bioreactors.[17] Typically, a decrease in productivity is observed with high temperatures although some thermophile microalgae can grow at temperatures up to 80°C (176°F), for example, in deserts or hot springs.[18]

Culture temperatures under natural solar radiation in summer in the south of France were measured at 130°F under a peak radiation of 1,000 W/m^2 and 121°F under 750 W/m^2 between 1 p.m. and 3 p.m., which exceeds the high temperature threshold.[19] The researcher estimated several hundreds of hours yearly for overheating and lower temperature regimes. Different levels of reflectivity and exposure duration of the microalgae window were investigated. Reduction in culture temperature is achieved as the surface reflectivity increases and exposure time decreases.[20] *Spirulina*, for example, show maximum growth rates of 4.5 g dry mass/l/day under 30°C and pH-9. Temperature affects not only growth rates but also biochemical compositions. *Spirulina platensis* strain grown at 40°C resulted in a decrease in protein content (22%) but an increase in lipids (43%) and carbohydrates (30%).[21]

Typical ways to mitigate heat stress in the bioreactor involve providing shading, spraying with water, and submerging in water. On the other hand, excessive low temperatures during winter can also reduce biomass productivity. Therefore, a year-round heat exchanger to keep the culture temperature within an optimum range is required. A balance in operation is required because heat stress in the culture results in decreased productivity while an active temperature-regulating process using a heat exchanger increases operational costs. Room air circulation in lieu of a heat exchanger in the culture results in good temperature ranges and reduces the dependence on an active heat exchanger. In addition, reduction in operating costs can be achieved by the use of heat- and/or cold-tolerant microalgae while keeping a high growth rate over such wide temperature swings. The following figures illustrate temperature ranges impinging on different building surface orientations and site locations.

Aeration: Aeration and mixing are other important elements for increasing algae productivity. They help light distribution, nutrient dissolution, regulation of fluid flow, and suspension of microalgae cells. Bioreactors have higher cell densities than natural environments, and air bubbling helps cells float instead of settling down at the bottom of the bioreactor. Air bubbles and mixing dynamics such as eddies and turbulence flow cause shear stress on microalgae cells and affect productivity and cell morphology. Green microalgae show the greatest tolerance against shear stress followed by cyanobacteria and red algae.[22] Aeration helps maintain even mixing of cells in light and dark zones and provides access to light, nutrients, and CO_2.[23] With photosynthesis, carbon is an essential element of microalgae composition and constitutes up to 50% of the dry biomass. Carbon dioxide as a sole inorganic carbon source is converted into organic compounds and O_2. Carbon dioxide is converted to high-value compounds such as carbohydrates, lipids, and protein. A mole of CO_2 has 12 g of carbon out of 44 g molar mass. Therefore, carbon uptake from microalgae can be calculated as follows:

$$\frac{44\left(\frac{g}{mol}CO2\right)}{12\left(\frac{g}{mol}Carbon\right)} \times \frac{0.5g\ Carbon}{1g\ algal\ biomass} = 1.8g\ CO2\ per\ 1g\ algal\ biomass$$

This equation indicates that, assuming carbon comprises 50% of the biomass, 1 g of dry mass sequesters 1.8 g of CO_2. Biomass productivity increases linearly up to 5% CO_2 concentration and a higher lipid content was obtained with 15% CO_2 aeration.[24] Microalgae productivity under room air (0.03% CO_2) reaches the same maximum growth rate as the microalgae grown under 5% CO_2 circulation, but it takes more in number of cultivation days to reach the same productivity.[25] Carbon dioxide bubble size and the residence time in the culture medium affect CO_2 fixation efficiency by microalgae cells. Slower bubble aeration and smaller diameters promote CO_2 diffusion. Bubble velocities higher than 0.05 m/s creates slug and annular flows that shorten the residency time in the media and discourage microalgae growth. High-speed aeration causes high shear stress to cells and causes unsuitable mixing for nutrients, lights, and CO_2.[26] Bubble sizes of 1–7 mm diameter are recommended for higher interfacial area and maximum residency of CO_2, whereas bubbles with sizes less than 1 mm in diameter scatter light and reduce growth rates (Figure 8.1).[27]

pH: The pH of the microalgae media is one indicator for available dissolved carbon affecting productivity. Elevated dissolved CO_2 levels turn the media acidic, which reduces photosynthesis. The optimum growth pH of the media varies with the strains and cultivation period. Most microalgae grow well in a pH range of 6–8.76,[28] but *Chlorella vulgaris* grows well at a pH of 9–10.[29] For *C. vulgaris*, the pH is nearly neutral at the beginning of cultivation, quickly increases to alkaline

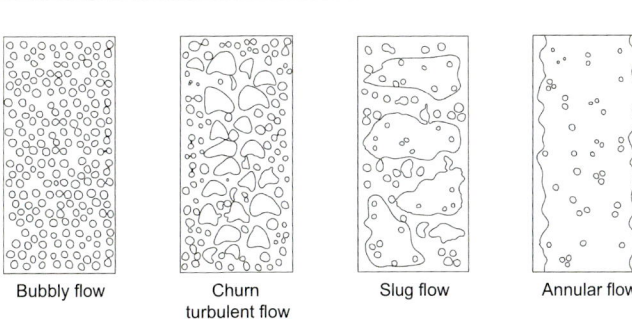

Increasing velocity

| Bubbly flow | Churn turbulent flow | Slug flow | Annular flow |

▲ Figure 8.1

Bubble flows in bubble columns: bubble size increases as superficial gas velocity increases. Velocities higher than 0.05 m/s cause slug and annular flows which are not suitable for cultivation (reproduced from Kommareddy et al., 2013).[30]

during the active phase of growth, and slightly decreases toward the end of the cultivation period.[31] Some other strains such as cyanobacteria are resistant to lower pH values, but increasing the pH and salinity is harmful for algae cells.[32] When microalgae reach the maximum growth stage (i.e., stationary stage), typically the pH rises. Depending on the acidic characteristics of the cultures, different strains can be selected for maximum productivity.

Nutrients: Nutrients are important for the speed of microalgae growth, and the accumulation of chemical compounds (e.g., carbohydrates, lipids).[33] The essential macro nutrients for microalgae are nitrogen, phosphorous, and inorganic carbon. The micronutrients are Mo, K, Co, Fe, Mg, Mn, B, and Zn affecting biomass, chemical composition, and enzymatic activities. Macronutrient phosphorous is important to microalgae growth, helping "cellular metabolism ranging from energy storage, to cellular structure, to the very genetic material that encodes all life on the planet."[34] Nitrogen deficiency reduces the algae growth but increases the production of carbohydrates and lipids.[35] Nutrients for commercial microalgae cultivation require a substantial supply of macronutrients. Wastewater, on the other hand, contains excessive phosphorus and nitrogen, which causes eutrophication when discharged into natural environments. The chemical and physical processes of treating wastewater are not economical, and the use of microalgae in nutrient removal from wastewater has shown a promising biological economic solution.[36] In addition, CO_2 is an important carbon source for microalgae growth and accumulation of fatty acids, whereas flue gases from power plants contain 10–15% (v/v) CO_2.[37] Use of nutrients, nitrogen oxide, sulfur dioxide, and CO_2 from wastewater and flue gas has shown increases in the growth rate and lipid content of *Nannochloropsis* sp. cells.[38, 39]

Algae cultivation: Industry-scale cultivation is essential to augment the multiple benefits of microalgae and tackle societal, environmental, and economic challenges. Since they do not rely on arable land, sites for microalgae cultivation are ubiquitous such as homes, offices, community gardens, rooftops, public spaces, and building surfaces. Microalgae need nutrients and CO_2 for growth and assimilation of chemical compounds. Wastewater and flue gas are proven to

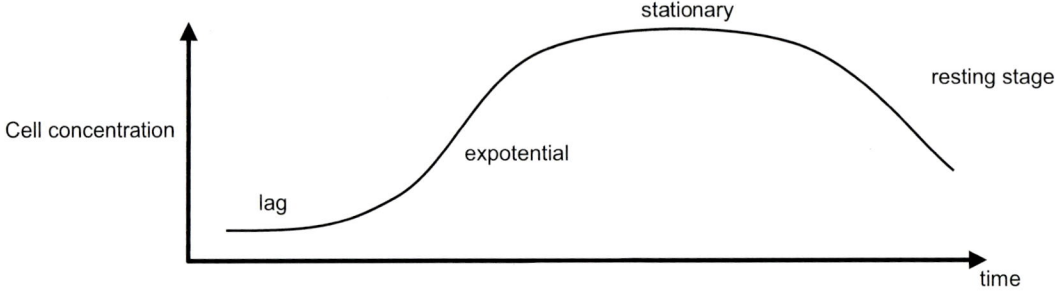

▲ Figure 8.2

Microalgae growth curve consists of lag, exponential, stationary, and dead phases.

nourish microalgae and turn mass cultivation into a more economical and environ-mentally friendly operation. Microalgae cells go through different growing lifecycle phases—lag, exponential, stationary, and dead (Figure 8.2). These growing phases are affected by various factors such as strains, lighting conditions, temperature, nutrients, pH, and so on.

8.2 Bioreactor Design and Operation

Cultivation system: Bioreactor operation plays an important role in cell cultivation. Raceway ponds are the most common and economic cultivation method today, accounting for over 95% of microalgae production worldwide, but they are prone to contamination, which increases the risk for culture survival.[40] For this reason, strain selection for open pond cultivation must consist of robust, fast-growing strains that can tolerate extreme conditions. Recommended algae for raceway ponds include *Tetraselmis suecica* (productivity up to 9 g/m^2/day), *Nannochloropsis* sp. (productivity up to 13 g/m^2/day), and *Spirulina* sp. (productivity up to 21 g/m^2/day).[41] Major advantages of raceway ponds are their low construction cost, low maintenance, and low energy requirements during operation.

Photobioreactors, on the other hand, cultivate microalgae in a protected environment and offer high productivity with quality control. A photobioreactor is primarily constructed with one of three configurations: vertical, horizontal, or flat. Vertical photobioreactors perform better than horizontal ones, and flat panel pho-tobioreactors have the highest average photosynthetic efficiency when compared to other bioreactors.[42] The horizontal reactor receives high solar energy, although much of this extra energy is dissipated as heat, resulting in a decrease in produc-tivity. In vertical photobioreactors, on the other hand, microalgae achieve higher photosynthetic efficiency and areal productivity because solar intensity is relatively moderate compared to the horizontal counterpart with less land area required for installation.[43]

Tubular photobioreactors are the most common closed-system design for microalgae cultivation at an industrial scale. Because tubular photobioreactors are protected from outdoor environments, they are less susceptible to contamina-tion than raceway ponds, enhancing biomass productivity and quality. However, the system is susceptible to biofouling in which microalgae cells are adhered to

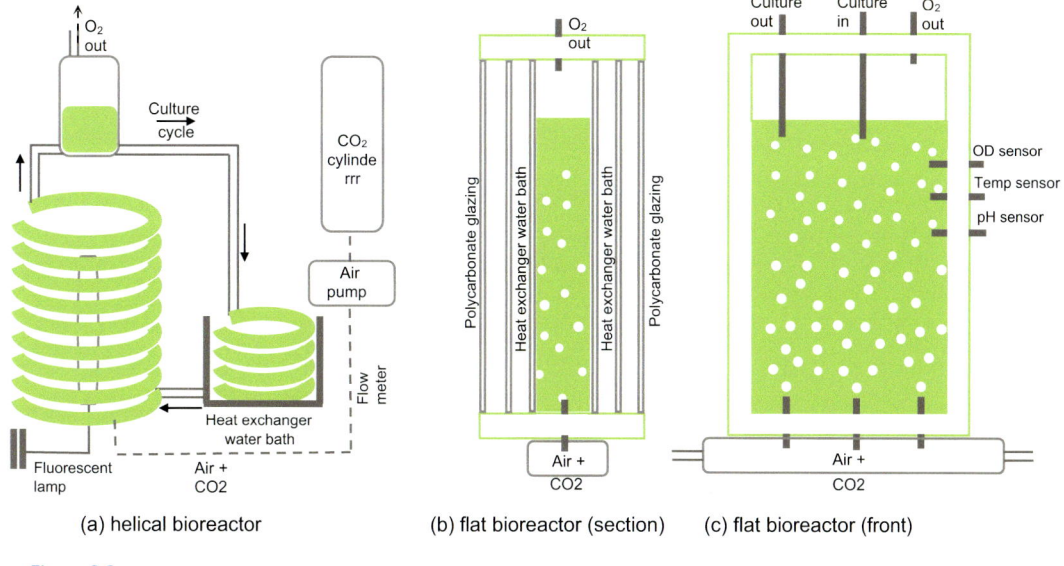

▲ Figure 8.3

Bioreactor design with upright helical photobioreactor (a) and flat vertical bioreactor (b, c); it consists of air pump, storage tank and pump, sensors (OD, temp, pH) (reproduced from Wang et al.[44] and Singh et al.[45]).

the wall of the system and light penetration is reduced to the rest of the culture, requiring periodic maintenance.[46] Tubular photobioreactors consist of production loops and a mixing/retention tank. Production loops are bioreactors made of transparent glass or plastic tubes. The tube diameter is around 10 cm (~4 inches), and the surface-to-volume ratio is 80:1 with a 20–400 m length. The culture inside the production loop is circulated by pumps and air streams. Due to light penetration and mixing of nutrients and gases, a tube diameter of 5–9 cm and a tube length of 100–150 m is considered optimal.[47] Power requirements for tubular photobioreactors range from 10 to 100 W/m^2.[48] When the production loops are exposed to sunlight, the bioreactors absorb substantial heat. Cooling is necessary, and various methods are available: spraying water, self-shading by overlapping the tubes, water baths, and heat exchangers.[49] Although the system is a closed loop running in a protected environment, microalgae are exposed to contamination especially when water and media are first introduced. Therefore, photobioreactors operated with wastewater and natural waterways require careful attention to mitigate external contamination.[50] Size and operational specifications for mixing and retention tanks will vary depending on strains, culturing requirements, and the production mode (e.g., batch production, semi-continuous production, continuous production). Figure 8.3 illustrates bioreactor typologies and key components for operation. Table 8.2 summarizes PBR design parameters that affect microalgae growth.

Production method: Microalgae are grown in a bioreactor by three methods: batch production, semi-continuous production, and continuous production. Batch production is an operational environment in which microalgae are cultivated for a long period until they reach the stationary phase. During the batch cultivation period, an aeration system continuously feeds CO_2 and major nutrients can

PBR design parameters affecting maximum microalgae growth

PBR design parameters	Description
Materials	Photosynthesis—Transparent materials, for example, UV-resistant plastic, glass, ETFE
	Excessive heat mitigation—IR reflective coating or film
Depth for light penetration	Up to 10 cm (4")
Geometry	High illuminated surface-to-volume ratio
	Modular system allowing for easy assembly and disassembly
Architectural aesthetics	Flat, tubular, helical, stranded, woven, inflated, interlocking, divided

be added as needed. When the cell density reaches target performance, the majority of the culture batch is harvested, and a new production cycle starts. *Nannochloropsis* sp. and *Chlorella* sp. in a batch cultivation model yield biomass productivity of 3.83 g/L/day and 3.76 g/L/day, respectively.[51] For the semi-continuous production mode, microalgae grow in a batch culture condition at the beginning. When cell density reaches maximum growth, a portion of the culture volume is harvested, and fresh media is added. The harvesting cycle can be every day or every other week depending on the target cell density and lipid content. The semi-continuous production mode aerated with 15% CO_2 increases *Chlorella* sp. biomass productivity to 17.2 g/L/day.[52] Lastly, the continuous production mode is operated based on the continual withdrawal of culture and immediate replacement of fresh media at the same volume. *Chlorella minutissima* and *Dunaliella tertiolecta* under continuous cultivation mode reach 137 mg/L/day and ~91 mg/L/day, respectively.[53] The semi-continuous production can be used for industry-scale production while keeping a high growth rate due to its ability to control nutrition and other growth parameters.[54]

Control system: Microalgae cultivation requires close monitoring and control of environmental factors to ensure the efficient operation of the system as concerns culturing, enrichment, microalgae collection, harvesting, and bioproduct processing. The target environmental conditions are monitored and controlled by measuring data using various sensors such as temperature sensors, photometers, pH sensors, O_2 sensors, turbidity sensors, flow meters, and so on. These sensors detect system changes in culture temperature, light intensity, pH of the medium, nutrients, salinity, and so on. When environmental conditions are out of the recommended range, inflow and outflow of media, energy, gas, other materials, and so on will be adjusted by an automatic control and monitoring system to meet the target. A programmable logic controller can be developed for maximum culture productivity. Figure 8.4 exemplifies a programmable control logic for maximum biomass productivity.

Popular microalgae in biofuel production: *Chlorella* (green microalgae) is a circular green alga with 2–10 μm diameter cells, found naturally in freshwater, saltwater, and soil. Due to their fast growth rate and strong tolerance to growing environments, *Chlorella* is one of the most popular microalgae in scientific research in biofuels, with some studies achieving a doubling of biomass in 3.5 days.[55] Their relatively high tolerance for nitrates, phosphates, and CO_2 concentrations make them one of the best microalgae for use in wastewater and flue gas remediation. *Chlorella* has been shown to reduce phosphorus and nitrogen in municipal

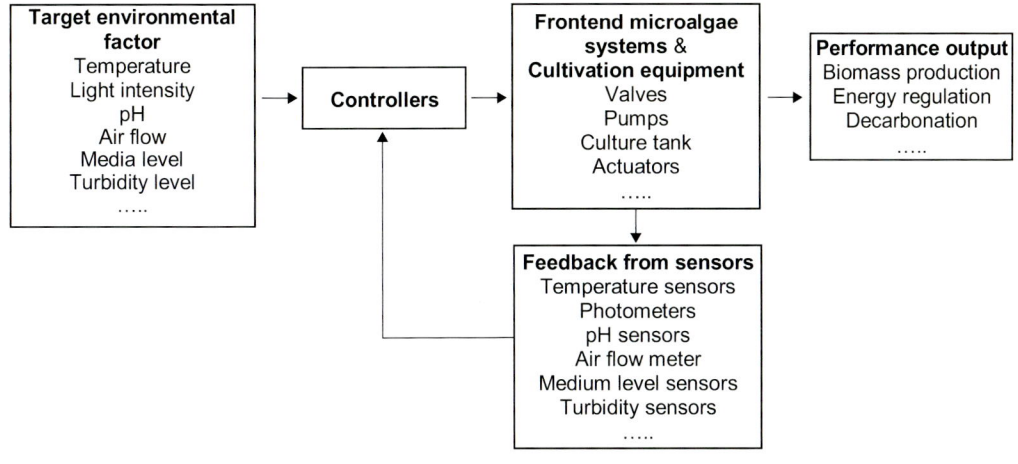

An example of a programmable control logic design for microalgae building applications.

wastewater by 90% and 83%, respectively,[56] and shows an exemplary growth rate with up to 50% CO_2.[57] *C. vulgaris* achieved maximum growth of 5 g/L in a nitrate-rich media and 10 g/L with an increased nitrate level.[58] *Chlorella*'s thick cell wall makes it more resistant to damage from heavy metals that may be present in wastewater.[59] *C. vulgaris* was ranked the highest among biofuel potential strains, and its high growth rate and high lipid content favors *Chlorella* for biofuel production; however, challenges exist in the post-cultivation process in that their small size and round shape present difficulties with harvesting them in a cost-effective manner.[60] *Chlorella* is also used in biofertilizers for agriculture in enhancing growth yield and weight gain of a wheat variety by 140% and 40%, respectively, after treatment with a *Chlorella* fertilizer.[61] Figure 8.5 illustrates how *Chlorella* is harvested and how each chemical compound is converted into a final end product such as biofuel, feeds, biofertilizer, and bioactive compounds.

Dunaliella (green microalgae): First discovered in the salterns of France in 1838, *Dunaliella* are nearly ubiquitous in salty waters around the globe.[62] *Dunaliella* is considered a model organism for the study of salt adaptation in algae.[63] With cell sizes of 5–25 μm in length and 3–13 μm in width, their cells can be ellipsoid, oval-shaped, near-spherical, pear-shaped, or tapered at both ends and are either radially symmetrical, bilateral, or slightly asymmetrical.[64] The cells accumulate beta-carotene and turn to yellow and ultimately red color as they grow.[65] Some *Dunaliella* are motile, while others are not, and the elastic plasma membrane enclosing the cell offers morphological variability depending on environmental conditions.[66] *Dunaliella* is a fast-growing microalga, and a study conducted by the U.S. Department of Energy's Aquatic Species Program in the mid-1980s measured one strain (ASU0038) as achieving a 2.58 doubling rate per day.[67] *Dunaliella* is also shown to be effective bioremediators for wastewater treatment and flue gas fixation. *Dunaliella salina* removed 57.5% of the total nitrogen and 69% of the total phosphorus from tertiary-treated municipal wastewater.[68] It also shows growth performance and high tolerance to CO_2 concentrations up to 24%.[69] *Dunaliella* is among the four most common microalgae grown in industry, alongside *Spirulina*, *Chlorella*, and *Haematococcus*, mainly as sources of β-carotene, glycerol, and

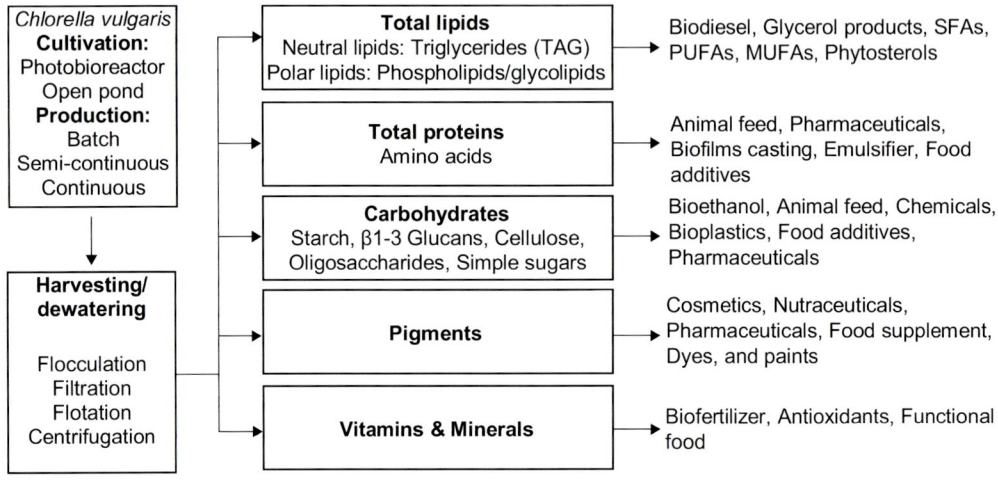

▲ Figure 8.5

Summary of cultivation process, chemical compounds, and potential uses of Chlorella in different industries (reproduced from Carl Safi, et al.).[70]

protein that can be easily extracted due to the absence of cell walls.[71] High growth rates and high oil content along with their ability to uptake nutrients in wastewater make *Dunaliella* a good candidate for future renewable fuel stock.[72] Table 8.3 shows the growth rate of various microalgae under different culturing conditions. Figure 8.6 illustrates a simulation set-up of a reference building to quantify photosynthetic photon flux density (PPFD) and temperature values impinging on the microalgae building surface across seasons.

Summary

The built environment is responsible for substantial energy use and pollutant emissions. A fast growth rate and high lipid content make microalgae a promising renewable resource for bioenergy production. They also have attracted attention for bioremediation applications in soil, water, and air quality. They effectively sequester CO_2 and bioremediate contaminants attributed to anthropogenic activities. Industry scalization is necessary for augmenting microalgae benefits, but mass cultivation is still economically challenging in part due to lack of real data from commercial facilities. To make cultivation cost-effective requires continued development in biotechnical engineering and system optimization. Productivity rates depend on various environmental factors, including light density, light/dark cycles, temperature, CO_2, nutrients, pH, and so on. Manipulation between a deficiency and an abundance of growth factors such as sunlight, nutrients, and CO_2 can yield high biomass productions and concentrations of lipids that can contribute to process economics. A semi-production mode effectively controls nutrition and other growth parameters and is recommended for large-scale industrial production. Nutrients can be supplied from wastewater and flue gas which can offset operational cost. A successful application can improve environmental and economic performance and contribute to healthy and livable environments.

▼ Table 8.3

Lab experiments for various microalgae growth under different environmental conditions

Microalgae	Temp (°C)	Nutrient/Media	Illumination (μmol/m²/s)	Photoperiod	pH	CO_2	Growth rate	Source
Botryococcus braunii	25	Modified BG-11 medium	150 ± 10	Continuous	N/A	2%, 5%, 10%, and 20%	2.31 g/L on day 25	Ge et al. 2011
Chlorella pyrenoidosa	25 ± 2	BG-11 medium	6,000 lx	N/A	8 (initially)	N/A	2.03 g/L	Wu & Maio 2014
Chlorella sorokiniana	20–25	BBM, BG-11, and piggery wastewater	150	N/A	BBM: 7.4 ± 0.3; BG-11: 7.7 ± 0.3	0.1 vvm with 2.5% CO_2	2.3–2.5 g/L	Marjakangas et al. 2015
Chlamydomonas reinhardtii	25 ± 1	Tris-acetate-phosphate	~100	Continuous	N/A	2% CO_2 v/v air at 100 mL/min	0.45 mg/mL~ 0.9 mg/mL	Meng et al. 2020
Chlamydomonas sp.	20–25	BBM, BG-11, and piggery wastewater	150	N/A	BBM: 7.4 ± 0.3; BG-11: 7.7 ± 0.3	0.1 vvm with 2.5% CO_2	2.4–2.8 g/L	Marjakangas et al. 2015
Chlorella sp.	24 ± 1	BG-11, BBM, Fog's, and M4N	~2500 lx	16:8 light:dark	7	N/A	274.46 mg/L/day	Sharma et al. 2016
Chlorella vulgaris	25	Varied	150	Continuous	7	5%	0.217 g/L/day (N-); 0.668 g/L/day (N+)	Breuer et al. 2012
Chlorella vulgaris	25	BG-11	60	N/A	N/A	N/A	5.050 x 10^6 cell/mL	Sajadian et al. 2018
Chlorella vulgaris	20–25	BBM, BG-11, and piggery wastewater	150	N/A	BBM: 7.4 ± 0.3; BG-11: 7.7 ± 0.3	0.1 vvm with 2.5% CO_2	1.8–2.0 g/L	Marjakangas et al. 2015

▼ Table 8.3

(Continued)

Microalgae	Temp (°C)	Nutrient/Media	Illumination (μmol/m²/s)	Photoperiod	pH	CO_2	Growth rate	Source
Chlorella zofingiensis	25	Varied	150	Continuous	7	5%	0.508 g/L/day (N-); 0.792 g/L/day (N+)	Breuer et al. 2012
Chlorella zofingiensis	25 ± 1	BBM	60, 85, 115, 170	Continuous	N/A	1 vvm CO2 1% v/v	0.11–0.72 g/L/day	Oncel et al. 2011
Dunaliella	23	Erdschreiber	~180	12h/12h	7.8	N/A	2–8 × 10^6 cells/mL	Byrd et al. 2017
Dunaliella tertiolecta	25	Varied	150	Continuous	7.5	5%	0.341 g/L/day (N-); 0.487 g/L/day (N+)	Breuer et al. 2012
Dunaliella salina	28 ± 0.05	High salinity medium	100 (low) 800 (high)	Continuous	N/A	1%/10% v/v at 0.1 L/min	3.5 g/L after 5 days	Yuan et al. 2019
Fistulifera solaris	Variable ambient	Modified f/2 medium	Sunlight	Sunlight	N/A	N/A	6.45–23.30 mg/L/day	Matsumoto 2017
Haematococcus pluvialis	14, 15, 19, 24	Modified BBM	10, 50, 90	N/A	7	N/A	0.54–1.22 g/L dry at 3 weeks	Harker et al. 1995
Haematococcus pluvialis	25	OHM	20 ± 5	Continuous	N/A	N/A	0.243 g/L/day	Jeon et al. 2006
Haematococcus pluvialis	25 ± 1	BBM	60, 85, 115, 170	Continuous	N/A	1 vvm CO2 1% v/v	0.22–0.69 g/L/day	Oncel et al. 2010
Haematococcus pluvialis	25 ± 1	BG-11	30	Continuous	N/A	1.5% (v/v) CO2	N/A	Zhang et al. 2014
Haematococcus pluvialis	25 ± 1	BG-11	0–150	N/A	N/A	1.5% CO2 (v/v)	6.6 ± 0.01 g/m^2/day	Zhang et al. 2014

Species	Temp	Medium	Light	Photoperiod	pH	CO₂	Productivity	Reference
Isochrysis aff. Galbana	25	Varied	150	Continuous	7.5	5%	0.149 g/L/day (N-); 0.401 g/L/day (N+)	Breuer et al. 2012
Mayamaea sp. JPCC CTDA0820	variable (avg. <15)	Modified f/2 medium	Sunlight	Sunlight	N/A	N/A	12.45–17.47 mg/L/day	Matsumoto et al. 2017
Nannochloropsis	23 ± 1	Modified f/2 medium	80	N/A	N/A	5%	4.38 g/L/day	He et al. 2020
Nannochloropsis sp.	25	Varied	150	Continuous	7	5%	0.524 g/L/day (N+); 0.264 g/L/day (N+)	Breuer et al. 2012
Neochloris oleo-abundans	25	Varied	150	Continuous	7	5%	0.426 g/L/day (N-); 0.451 g/L/day (N+)	Breuer et al. 2012
Nannochloropsis oculate	21 ± 1	Commercial F/2	50, 100, 150	14/10 light/dark	7.5 to 9	CO_2 @5 s 20–40 min	0.12–0.35 g/L per day	Parsy et al. 2020
Nannochloropsis oceanica	14.5–35.7	Guillard's f/2	34–80	N/A	~6.5 / ~7.8	N/A	0.7 g/L/day	Ras et al. 2005
Nannochloropsis oceanica	14.5–35.7	Guillard's f/2	34–80	sun cycle+ artificial light	~6.5 / ~7.8	N/A	0.7 g/L/day	Ras et al. 2005
Nannochloropsis oculate	20 ± 1	f/2 medium	500	12:12 light:dark	~8–11	N/A	~20 × 10^6 cells/mL	Tamburic et al. 2014
Nannochloropsis oculate	15–25	f/2 medium	500	12:12 light:dark	~8–11	filtered ambient air	~24 × 10^6 cells/mL	Tamburic et al. 2014
Phaeodactylum tricornutum	25	Varied	150	Continuous	7.5	5%	0.122 g/L/day (N-); 0.486 g/L/day (N+)	Breuer et al. 2012

(Continued)

Microalgae	Temp (°C)	Nutrient/Media	Illumination (μmol/m²/s)	Photoperiod	pH	CO$_2$	Growth rate	Source
Porphyridium cruentum	25	Varied	150	Continuous	7.5	5%	0.308 g/L/day (N-); 0.679 g/L/day (N+)	Breuer et al. 2012
Scenedesmus obliquus	25	Varied	150	Continuous	7	5%	0.719 g/L/day (N-); 0.767 g/L/day (N+)	Breuer et al. 2012
Scenedesmus obliquus	25 ± 2	BG-11 medium	6000 lx	N/A	8 (initially)	N/A	2.70 g/L in nitrate content of 0.6 g/L.	Wu & Maio 2014
Scenedesmus sp.	10, 20, 25, 30	Modified 50% BG-11	~55–60	14/10 light/dark	N/A	N/A	313.3 g biomass /(gP) at 20°C	Xin et al. 2011
Spirulina platensis	23	Zarrouk's media	1250 lumens	12h/12h	N/A	N/A	N/A	Longtin et al. 2021
Spirulina	32	Modified Zarrouk's	85–430	N/A	N/A	N/A	53.2–67.7 mg/L/day	Delrue et al. 2017
Spirulina	15.3–27.6	Modified Zarrouk's	85–430	daily cycle	N/A	N/A	averaged 55 mg/L/day	Delrue et al. 2017

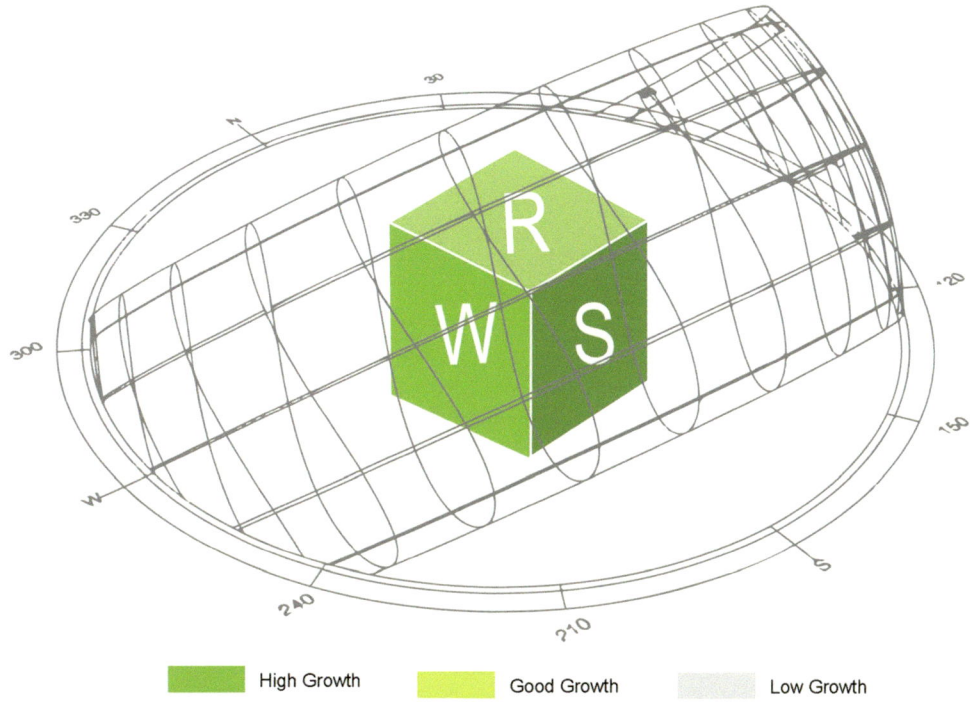

High Growth   Good Growth   Low Growth

Simulated data for PPFD (photosynthetic photon flux density) and temperatures impinging on different building envelop orientations (Roof, East, West, South, North).

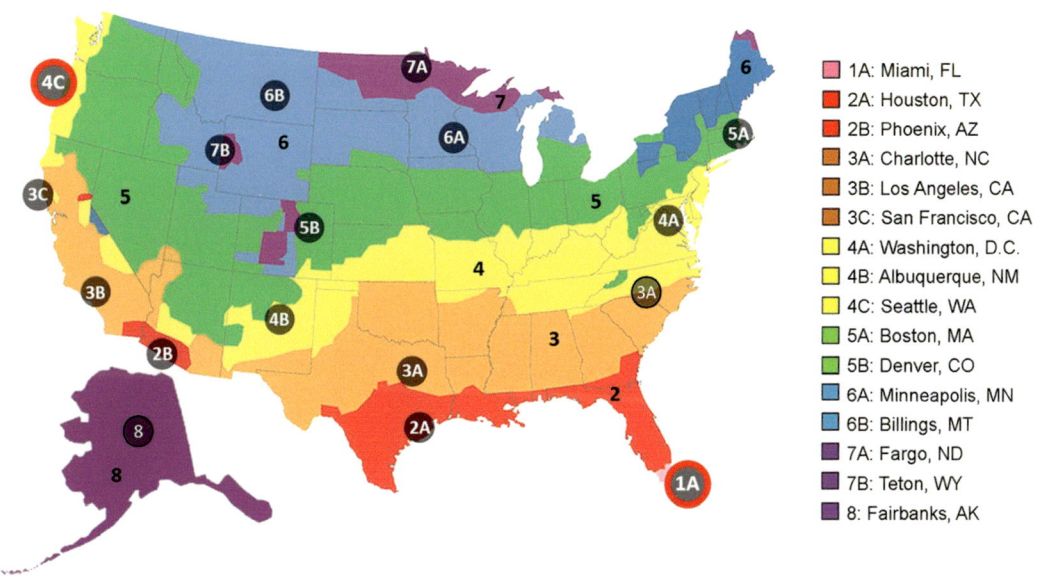

1A: Miami, FL
2A: Houston, TX
2B: Phoenix, AZ
3A: Charlotte, NC
3B: Los Angeles, CA
3C: San Francisco, CA
4A: Washington, D.C.
4B: Albuquerque, NM
4C: Seattle, WA
5A: Boston, MA
5B: Denver, CO
6A: Minneapolis, MN
6B: Billings, MT
7A: Fargo, ND
7B: Teton, WY
8: Fairbanks, AK

Sixteen cities with different climate characteristics for PPFD and temperature simulations

▲ Figure 8.6

A simulation set-up to calculate PPFD and temperature ranges falling on the microalgae building enclosure; Figures 8.7.1– 8.7.32 showing simulated PPFD and temperature data affecting microalgae growth.

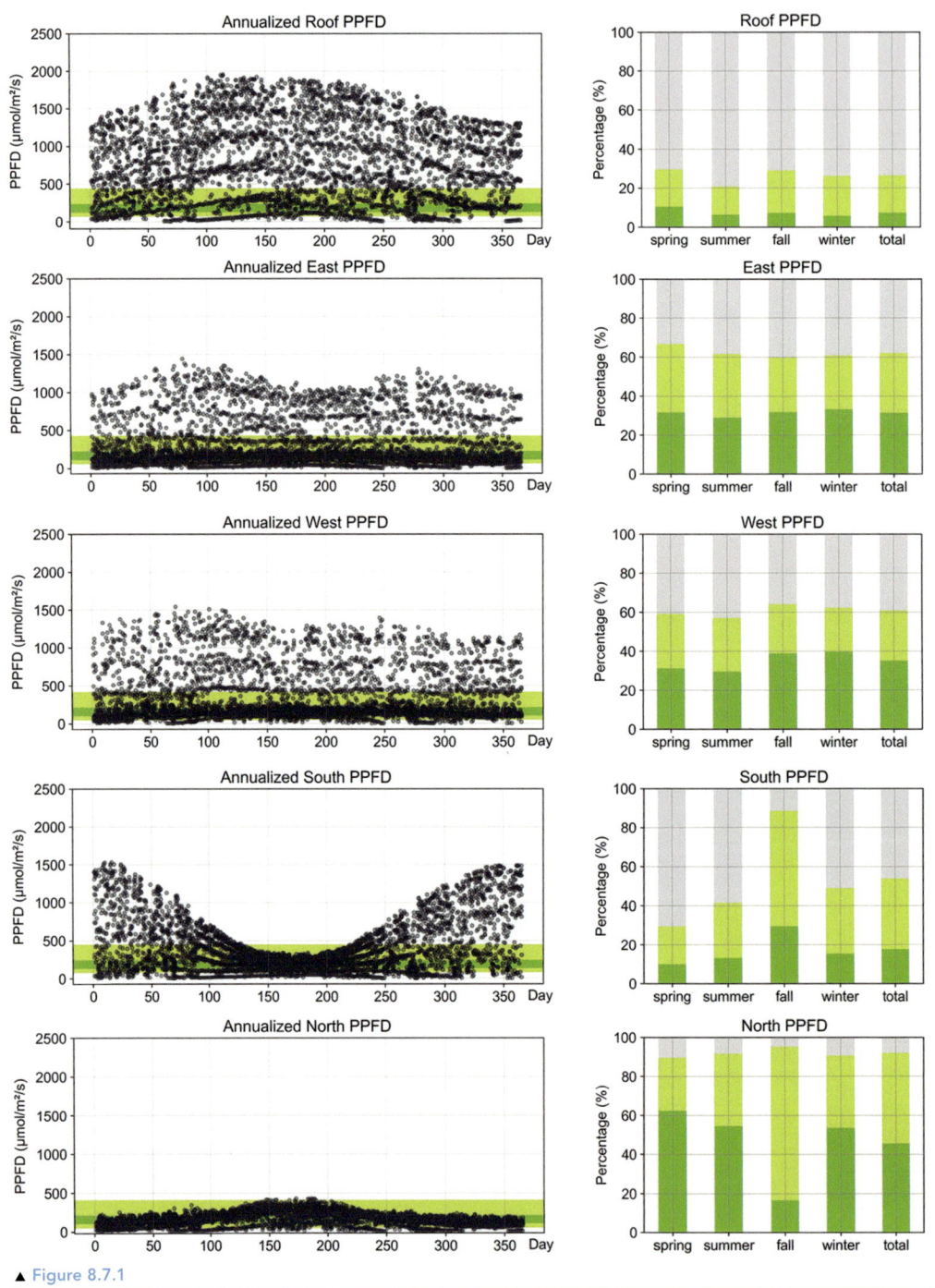

Simulated PPFD in Miami, FL
Climate zone 1A in accordance with ASHRAE 90.1

▲ Figure 8.7.1

PPFD ranges for different building envelop locations (roof, E, W, S, N) and the percentage of the data points that belong to good growing PPFD ranges in Miami, Florida (climate zone 1A per ASHRAE 90.1).

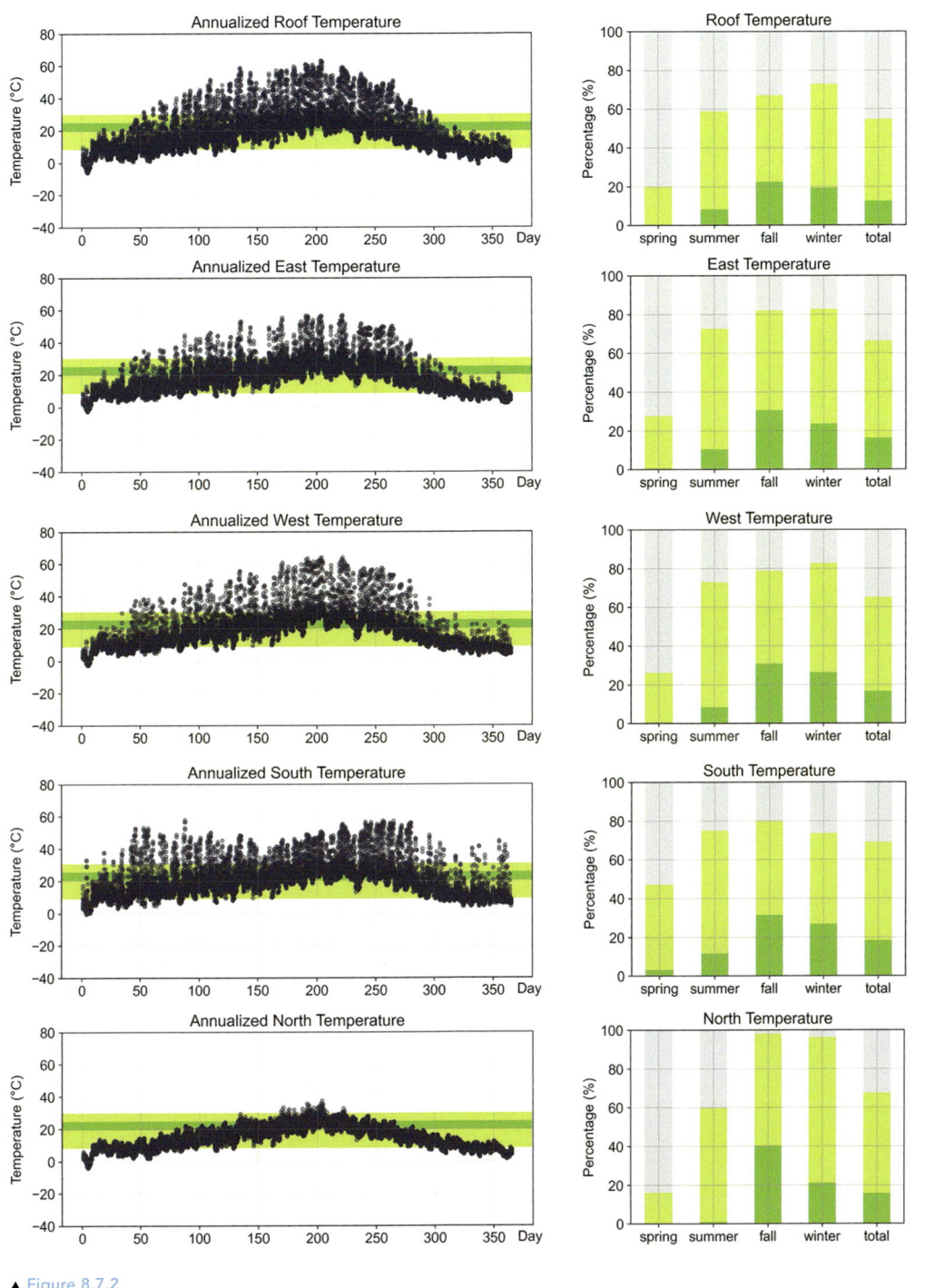

▲ Figure 8.7.2

Temperature ranges for different building envelop locations (roof, E, W, S, N) and the percentage of the data points that belong to good growing temperature ranges in Miami, Florida (climate zone 1A per ASHRAE 90.1).

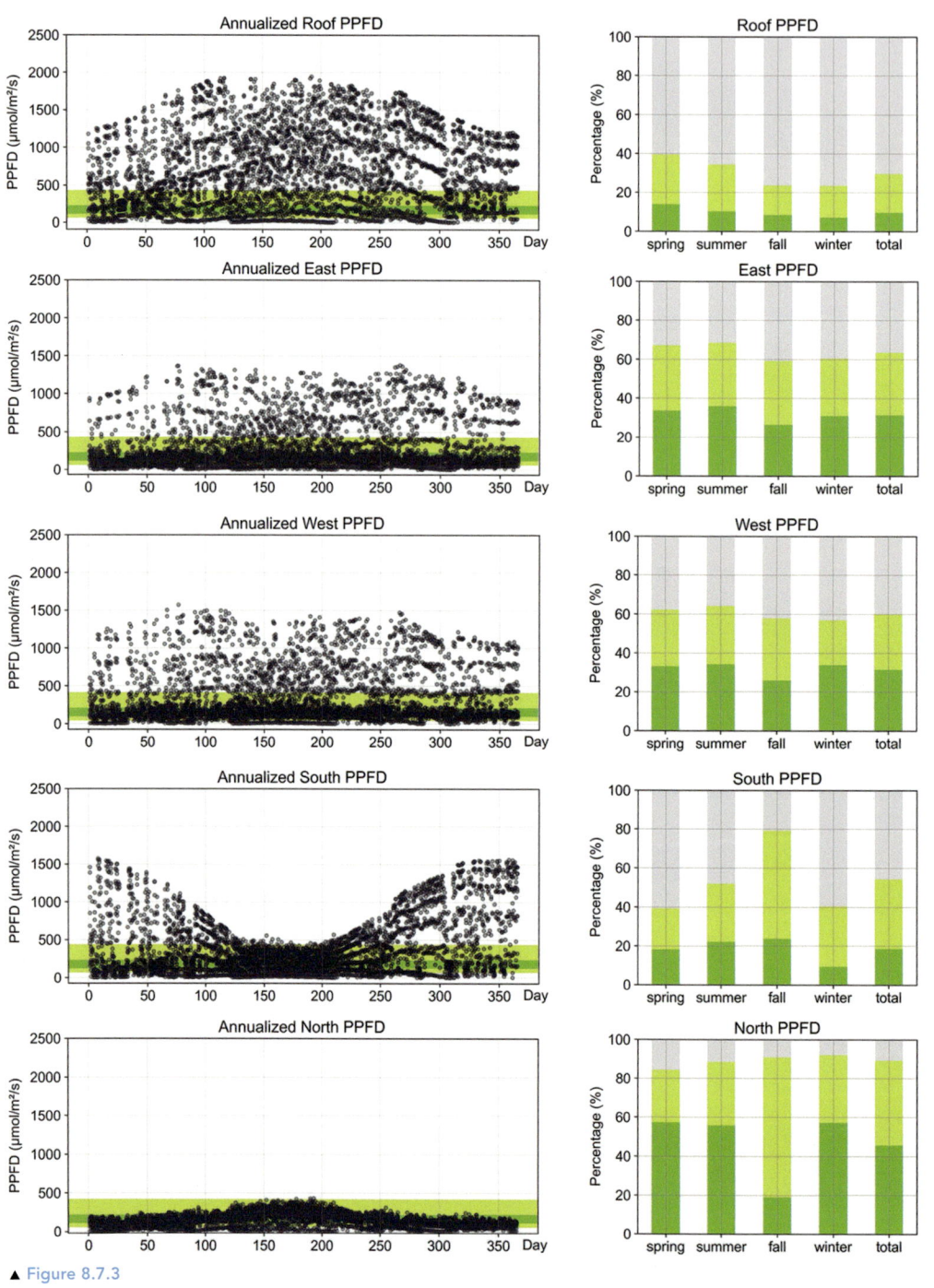

Simulated PPFD in Houston, TX
Climate zone 2A in accordance with ASHRAE 90.1

▲ Figure 8.7.3

PPFD ranges for different building envelop locations (roof, E, W, S, N) and the percentage of the data points that belong to good growing PPFD ranges in Houston, Texas (climate zone 2A per ASHRAE 90.1).

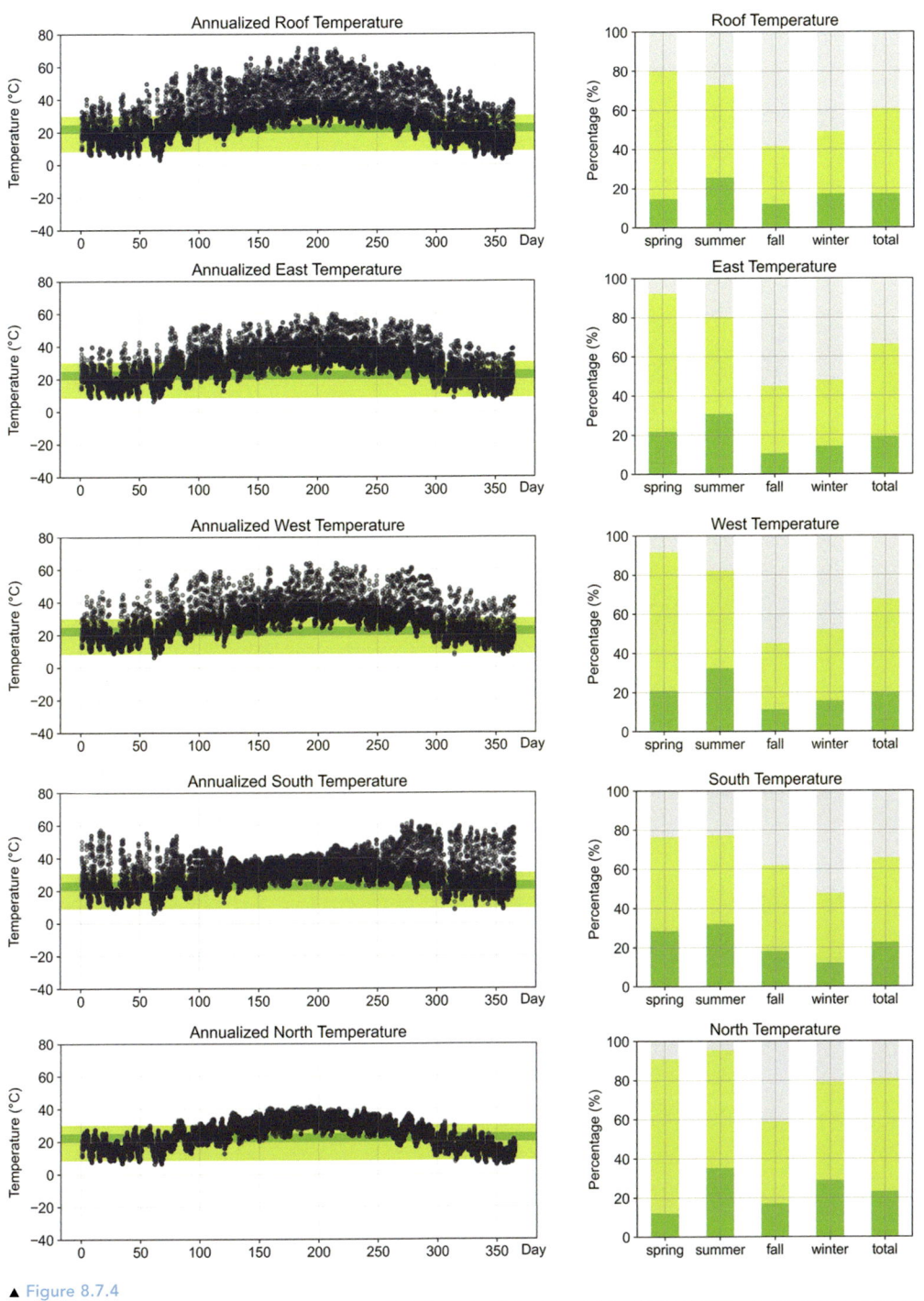

▲ Figure 8.7.4

Temperature ranges for different building envelop locations (roof, E, W, S, N) and the percentage of the data points that belong to good growing temperature ranges in Houston, Texas (climate zone 2A per ASHRAE 90.1).

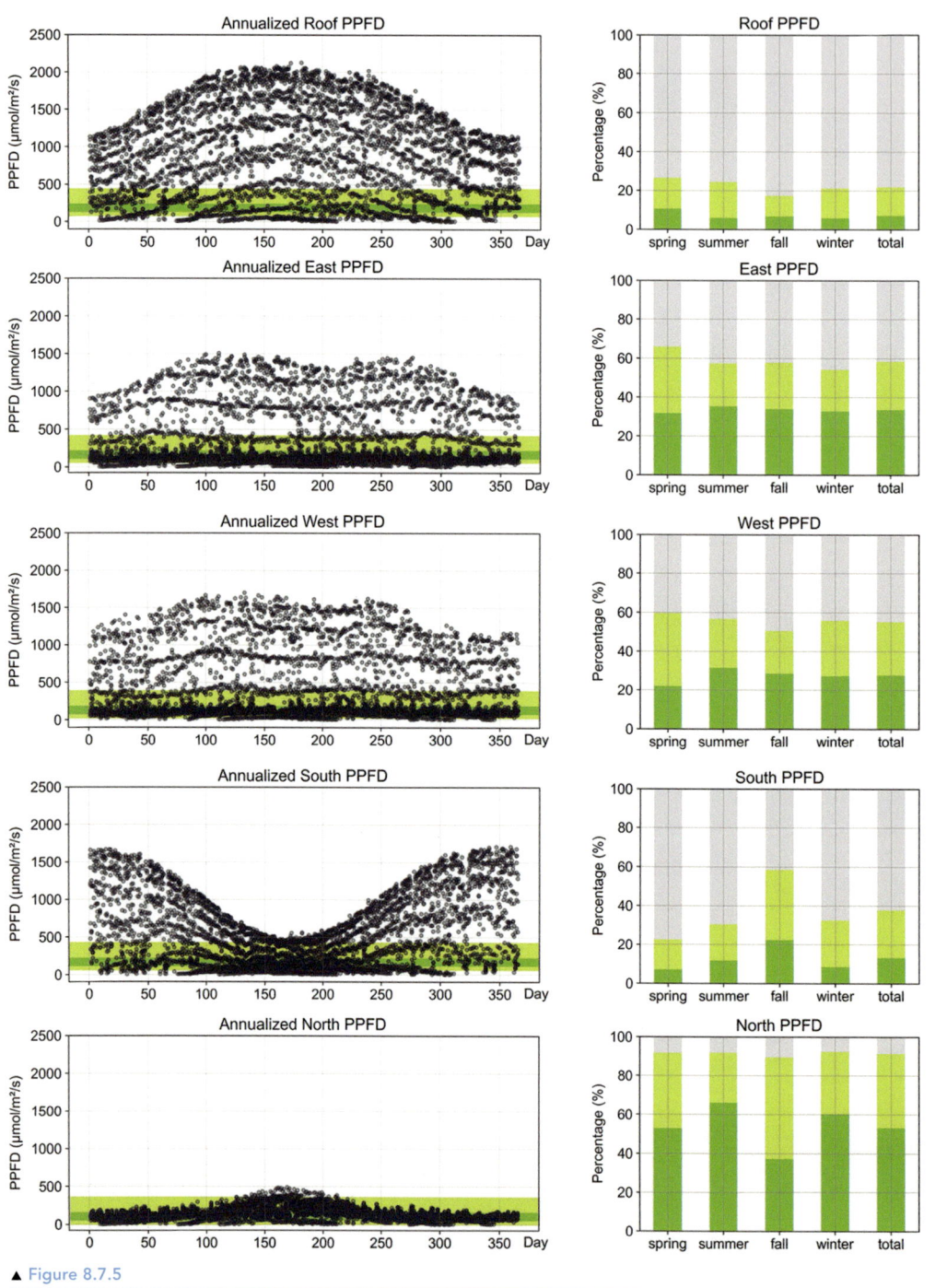

Simulated PPFD in Pheonix, AZ
Climate zone 2B in accordance with ASHRAE 90.1

▲ Figure 8.7.5

PPFD ranges for different building envelop locations (roof, E, W, S, N) and the percentage of the data points that belong to good growing PPFD ranges in Phoenix, Arizona (climate zone 2B per ASHRAE 90.1).

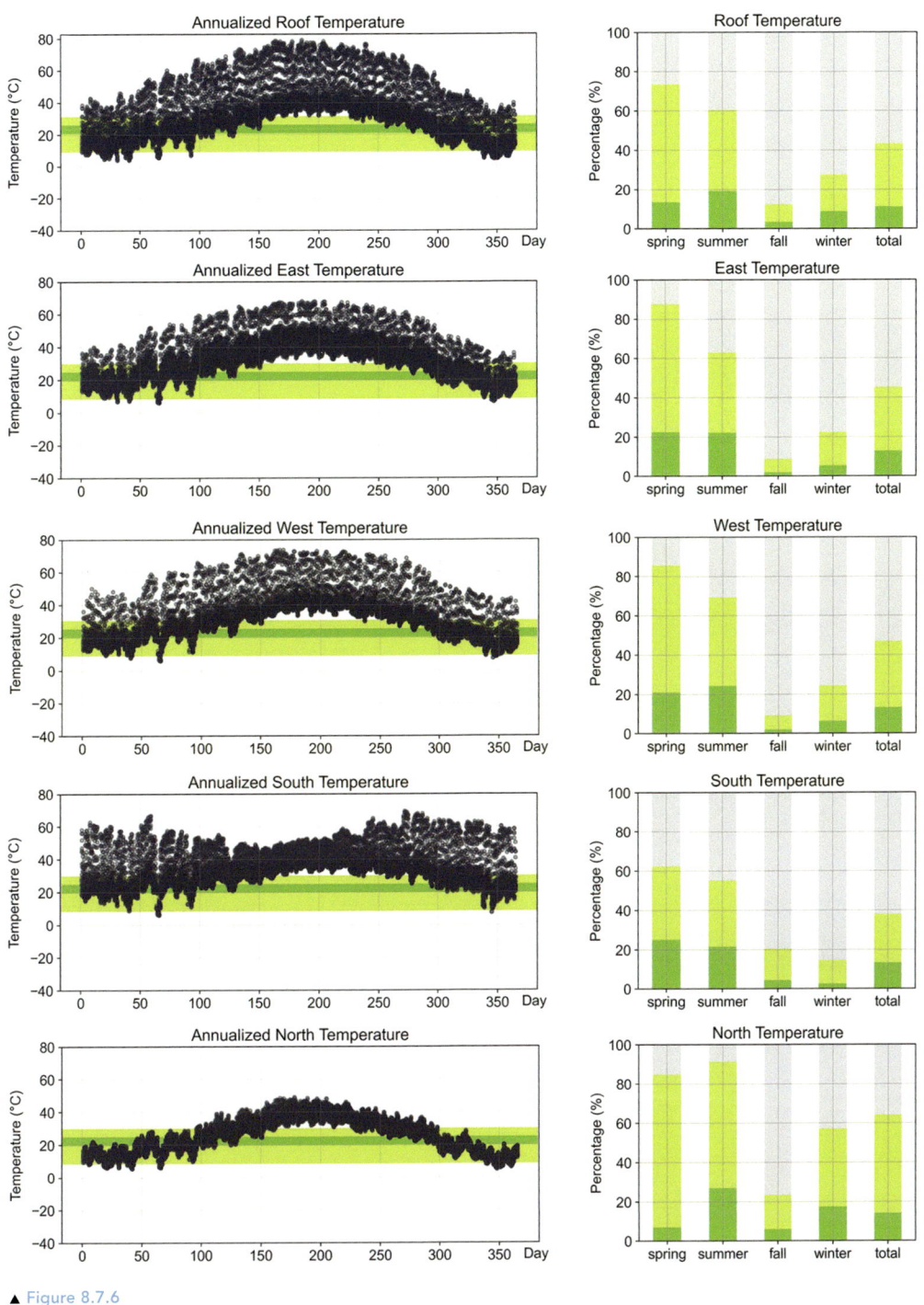

▲ Figure 8.7.6

Temperature ranges for different building envelop locations (roof, E, W, S, N) and the percentage of the data points that belong to good growing temperature ranges in Phoenix, Arizona (climate zone 2B per ASHRAE 90.1).

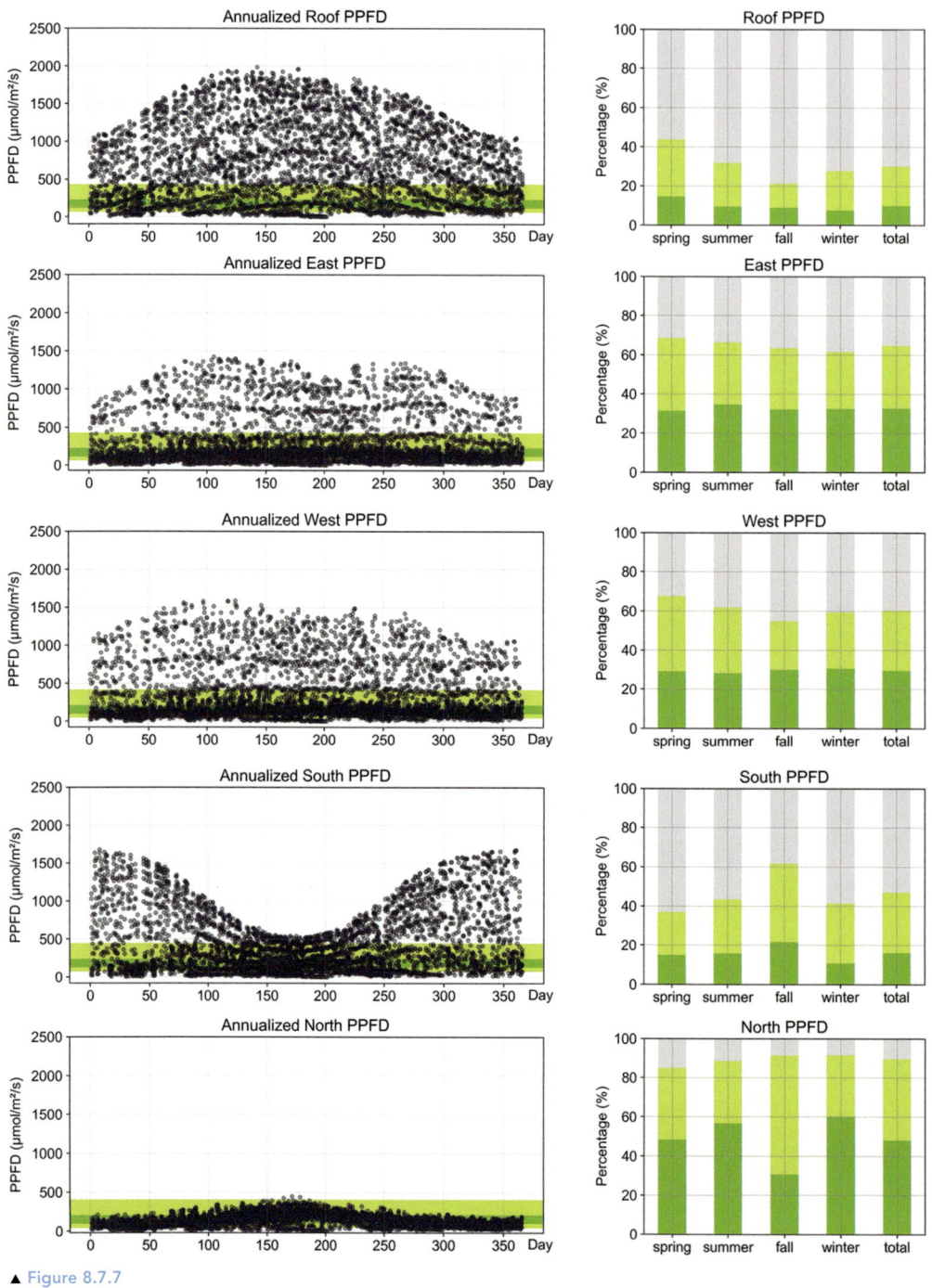

▲ Figure 8.7.7

PPFD ranges for different building envelop locations (roof, E, W, S, N) and the percentage of the data points that belong to good growing PPFD ranges in Charlotte, North Carolina (climate zone 3A per ASHRAE 90.1).

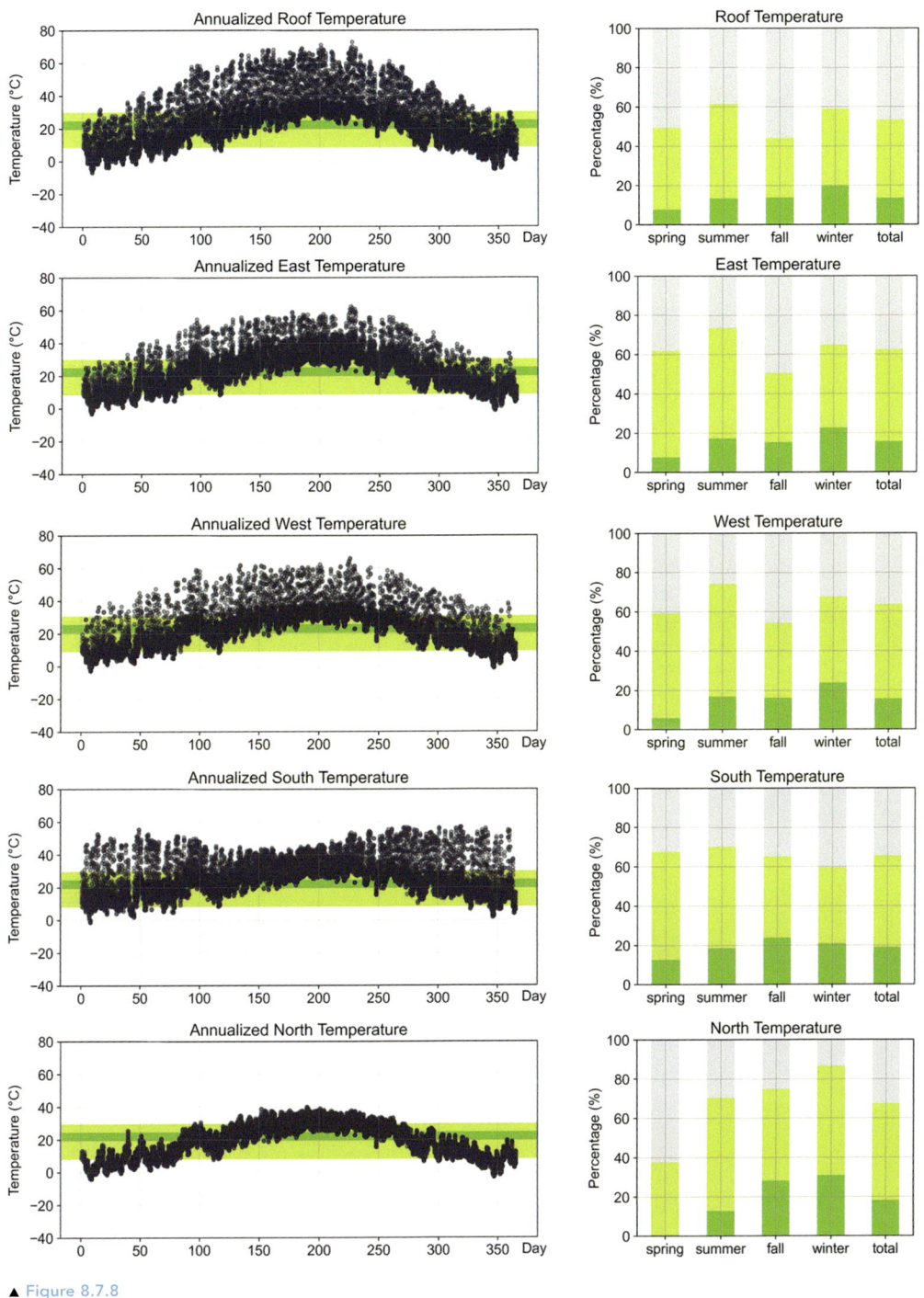

Simulated Temperatures in Charlotte, NC
Climate zone 3A in accordance with ASHRAE 90.1

▲ Figure 8.7.8

Temperature ranges for different building envelop locations (roof, E, W, S, N) and the percentage of the data points that belong to good growing temperature ranges in Charlotte, North Carolina (climate zone 3A per ASHRAE 90.1).

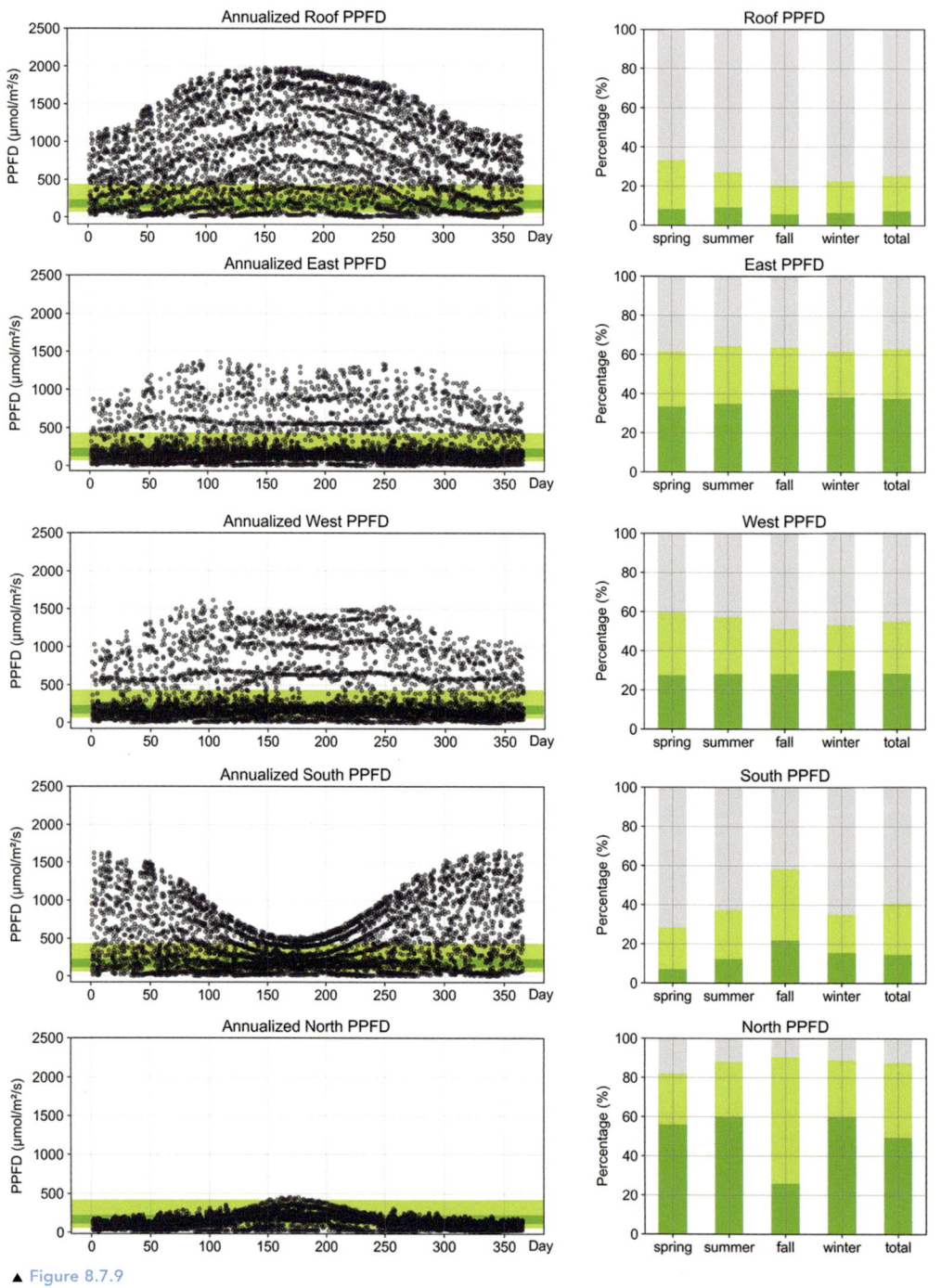

Simulated PPFD in Los Angeles, CA
Climate zone 3B in accordance with ASHRAE 90.1

▲ Figure 8.7.9

PPFD ranges for different building envelop locations (roof, E, W, S, N) and the percentage of the data points that belong to good growing PPFD ranges in Los Angeles, California (climate zone 3B per ASHRAE 90.1).

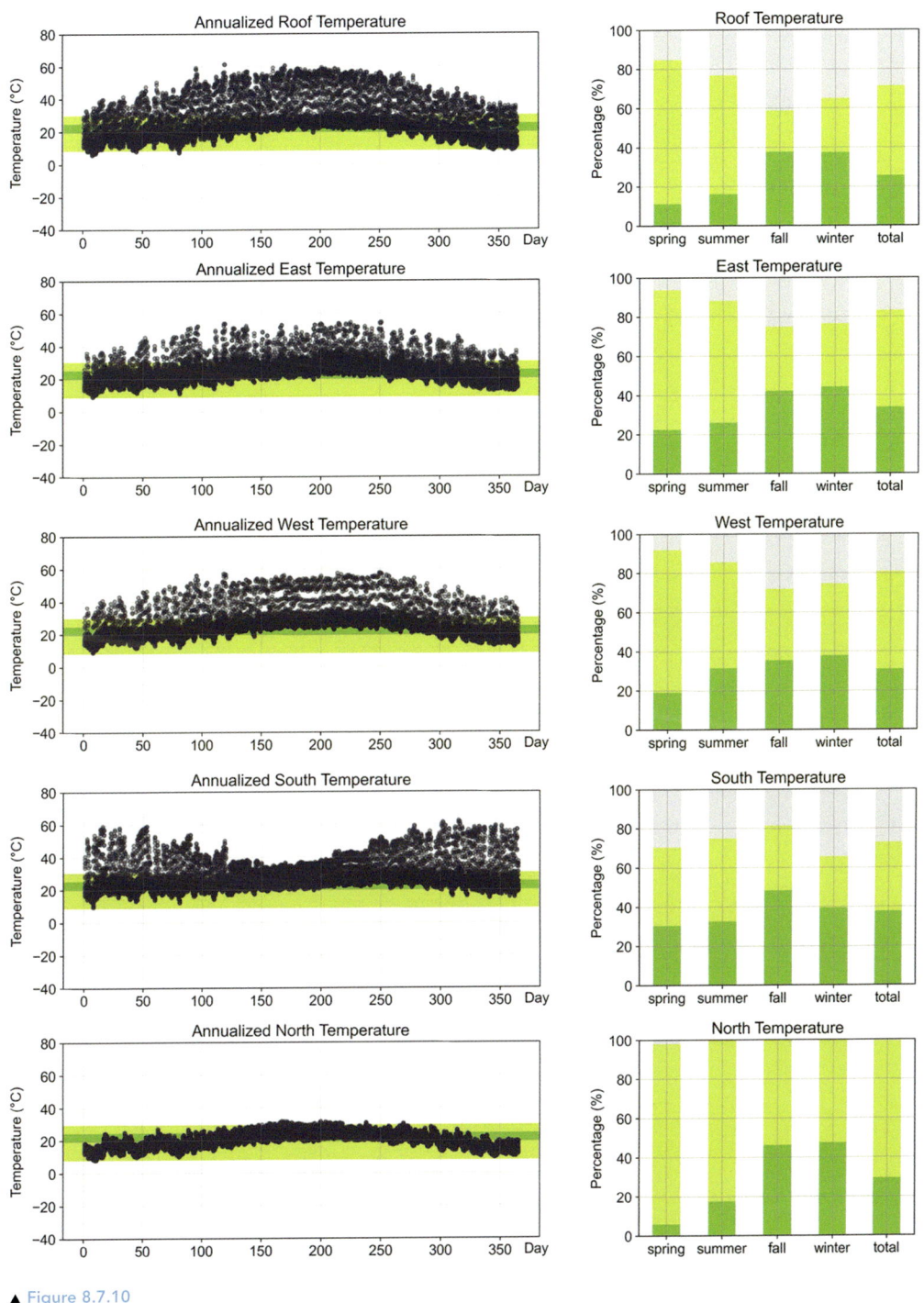

▲ Figure 8.7.10

Temperature ranges for different building envelop locations (roof, E, W, S, N) and the percentage of the data points that belong to good growing temperature ranges in Los Angeles, California (climate zone 3B per ASHRAE 90.1).

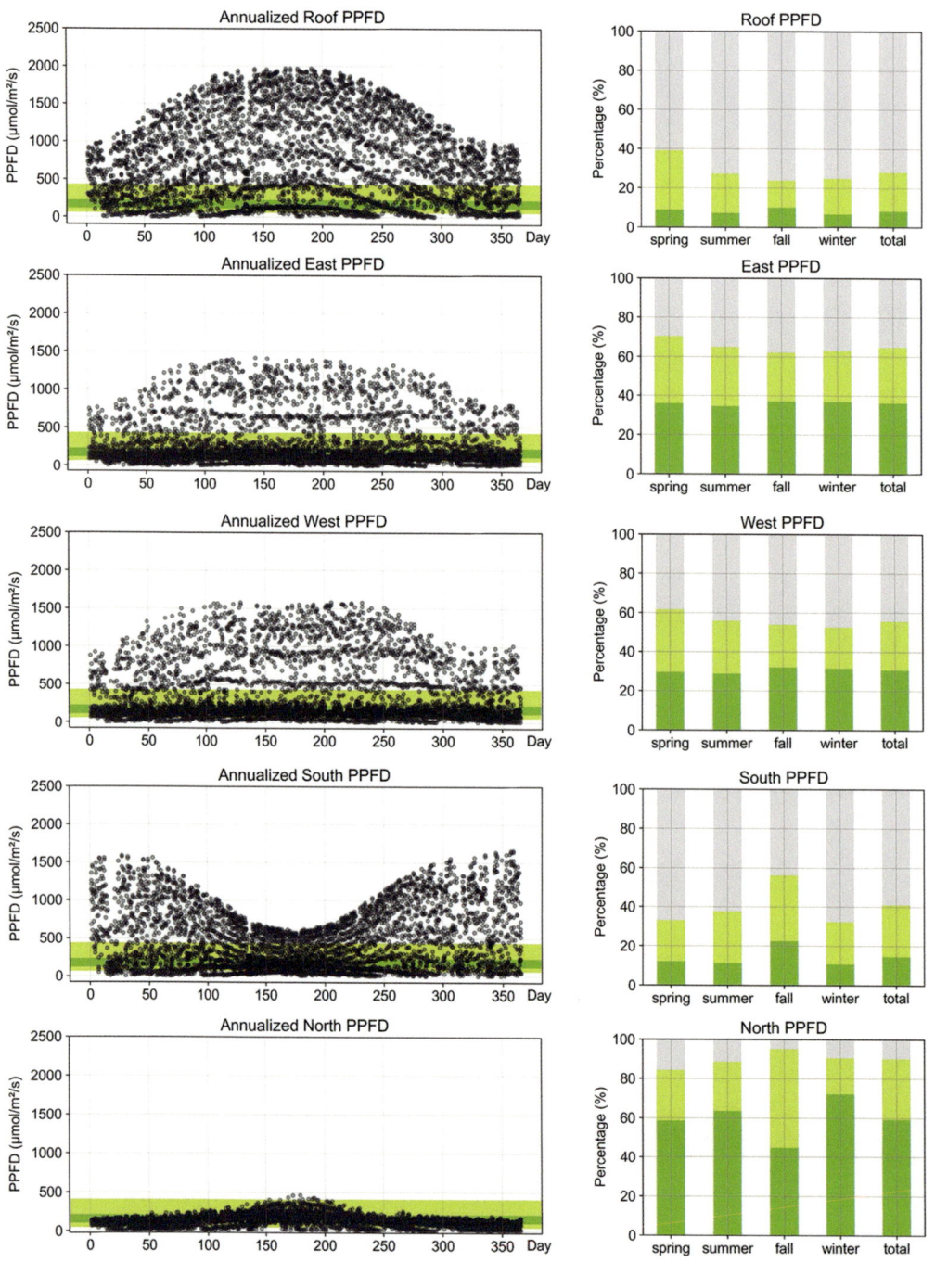

▲ Figure 8.7.11

PPFD ranges for different building envelop locations (roof, E, W, S, N) and the percentage of the data points that belong to good growing PPFD ranges in San Francisco, California (climate zone 3C per ASHRAE 90.1).

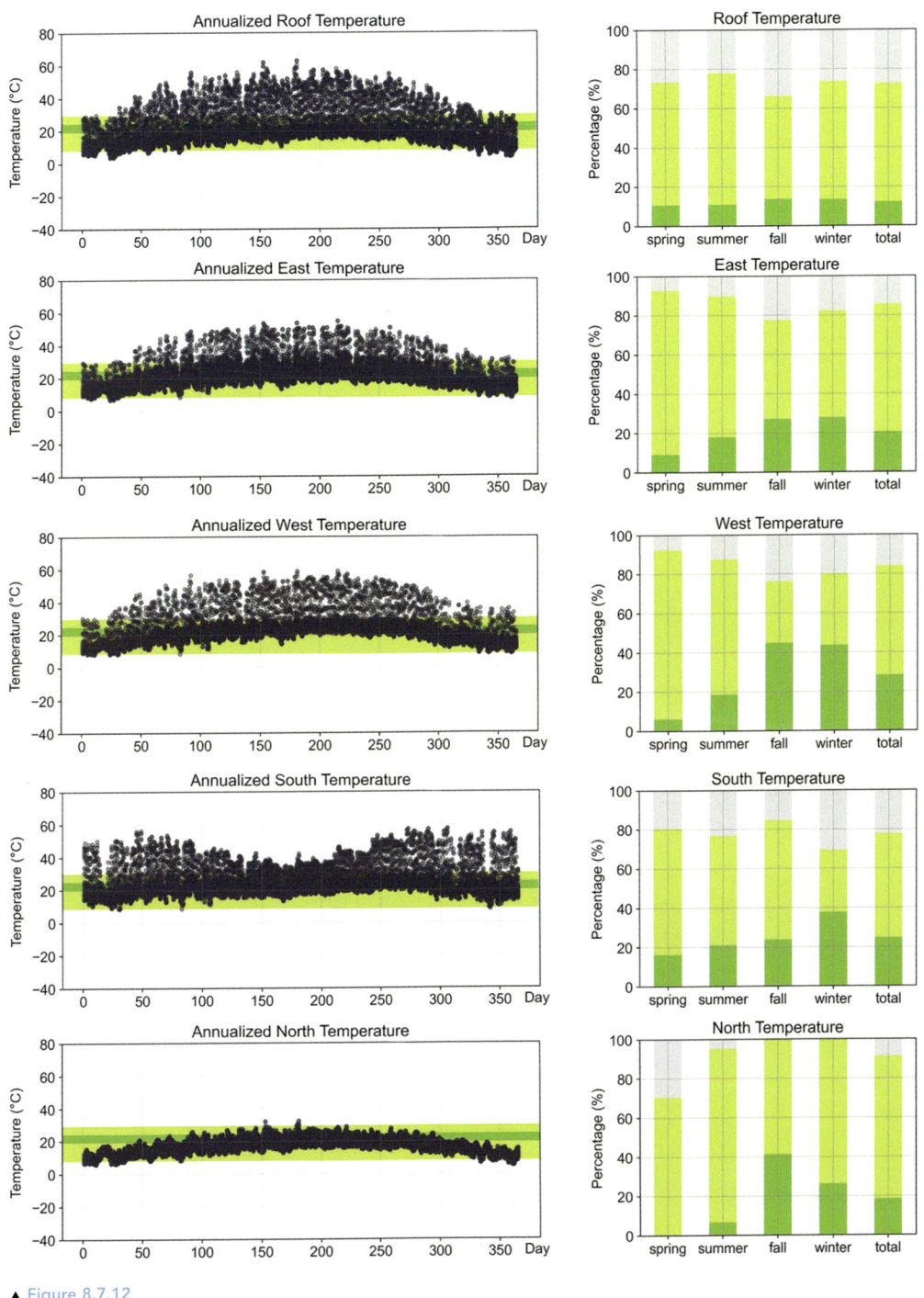

▲ Figure 8.7.12

Temperature ranges for different building envelop locations (roof, E, W, S, N) and the percentage of the data points that belong to good growing temperature ranges in San Francisco, California (climate zone 3C per ASHRAE 90.1).

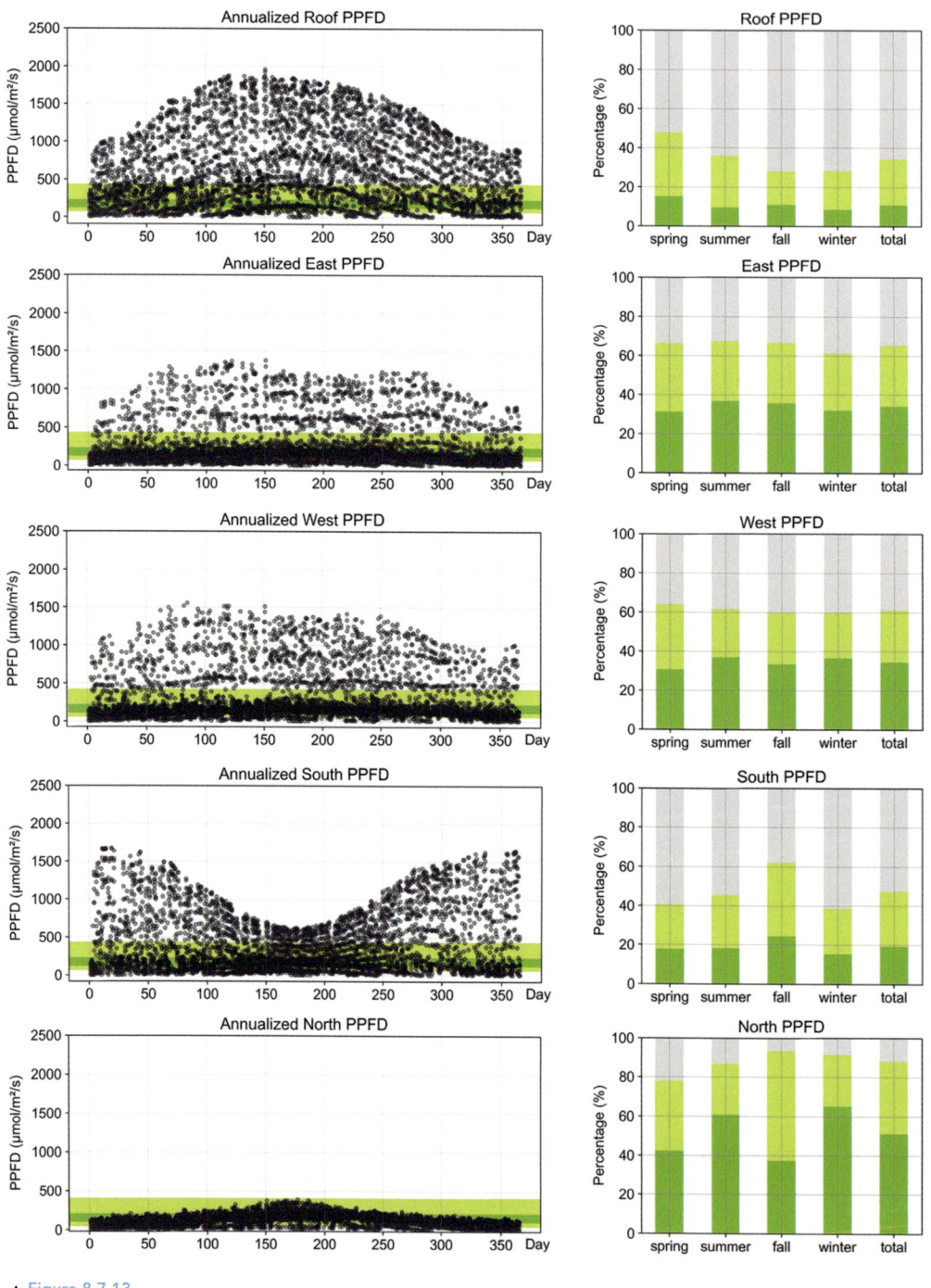

Simulated PPFD in Washington DC
Climate zone 4A in accordance with ASHRAE 90.1

▲ Figure 8.7.13

Temperature ranges for different building envelop locations (roof, E, W, S, N) and the percentage of the data points that belong to good growing temperature ranges in Washington, D.C. (climate zone 4A per ASHRAE 90.1).

Simulated Temperatures in Washington DC
Climate zone 4A in accordance with ASHRAE 90.1

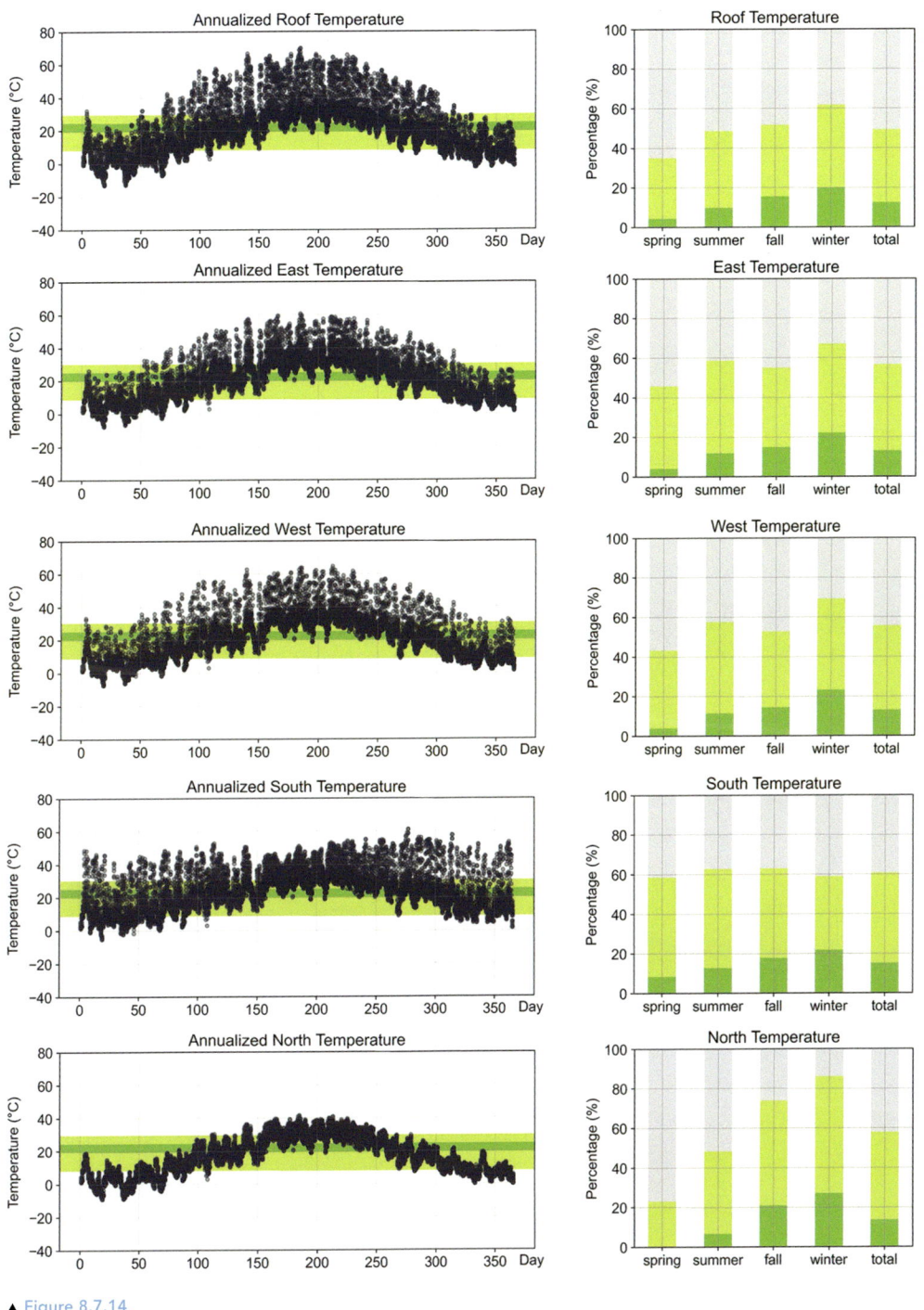

Temperature ranges for different building envelop locations (roof, E, W, S, N) and the percentage of the data points that belong to good growing temperature ranges in Washington, D.C. (climate zone 4A per ASHRAE 90.1).

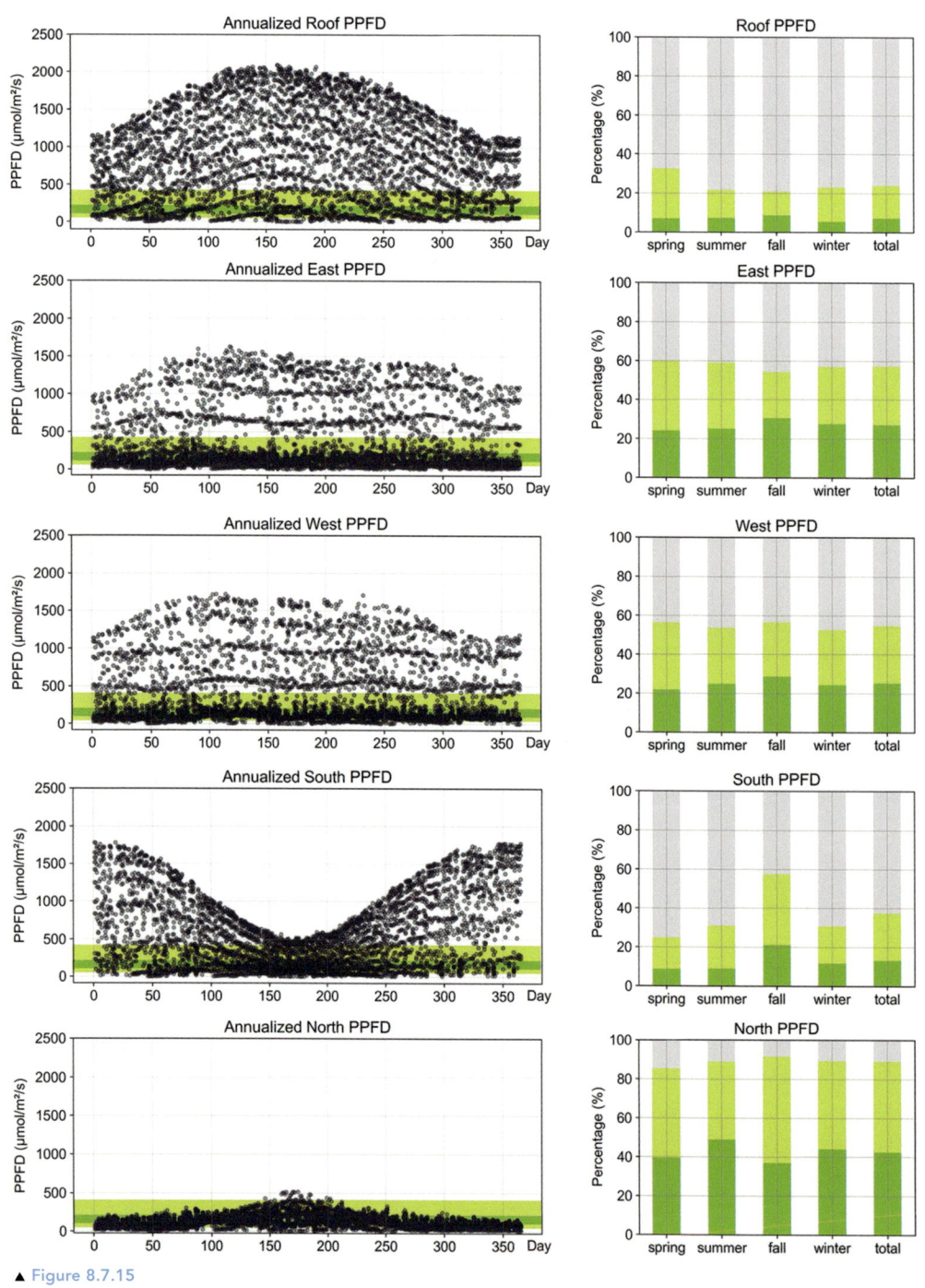

▲ Figure 8.7.15

Temperature ranges for different building envelop locations (roof, E, W, S, N) and the percentage of the data points that belong to good growing temperature ranges in Albuquerque, New Mexico (climate zone 4B per ASHRAE 90.1).

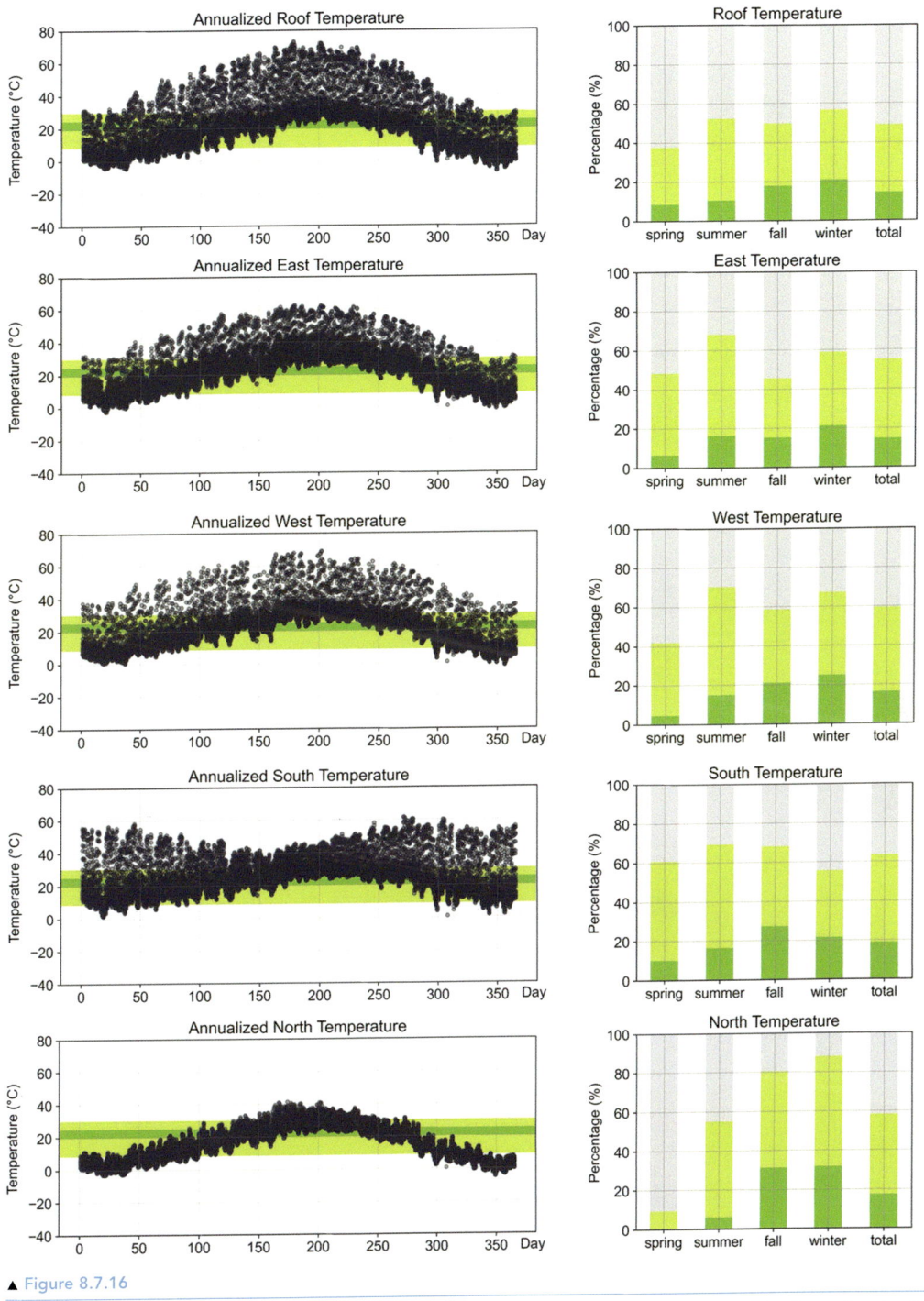

▲ Figure 8.7.16

Temperature ranges for different building envelop locations (roof, E, W, S, N) and the percentage of the data points that belong to good growing temperature ranges in Albuquerque, New Mexico (climate zone 4B per ASHRAE 90.1).

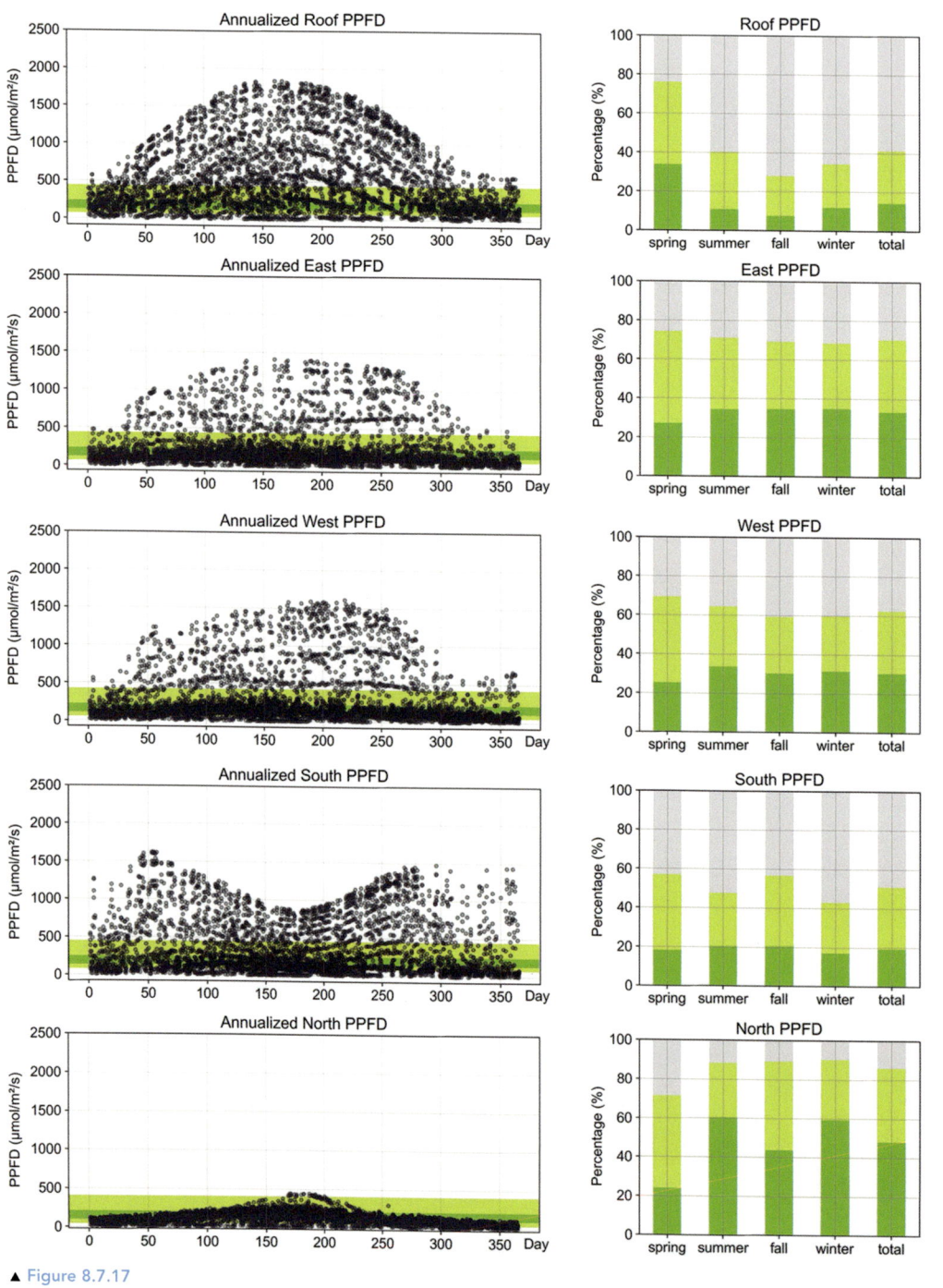

Simulated PPFD in Seattle, WA
Climate zone 4C in accordance with ASHRAE 90.1

▲ Figure 8.7.17

Temperature ranges for different building envelop locations (roof, E, W, S, N) and the percentage of the data points that belong to good growing temperature ranges in Seattle, Washington (climate zone 4C per ASHRAE 90.1).

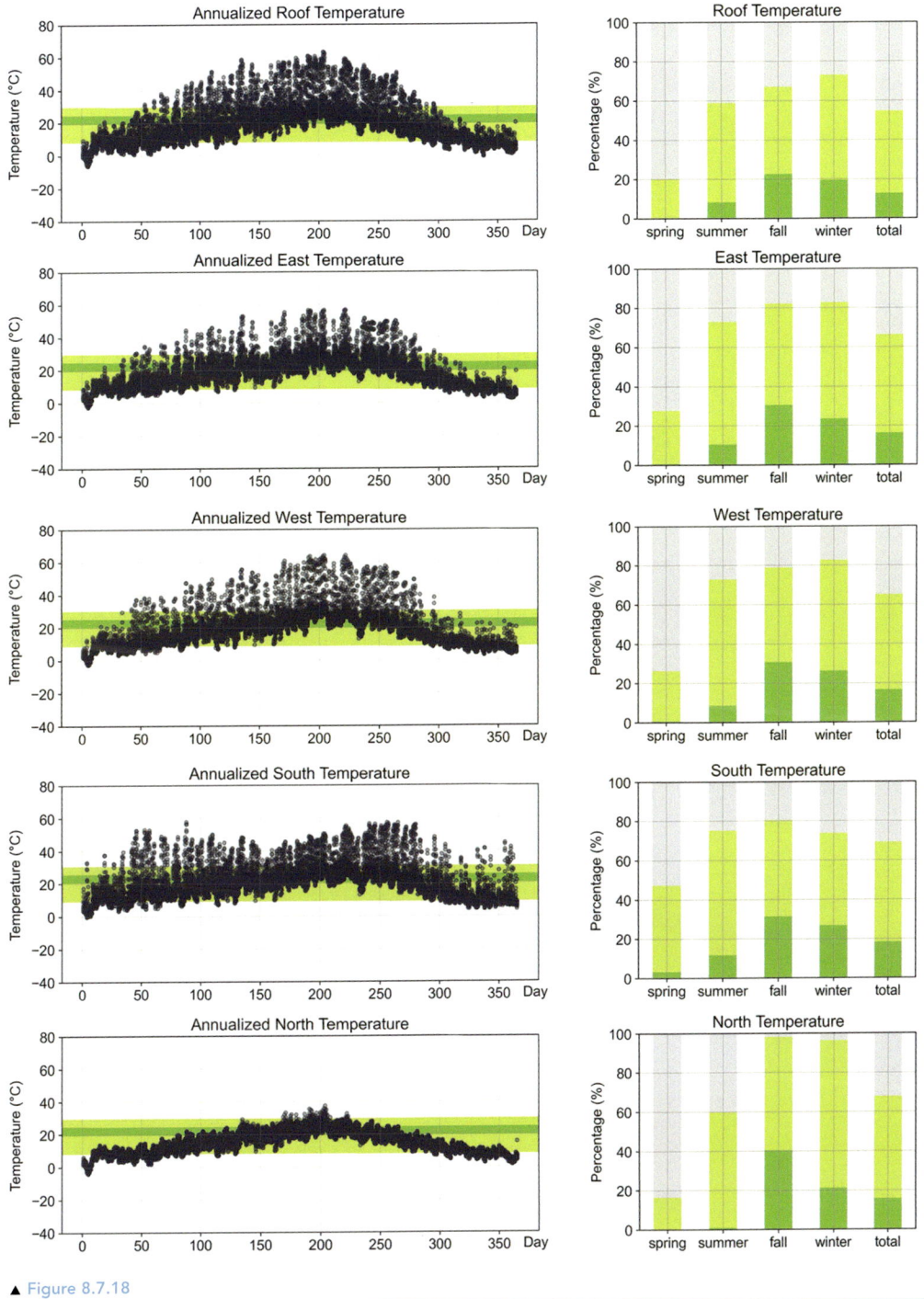

Simulated Temperatures in Seattle, WA
Climate zone 4C in accordance with ASHRAE 90.1

▲ Figure 8.7.18

Temperature ranges for different building envelop locations (roof, E, W, S, N) and the percentage of the data points that belong to good growing temperature ranges in Seattle, Washington (climate zone 4C per ASHRAE 90.1).

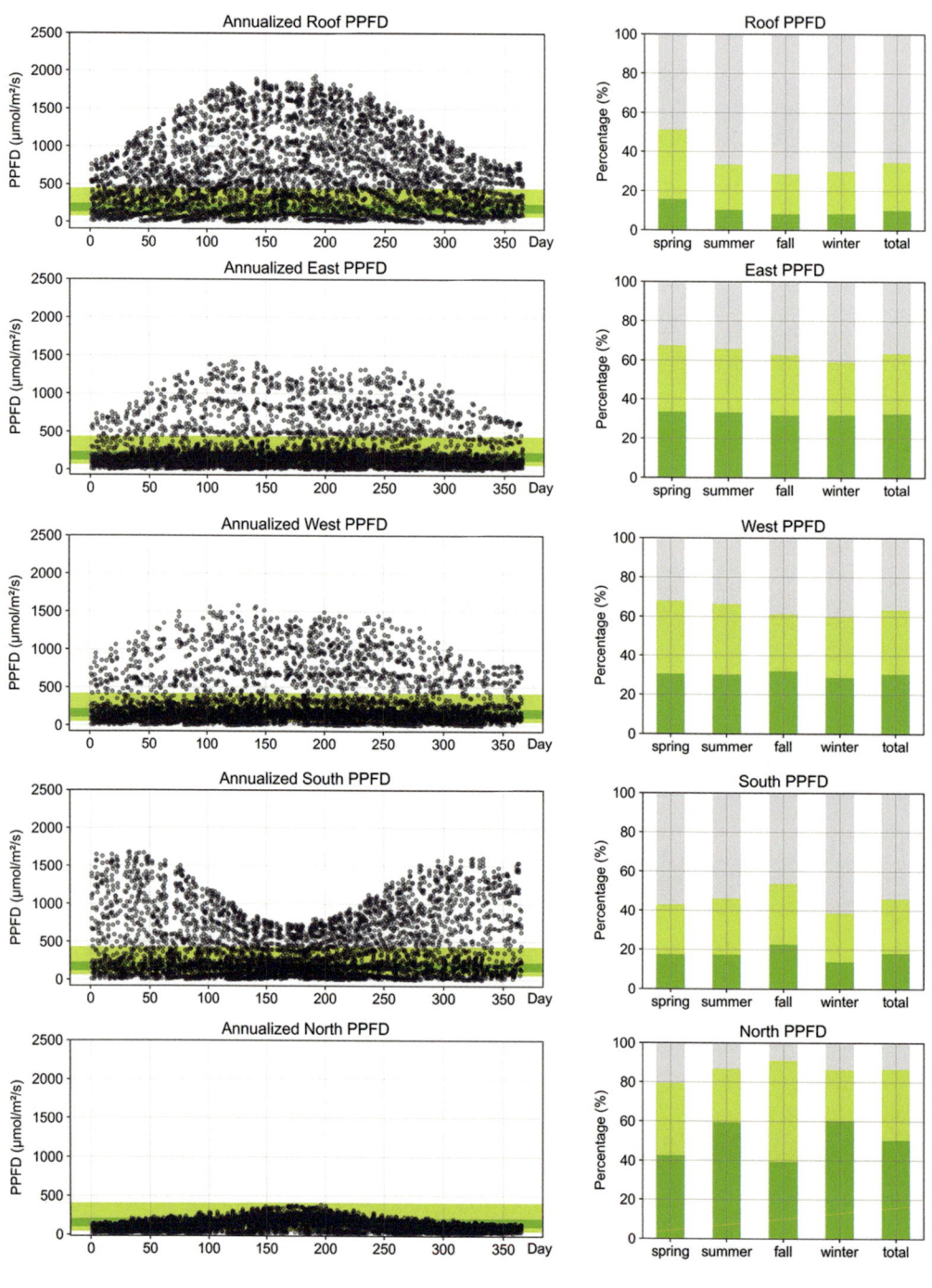

Simulated PPFD in Boston, MA
Climate zone 5A in accordance with ASHRAE 90.1

▲ Figure 8.7.19

Temperature ranges for different building envelop locations (roof, E, W, S, N) and the percentage of the data points that belong to good growing temperature ranges in Boston, Massachusetts (climate zone 5A per ASHRAE 90.1).

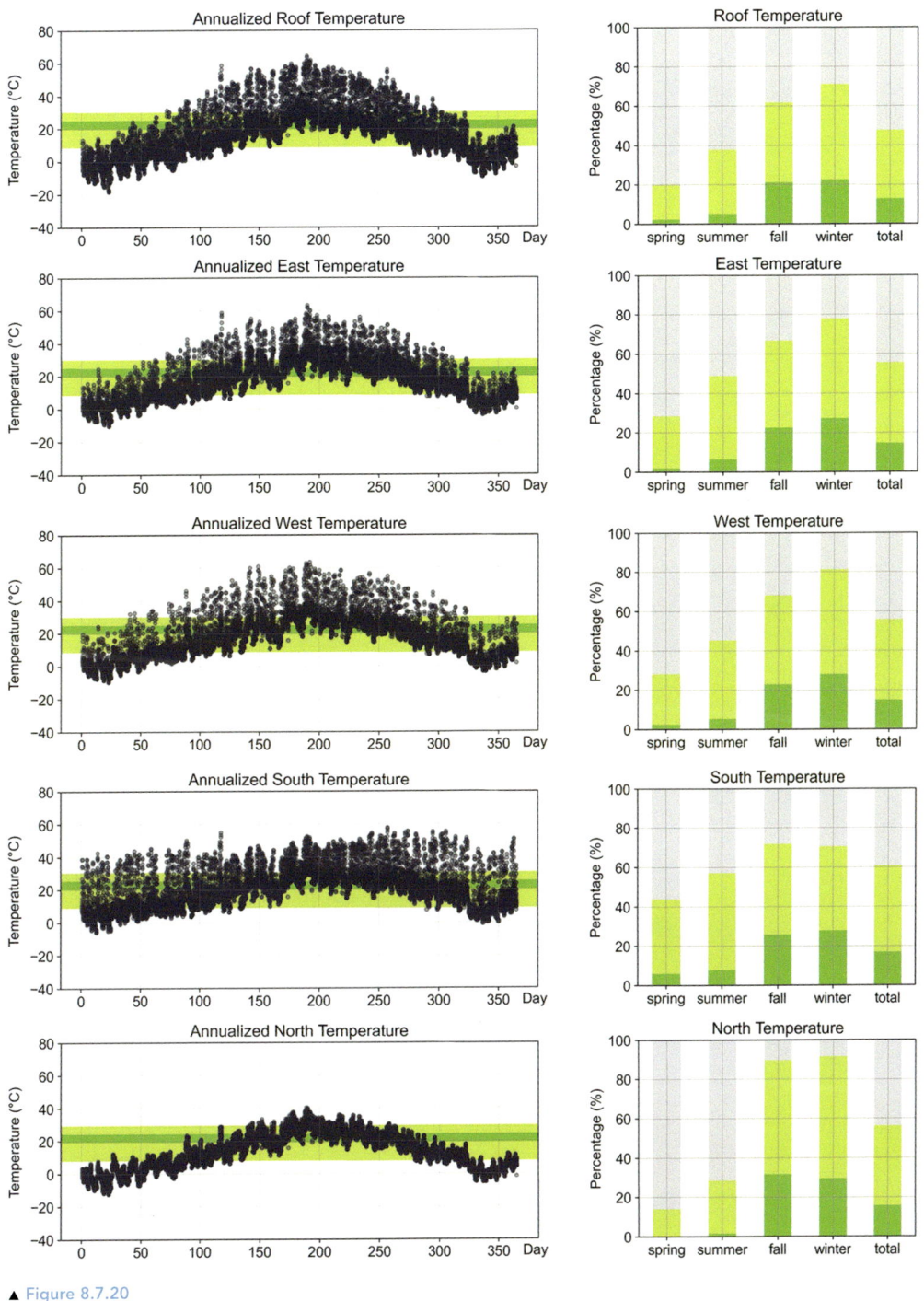

▲ Figure 8.7.20

Temperature ranges for different building envelop locations (roof, E, W, S, N) and the percentage of the data points that belong to good growing temperature ranges in Boston, Massachusetts (climate zone 5A per ASHRAE 90.1).

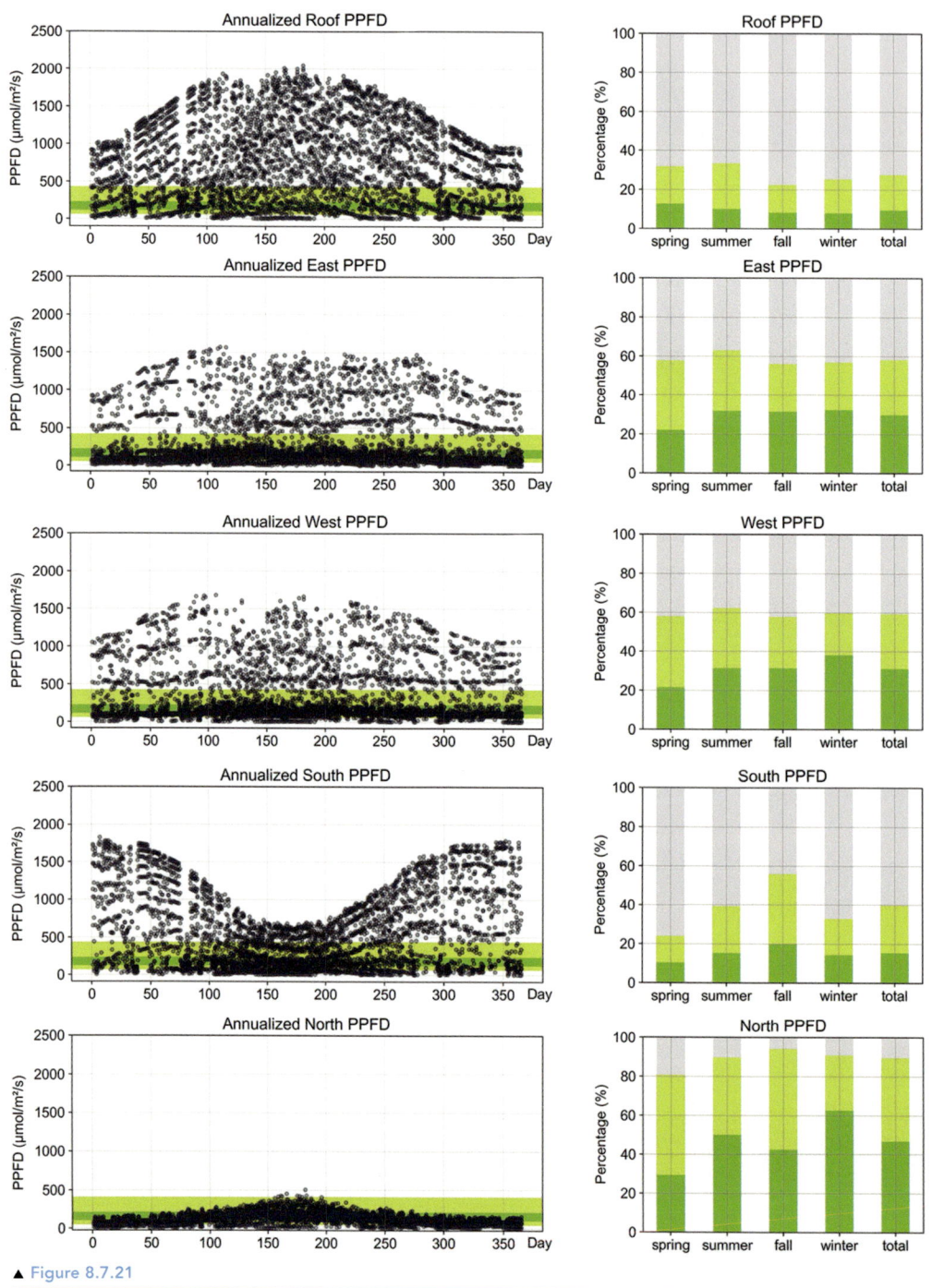

▲ Figure 8.7.21

Temperature ranges for different building envelop locations (roof, E, W, S, N) and the percentage of the data points that belong to good growing temperature ranges in Denver, Colorado (climate zone 5B per ASHRAE 90.1).

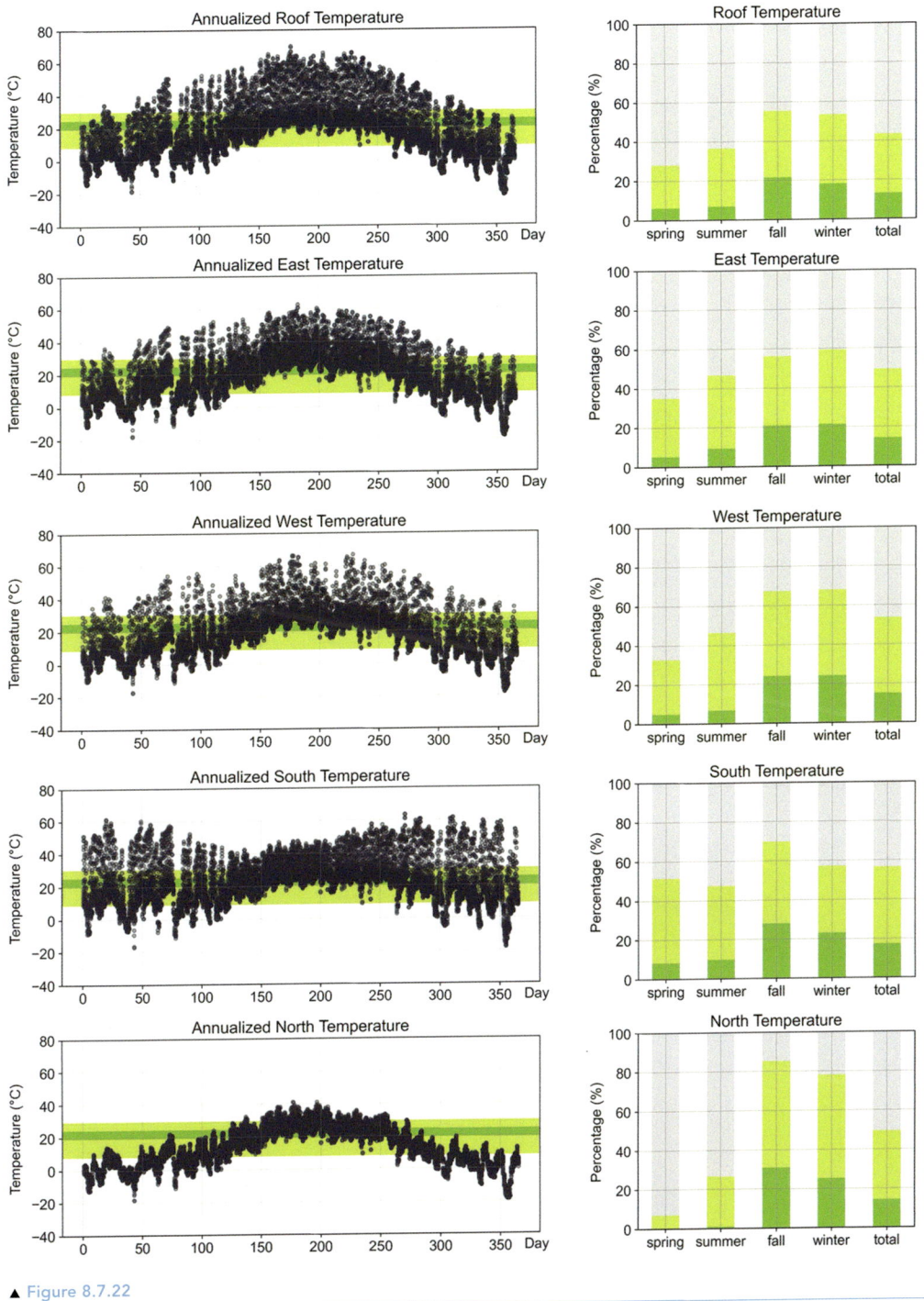

**Simulated Temperatures in Denver, CO
Climate zone 5B in accordance with ASHRAE 90.1**

▲ Figure 8.7.22

Temperature ranges for different building envelop locations (roof, E, W, S, N) and the percentage of the data points that belong to good growing temperature ranges in Denver, Colorado (climate zone 5B per ASHRAE 90.1).

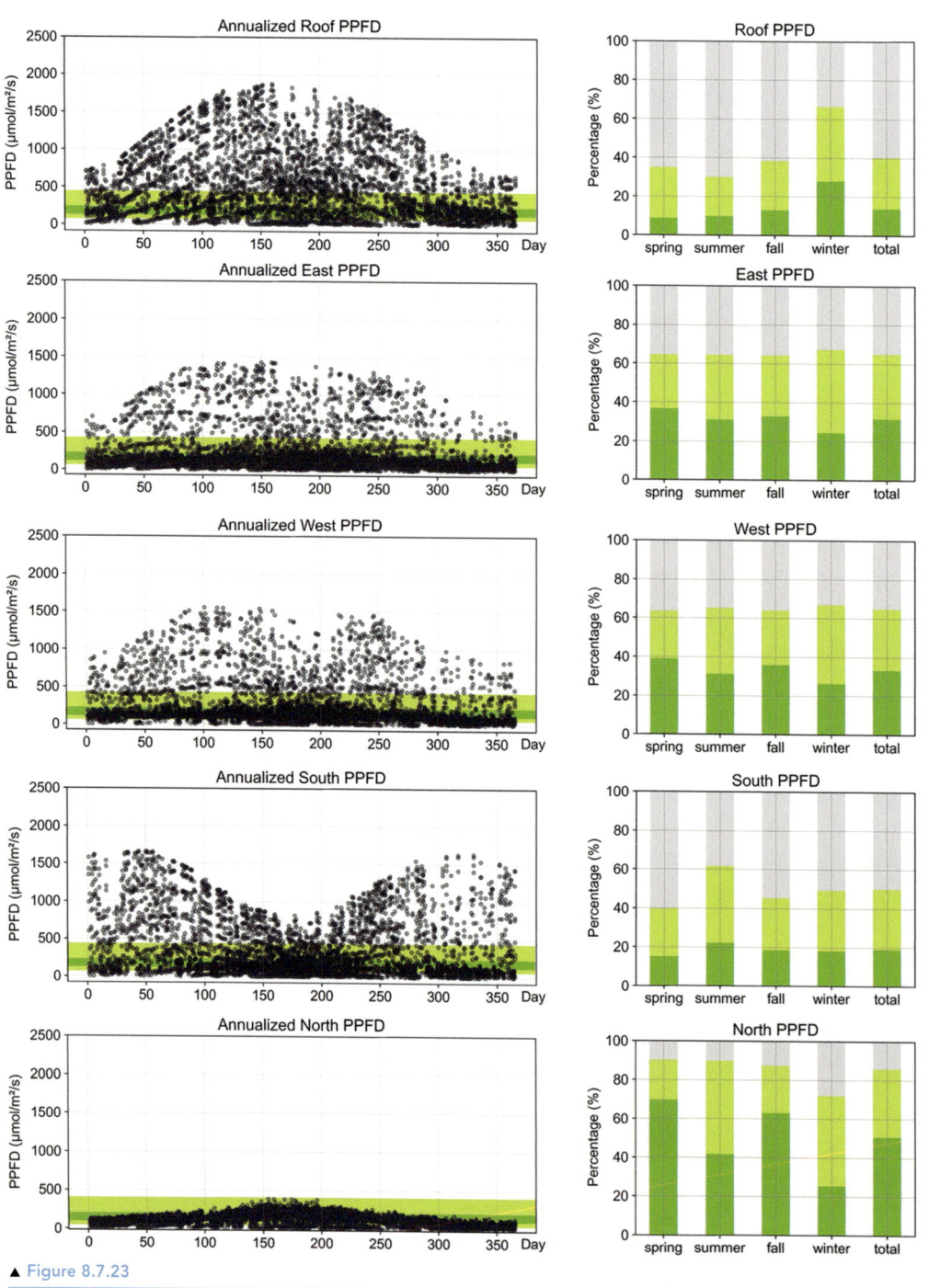

Simulated PPFD in Minneapolis, MN
Climate zone 6A in accordance with ASHRAE 90.1

▲ Figure 8.7.23

Temperature ranges for different building envelop locations (roof, E, W, S, N) and the percentage of the data points that belong to good growing temperature ranges in Minneapolis, Minnesota (climate zone 6A per ASHRAE 90.1).

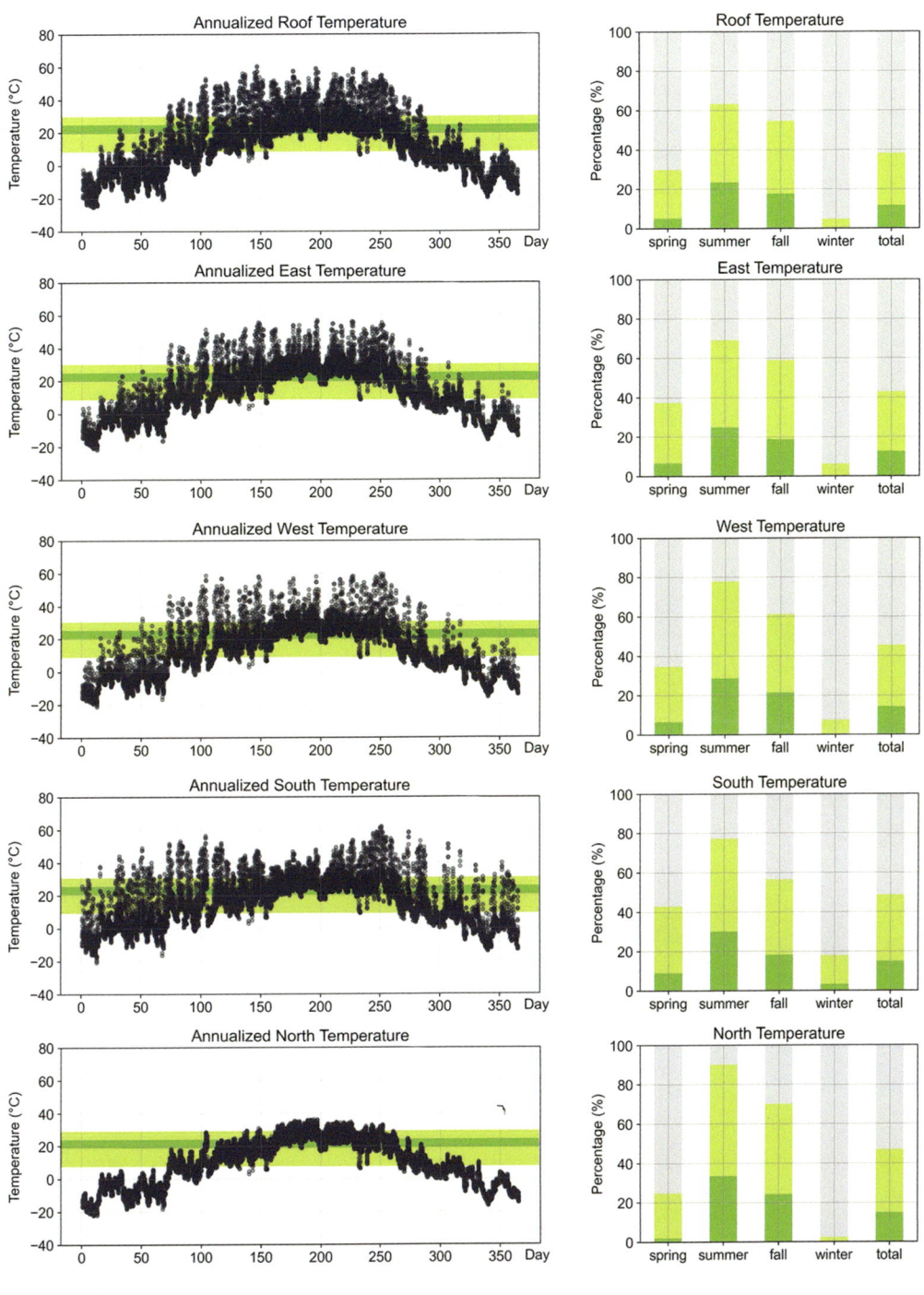

Simulated Temperatures in Minneapolis, MN
Climate zone 6A in accordance with ASHRAE 90.1

▲ Figure 8.7.24

Temperature ranges for different building envelop locations (roof, E, W, S, N) and the percentage of the data points that belong to good growing temperature ranges in Minneapolis, Minnesota (climate zone 6A per ASHRAE 90.1).

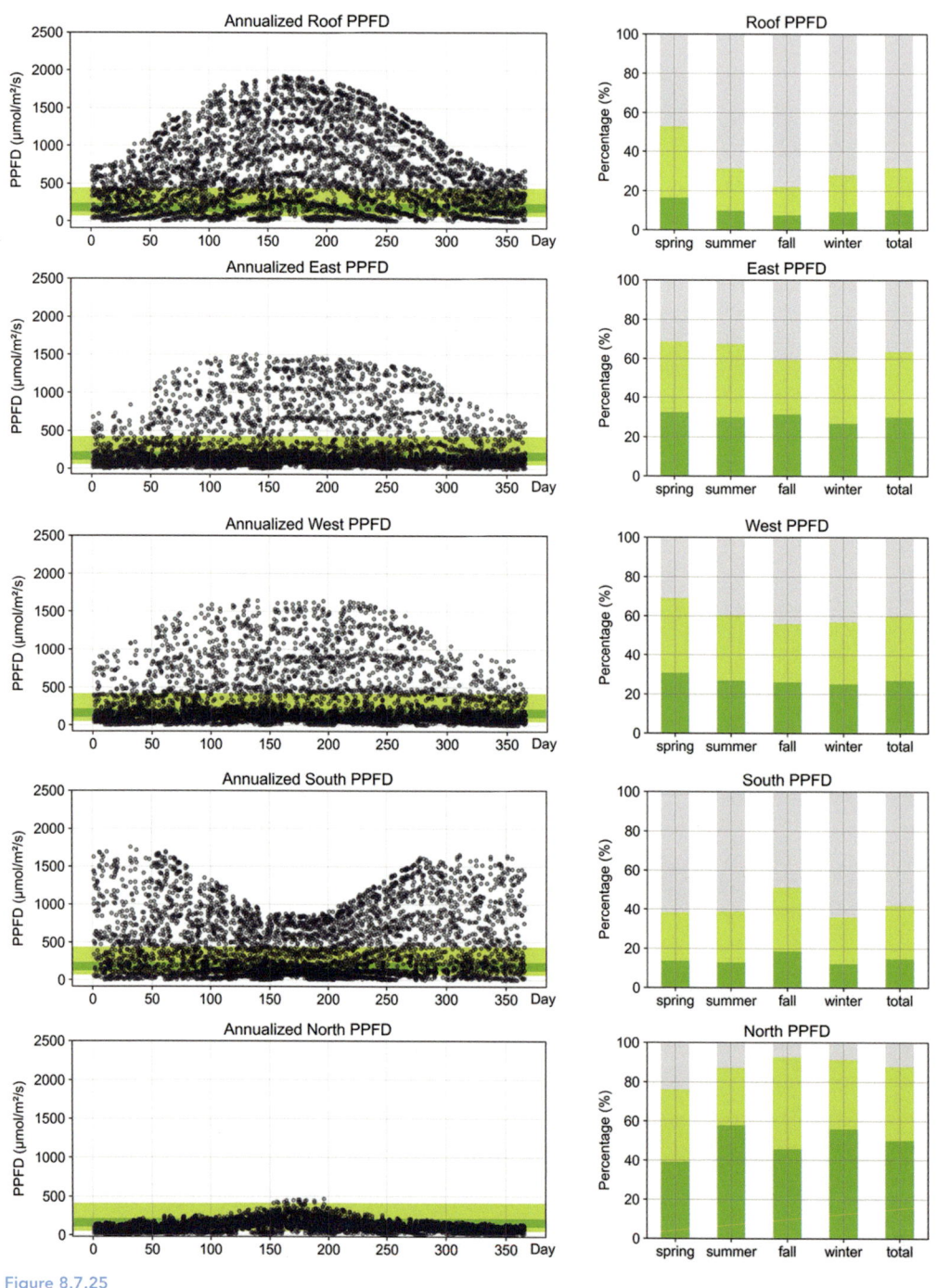

Simulated PPFD in Billings, MT
Climate zone 6B in accordance with ASHRAE 90.1

▲ Figure 8.7.25

Temperature ranges for different building envelop locations (roof, E, W, S, N) and the percentage of the data points that belong to good growing temperature ranges in Billings, Montana (climate zone 6B per ASHRAE 90.1).

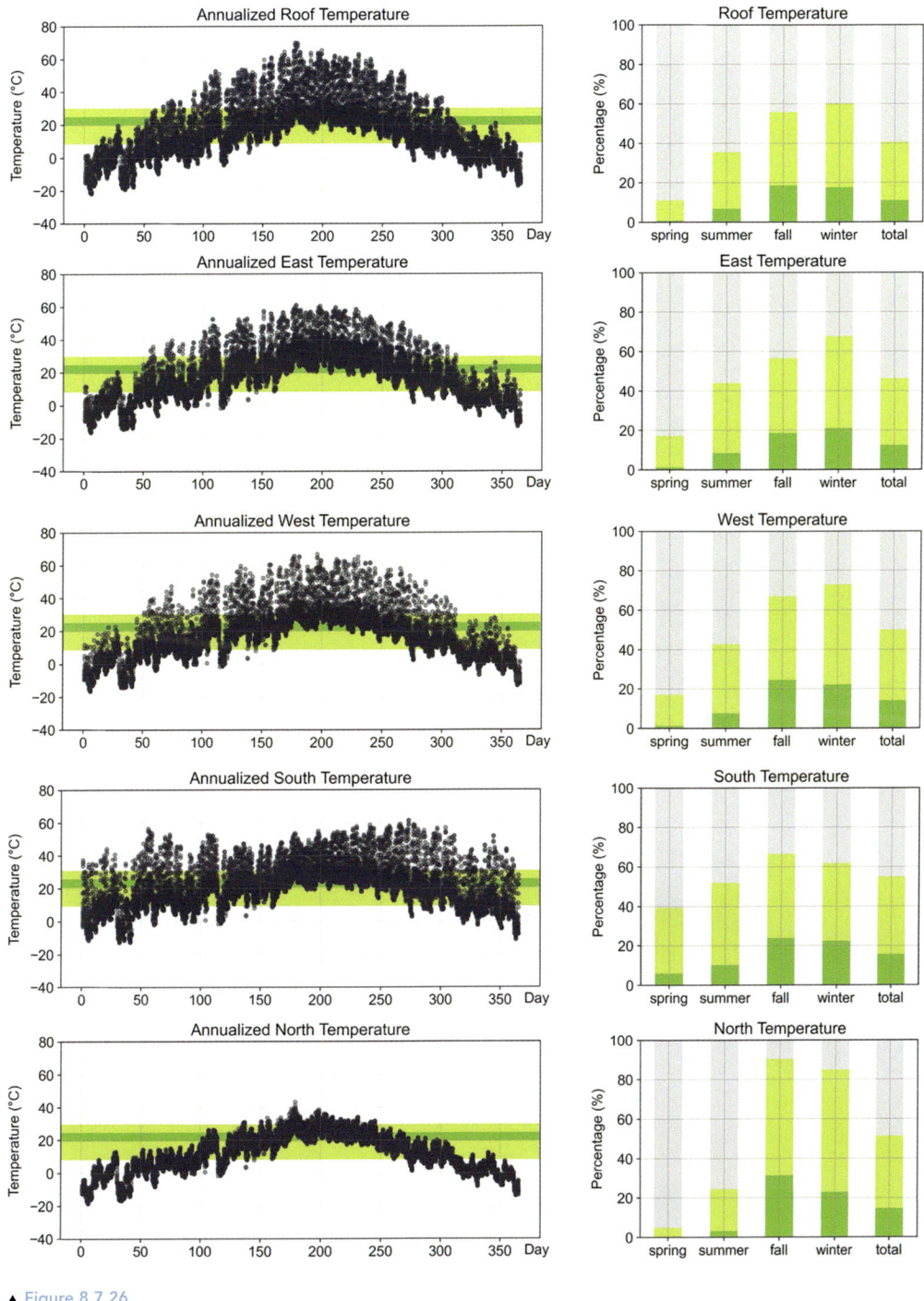

▲ Figure 8.7.26

Temperature ranges for different building envelop locations (roof, E, W, S, N) and the percentage of the data points that belong to good growing temperature ranges in Billings, Montana (climate zone 6B per ASHRAE 90.1).

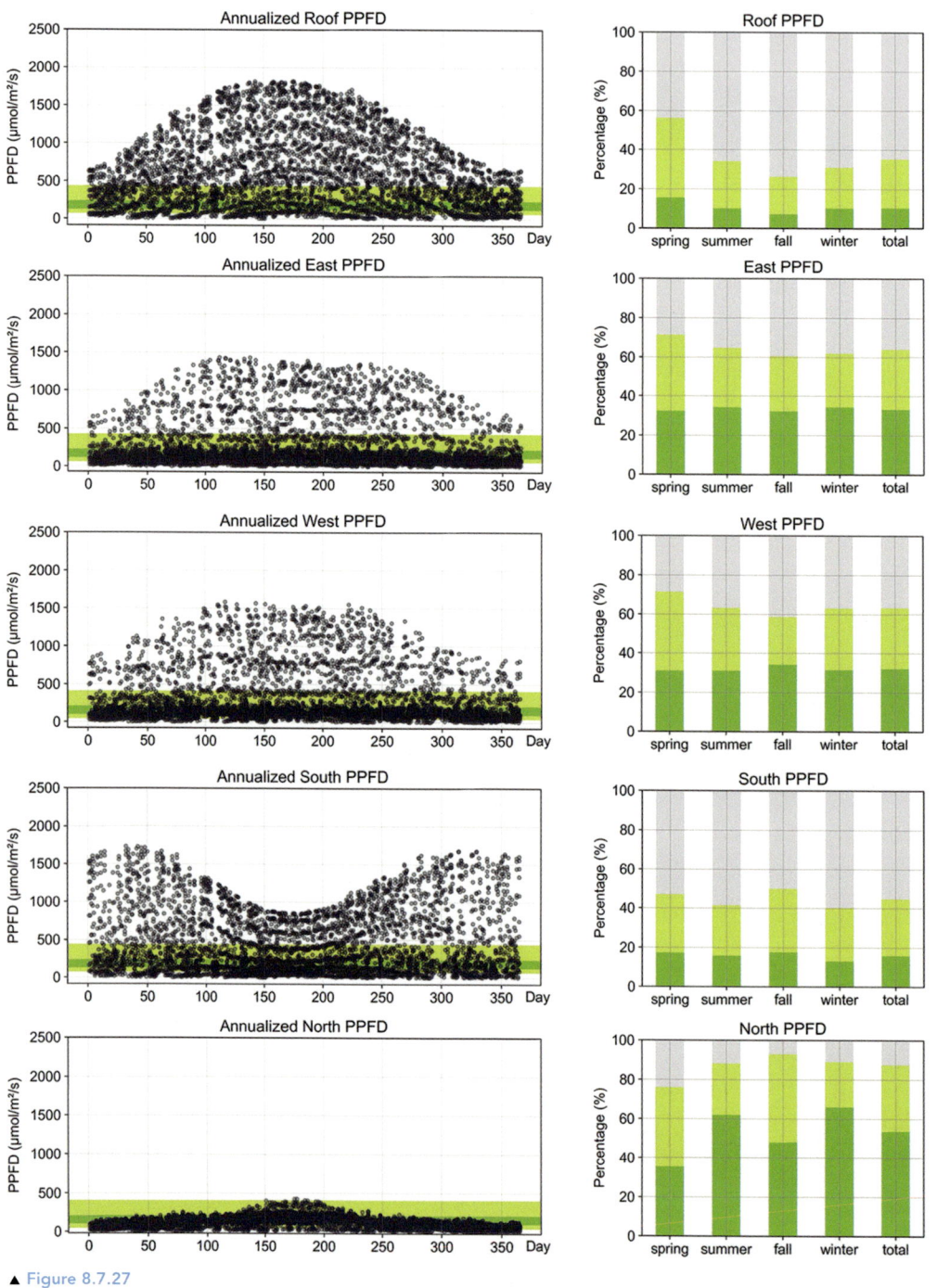

▲ Figure 8.7.27

Temperature ranges for different building envelop locations (roof, E, W, S, N) and the percentage of the data points that belong to good growing temperature ranges in Fargo, North Dakota (climate zone 7A per ASHRAE 90.1).

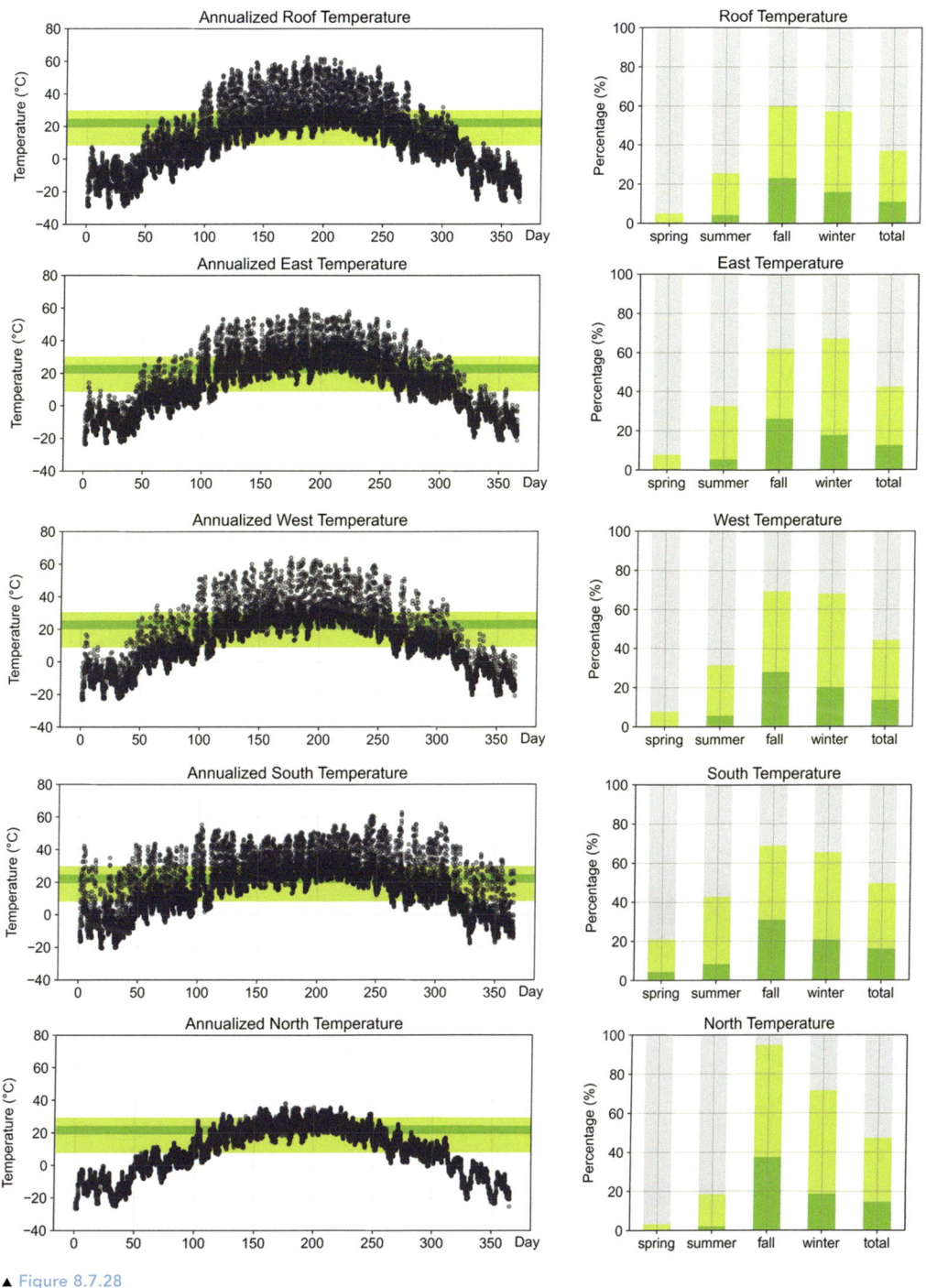

▲ Figure 8.7.28

Temperature ranges for different building envelop locations (roof, E, W, S, N) and the percentage of the data points that belong to good growing temperature ranges in Fargo, North Dakota (climate zone 7A per ASHRAE 90.1).

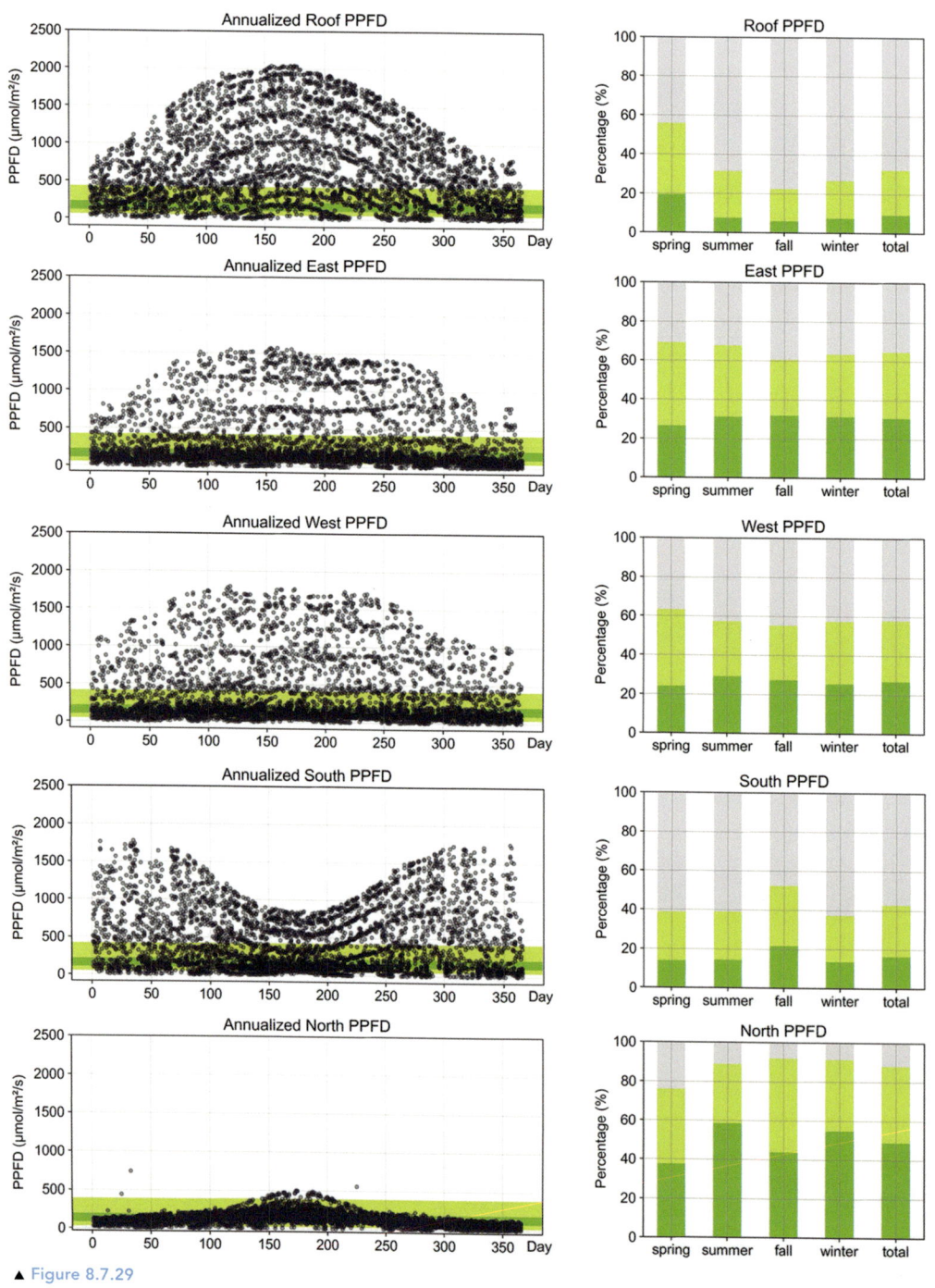

▲ Figure 8.7.29

Temperature ranges for different building envelop locations (roof, E, W, S, N) and the percentage of the data points that belong to good growing temperature ranges in Teton, Wyoming (climate zone 7B per ASHRAE 90.1).

PART III Microalgae Building Enclosure Design

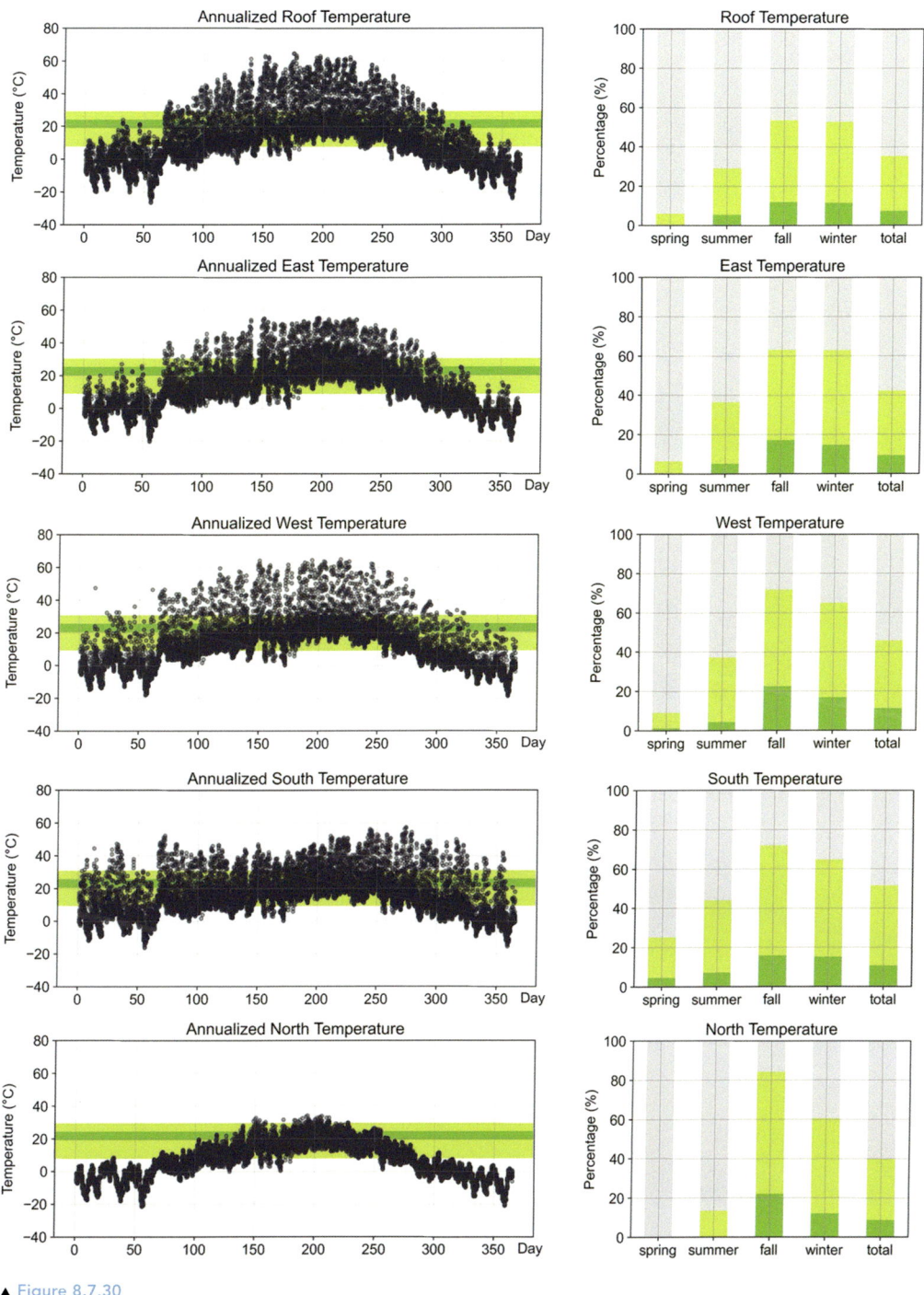

▲ Figure 8.7.30

Temperature ranges for different building envelop locations (roof, E, W, S, N) and the percentage of the data points that belong to good growing temperature ranges in Teton, Wyoming (climate zone 7B per ASHRAE 90.1).

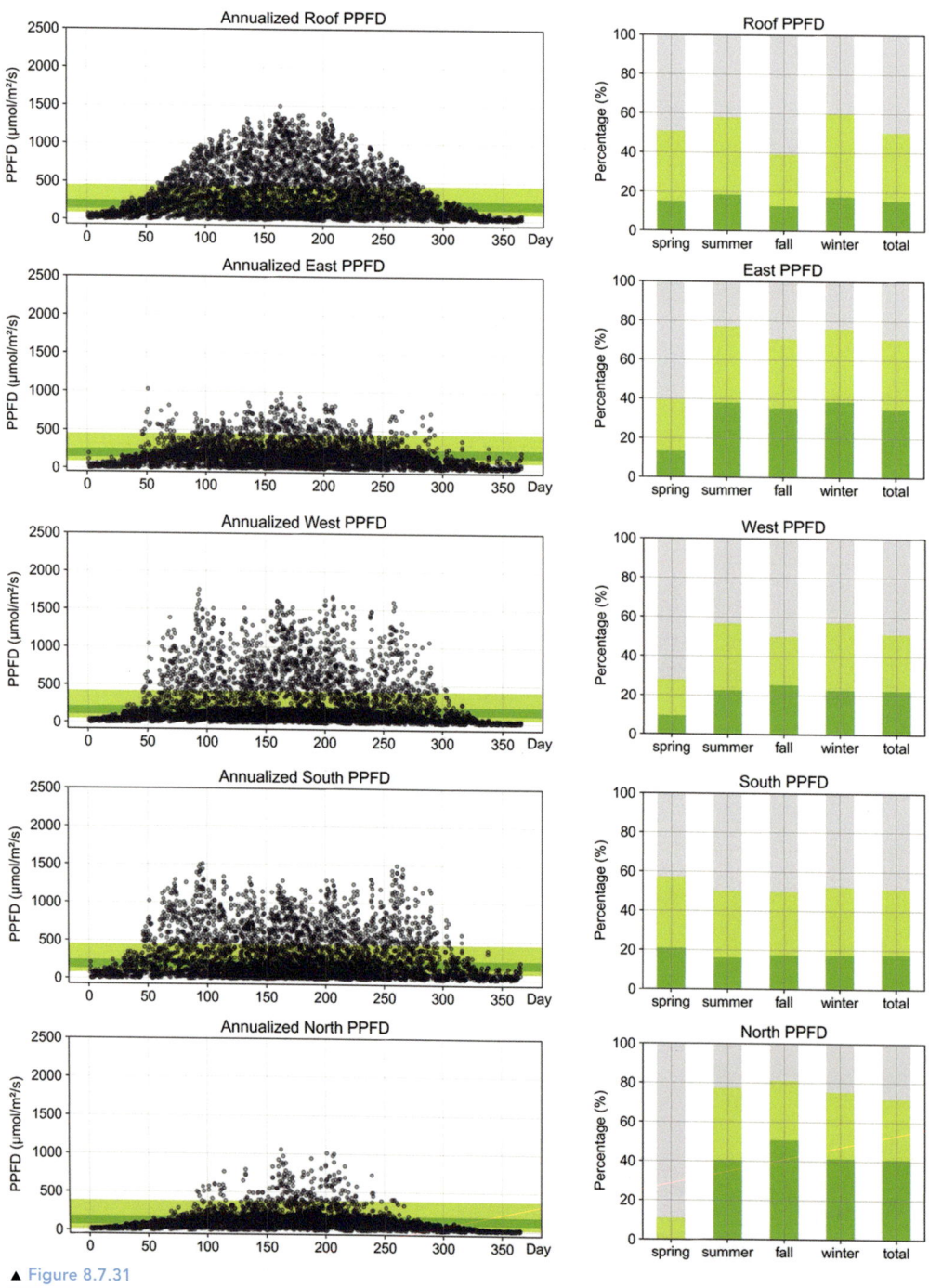

Simulated PPFD in Fairbanks, AK
Climate zone 8 in accordance with ASHRAE 90.1

▲ Figure 8.7.31

Temperature ranges for different building envelop locations (roof, E, W, S, N) and the percentage of the data points that belong to good growing temperature ranges in Fairbanks, Alaska (climate zone 8 per ASHRAE 90.1).

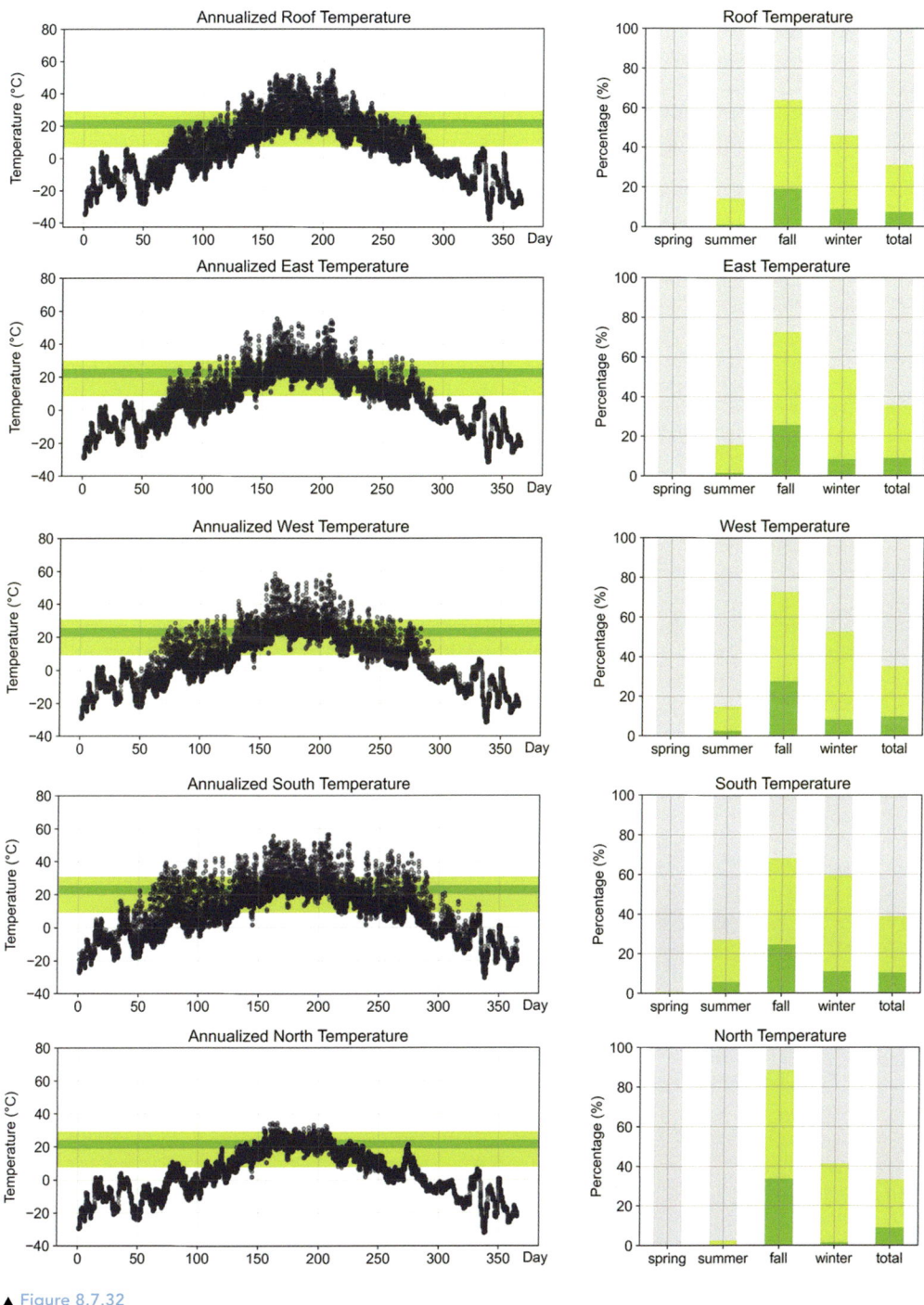

▲ Figure 8.7.32

Temperature ranges for different building envelop locations (roof, E, W, S, N) and the percentage of the data points that belong to good growing temperature ranges in Fairbanks, Alaska (climate zone 8 per ASHRAE 90.1).

Notes

1   U.S. Department of Energy (2020), Bioenergy Technologies Office 2019 R&D State of Technology, DOE/EE-2082, https://www.energy.gov/sites/prod/files/2020/07/f76/beto-2019-state-of-technology-july-2020-r1.pdf

2   Yunhua Zhu, Susanne B. Jones, and Daniel B. Anderson, *Algae Farm Cost Model: Considerations for Photobioreactors*. No. PNNL-28201. Pacific Northwest National Lab (PNNL), Richland, WA (United States), 2018, 8.

3   Tasneema Ishika, Parisa A. Bahri, Damian W. Laird, and Navid R. Moheimani, "The Effect of Gradual Increase in Salinity on the Biomass Productivity and Biochemical Composition of Several Marine, Halotolerant, and Halophilic Microalgae," *Journal of Applied Phycology* 30, no. 3 (2018): 1453–1464.

4   Gulab Singh and S.K. Patidar. "Microalgae Harvesting Techniques: A Review," *Journal of Environmental Management* 217 (2018): 499–508.

5   Ibid., 500.

6   P. Lavens, P. Sorgeloos, No. 361 *Manual on the Production and Use of Live Food for Aquaculture*, Food and Agriculture Organization (FAO) 1996.

7   Ritu Verma and Aradhana Srivastava, "Carbon Dioxide Sequestration and Its Enhanced Utilization by Photoautotroph Microalgae," *Environmental Development* 27 (2018): 95–106.

8   Ritu Verma, K.V.L. Kusuma Kumari, Aradhana Srivastava, and Arinjay Kumar, "Photoautotrophic, Mixotrophic, and Heterotrophic Culture Media Optimization for Enhanced Microalgae Production," *Journal of Environmental Chemical Engineering* (2020): 104149.

9   Ibid., 5.

10   Muhammad Imran Khan, Jin Hyuk Shin, and Jong Deog Kim, "The Promising Future of Microalgae: Current Status, Challenges, and Optimization of a Sustainable and Renewable Industry for Biofuels, Feed, and Other Products," *Microbial Cell Factories* 17, no. 1 (2018): 1–21.

11   Ibid., 5.

12   Ibid., 5.

13   Apogee instruments, "Conversion—PPFD to Lux," accessed July 1, 2021, https://www.apogeeinstruments.com/conversion-ppfd-to-lux/

14   Caner Koc, Gary A. Anderson, and Anil Kommareddy, "Use of Red and Blue Light-emitting Diodes (LED) and Fluorescent Lamps to Grow Microalgae in a Photobioreactor" (2013).

15   J. Masojídek, M. Sergejevová, K. Rottnerová, V. Jirka, J. Korečko, J. Kopecký, I. Zaťková, G. Torzillo, and D. Štys. "A Two-stage Solar Photobioreactor for Cultivation of Microalgae Based on Solar Concentrators," *Journal of Applied Phycology* 21, no. 1 (2009): 55–63.

16   Jeremy Pruvost, B. Le Gouic, O. Lepine, J. Legrand, and F. Le Borgne, "Microalgae Culture in Building-integrated Photobioreactors: Biomass Production Modelling and Energetic Analysis," *Chemical Engineering Journal* 284 (2016): 850–861.

17   Ibid., 851.

18   Yadiralia Covarrubias, Enrique A. Cantoral-Uriza, J. Sergio Casas-Flores, and J. Viridiana García-Meza, "Thermophile Mats of Microalgae Growing on the Woody Structure of a Cooling Tower of a Thermoelectric Power Plant in Central Mexico," *Revista Mexicana de Biodiversidad* 87, no. 2 (2016): 277–287.

19   V. Goetz, F. Le Borgne, Jeremy Pruvost, G. Plantard, and J. Legrand, "A Generic Temperature Model for Solar Photobioreactors," *Chemical Engineering Journal* 175 (2011): 443–449.

20   Ibid., 448.

21   L. Tomaselli, L. Giovannetti, A. Sacchi, and F. Bocci. "Effects of Temperature on Growth and Biochemical Composition in Spirulina Platensis Strain M2," in *Algal Biotechnology*, ed. T. Stadler...[et al.] (1988).

22 Chinchin Wang and Christopher Q. Lan, "Effects of Shear Stress on Microalgae—A Review," *Biotechnology Advances* 36, no. 4 (2018): 986–1002.

23 Ibid., 986.

24 Seyedeh Fatemeh Mohsenpour and Nik Willoughby, "Effect of CO2 Aeration on Cultivation of Microalgae in Luminescent Photobioreactors," *Biomass and Bioenergy* 85 (2016): 168–177.

25 Ibid., 171.

26 Anil R. Kommareddy, Gary A. Anderson, Stephen P. Gent, and Ghazi S. Bari, "The Impact of Air Flow Rate on Photobioreactor Sparger/Diffuser Bubble Size (s) and Distribution," in *2013 Kansas City, Missouri, July 21-July 24, 2013*(American Society of Agricultural and Biological Engineers, 2013), 1.

27 Ibid., 3, 4.

28 Ibid., 3.

29 Khan et al., "The Promising Future of Microalgae," 6.

30 Ibid., 6.

31 Mohsenpour et al., "Effect of CO_2 aeration," 173.

32 Ibid., 174.

33 Khan et al., "The Promising Future of Microalgae," 6.

34 Sonya T. Dyhrman, "Nutrients and Their Acquisition: Phosphorus Physiology in Microalgae," *The Physiology of Microalgae* (2016): 155–183.

35 Khan et al., "The Promising Future of Microalgae," 6

36 Ting Cai, Stephen Y. Park, and Yebo Li, "Nutrient Recovery from Wastewater Streams by Microalgae: Status and Prospects," *Renewable and Sustainable Energy Reviews* 19 (2013): 360–369.

37 Liling Jiang, Shengjun Luo, Xiaolei Fan, Zhiman Yang, and Rongbo Guo, "Biomass and Lipid Production of Marine Microalgae Using Municipal Wastewater and High Concentration of CO_2," *Applied Energy* 88, no. 10 (2011): 3336–3341.

38 Ibid., 3336.

39 Sheng-Yi Chiu Chien-Ya Kao, Tzu-Ting Huang, Chia-Jung Lin, Seow-Chin Ong, Chun-Da Chen, Jo-Shu Chang, and Chih-Sheng Lin, "Microalgal Biomass Production and On-site Bioremediation of Carbon Dioxide, Nitrogen Oxide and Sulfur Dioxide from Flue Gas Using Chlorella sp. Cultures," *Bioresource Technology* 102, no. 19 (2011): 9135–9142.

40 F.G. Acién, et al., "Photobioreactors for the Production of Microalgae," in *Microalgae-based Biofuels and Bioproducts: From Feedstock Cultivation to End-products* (Cambridge, MA: Woodhead Publishing Series in Energy, 2017): 34.

41 Ibid., 14-15.

42 Ibid., 34.

43 Jeroen H. de Vree, et al., "Comparison of Four Outdoor Pilot-scale Photobioreactors," *Biotechnology for Biofuels* 8 no. 215 (2015): 4, https://doi.org/10.1186/s13068-015-0400-2.

44 Bei Wang, Christopher Q. Lan, and Mark Horsman, "Closed Photobioreactors for Production of Microalgal Biomasses," *Biotechnology Advances* 30, no. 4 (2012): 904–912.

45 R.N. Singh and Shaishav Sharma, "Development of Suitable Photobioreactor for Algae Production—A Review," *Renewable and Sustainable Energy Reviews* 16, no. 4 (2012): 2347–2353.

46 F.G. Acién, et al., "Photobioreactors for the Production of Microalgae," 34.

47 Ibid., 23.

48 Ibid., 23.

49 Ibid., 15.

50 Ibid., 23.

51 Benjamas Cheirsilp and Salwa Torpee. "Enhanced Growth and Lipid Production of Microalgae under Mixotrophic Culture Condition: Effect of Light Intensity, Glucose Concentration and Fed-batch Cultivation." *Bioresource Technology* 110 (2012): 510–516.

52 Sheng-Yi Chiu, Chien-Ya Kao, Chiun-Hsun Chen, Tang-Ching Kuan, Seow-Chin Ong, and Chih-Sheng Lin, "Reduction of CO2 by a High-density Culture of Chlorella sp. in a Semicontinuous Photobioreactor," *Bioresource Technology* 99, no. 9 (2008): 3389–3396.

53 Haiying Tang, Meng Chen, K.Y. Simon Ng, and Steven O. Salley, "Continuous Microalgae Cultivation in a Photobioreactor," *Biotechnology and Bioengineering* 109, no. 10 (2012): 2468–2474.

54 Juliana Botelho Moreira, Jorge Alberto Vieira Costa, and Michele Greque de Morais, "Evaluation of Different Modes of Operation for the Production of Spirulina sp.," *Journal of Chemical Technology & Biotechnology* 91, no. 5 (2016): 1345–1348.

55 J.R. Malapascua, et al., "Photosynthesis and Growth Kinetics of *Chlorella vulgaris* R-117 Cultured in an Internally LED-illuminated Photobioreactor," *Photosynthetica* 57, no. 1 (2019): 103, http://doi.org/10.32615/ps.2019.031.

56 Jyoti Sharma, et al., "Microalgal Consortia for Municipal Wastewater Treatment— Lipid Augmentation and Fatty Acid Profiling for Biodiesel Production," *Journal of Photochemistry & Photobiology* 202, no. 111638 (2020): 2, https://doi.org/10.1016/j.jphotobiol.2019.111638.

57 K.D. Sung, et al., "Isolation of a New Highly CO_2 Tolerant Fresh Water Microalga *Chlorella* sp. KR-1," *Renewable Energy* 16 (1999): 1019, https://doi.org/10.1016/S0960-1481(98)00362-0.

58 Virthie Bhola, et al., "Effects of Parameters Affecting Biomass Yield and Thermal Behavior," 380.

59 Eduardo Bittencourt Sydney, et al., "Respirometric Balance and Carbon Fixation," 76.

60 Muhammad Aminul Islam, et al., "Microalgal Species Selection for Biodiesel Production Based on Fuel Properties Derived from Fatty Acid Profiles," *Energies* 6 (2013): 5676, http://doi.org/10.3390/en6115676.

61 Carl Safi, et al., "Morphology, Composition, Production, Processing, and Applications of *Chlorella vulgaris*," 274.

62 Jorge Olmos Soto, "*Dunaliella* Identification Using DNA Fingerprinting Intron-Sizing Method and Species-Specific Oligonucleotides," in *Handbook of Marine Microalgae*, ed. Se-Kwon Kim (Elsevier, 2015), 559–560.

63 Ibid., 562.

64 Ibid., 560.

65 Ibid., 559.

66 Ibid., 560.

67 National Renewable Energy Laboratory, *A Look Back at the U.S. Department of Energy's Aquatic Species Program: Biodiesel from Algae*, by John Sheehan, Terri Dunahay, John Benemann, and Paul Roessler, NREL/TP-580-24190 (Golden: U.S. Department of Energy, 1998), 44.

68 Ayesha Shahid, et al., "Cultivating Microalgae in Wastewater for Biomass Production, Pollutant Removal, and Atmospheric Carbon Mitigation; A Review," *Science of the Total Environment* 704 (2020): 135308, https://doi.org/10.1016/j.scitotenv.2019.135303.

69 Wim Brilman, et al., "Capturing Atmospheric CO_2 Using Supported Amine Sorbents for Microalgae Cultivation," *Biomass and Bioenergy* 53 (2013): 45, http://dx.doi.org/10.1016/j.biombioe.2013.02.042.

70 Carl Safi, et al., "Morphology, Composition, Production, Processing, and Applications of *Chlorella vulgaris*," 274.

71 Eduardo Bittencourt Sydney, et al., "Respirometric Balance and Carbon Fixation," 75.

72 Michael H. Huesemann and John R. Benemann, "Biofuels from Microalgae: Review of Products, Processes and Potential, with Special Focus on Dunaliella sp.," *The Alga Dunaliella* (2019): 445–474.

Chapter 9

Bioclimatic Design Overview

Bioclimatic building design considers climate characteristics in specific site locations and building orientations. Bioclimatic strategies capitalize on surrounding resources, such as sunlight, wind, rain, and landscapes, and minimize building energy consumption especially in heating, cooling, lighting, and ventilation while offering health and comfort for users. As discussed in the previous chapters, Chapters 4–7, microalgae can be integrated with different building systems and infrastructures such as building enclosures, roofs, interior systems, furniture, urban farming, waste treatment facilities, community gardens, and so on. This section focuses on microalgae building enclosures and delves into their potential environmental, economic, and social benefits. The bioclimatic design strategies for microalgae enclosures start with finding a balance between vision area for occupants' view-out and microalgae area for maximum cultivation. The microalgae area offers energy attributes of U-factor (i.e., heat transmission), SHGC (solar heat gain coefficient), and VLT (visible light transmittance), whereas vision area provides view-out and daylighting related to occupant satisfaction. The cell density and color changes from the microalgae area provide shading in summer as microalgae cells grow faster under intense sunlight. The solar radiation in winter slows down the cell growth, allowing winter sun penetration into the room. The provision of view-out area within the microalgae system admits year-round daylighting. Microalgae systems utilize CO_2 inside the room and, in return, reduce CO_2 and generate oxygen for the occupants' health and well-being. This basic biological performance establishes the potential impact on overall energy performance, microalgae growing environments, and occupant comfort.

9.1 Microalgae Architecture Bioclimatic Design

Bioclimatic design is a climate-sensitive design to achieve comfort improvement and energy load reduction through climate responsive design strategies. A building following bioclimatic design strategies helps occupants to accommodate extended boundaries of thermal comfort zones. For example, the ability to open a window and access shading blinds gives wider temperature options for occupants. Tree-planting, light-colored surfacing, or solar shading help temper microclimates around the building. A bioclimatic approach analyzes climatic characteristics from the viewpoint of occupant comfort using bioclimatic charts in which the human thermal comfort zone is identified. Bioclimatic design starts from the analysis of

DOI: 10.4324/9780367814410-12

climatic characteristics based on climatic indices and variables. These variables include temperature, humidity, precipitation, solar radiation, wind speed, cooling/heating degree days, temperature swing, and effective temperature. The solar radiation consists of direct and diffused light throughout the year with maximum and minimum irradiance in a certain period. Direct radiation represents a major part of solar radiation. Daily ambient temperatures oscillate between high and low temperatures throughout the year. Wind speed and blowing directions affect ambient temperatures and microclimates around the building. The summer season is very hot, but the availability of prevailing winds on the site can temper thermal stress. The climate is also characterized by the amount of rainfall that affects relative humidity. Climatic characteristics are correlated with topological information such as latitude, longitude, altitude, and proximity to a coast.

The boundaries of comfort zones that an occupant can tolerate vary depending on climatic characteristics and design strategies. The boundary of thermal comfort zones is typically extended as bioclimatic design strategies are introduced. Bioclimatic design research has long been studied by design science researchers such as Olgyay, Giovoni, Milne and Giovoni, Dekay and Brown, Szokolay, and so on. The promise of bioclimatic design is that when a building is designed in a way that utilizes local climates for building operation, users tend to adapt to local climates and reduce reliance on active HVAC (heating, ventilation, and air conditioning) systems. One metric to evaluate the effectiveness of the bioclimatic building is the extension of the boundary of the indoor climate zone to determine an occupant's comfort level in a wide range of environmental conditions.

Olgyay developed the Mollier diagram in 1923 based on outdoor temperatures and relative humidity.[1] When the crossover of the temperature and relative humidity does not situate within the thermal comfort zone, air movement and solar radiation can assist in adjusting an uncomfortable case to the thermal comfort zone. However, Olgyay's bioclimatic chart evaluates outdoor climates and does not incorporate the indoor environment of a building. Giovoni's bioclimatic chart in 1969 incorporates both thermal comfort and bioclimatic design strategies along with temperature and humidity data based on indoor environments.[2] Bioclimatic design is important for improving user comfort when outdoor climates are not favorable for thermal comfort. Milne and Giovoni's bioclimatic chart in 1981 added more climate-responsive strategies. As a result, the boundaries of the thermal comfort zones were expanded and greater thermal comfort zones were achieved.[3] The Milne and Giovoni chart has been used by successive researchers and for the ASHRAE guidelines. ASHRAE defines thermal comfort as "that condition of mind that expresses satisfaction with the thermal environment."[4] Thermal comfort according to ASHRAE 55 is represented by the predicted mean vote (PMV). This index is a seven-point thermal sensation scale based on the mean value of votes of a group of occupants. This indoor thermal comfort can be predicted based on environmental and personal factors including air temperature, radiant temperature, air speed, humidity, clothing insulation, and metabolic rate.

Two levels of bioclimatic design strategies can be considered for microalgae enclosure design. One strategy is to reduce building energy consumption and maintain occupant satisfaction in the areas of thermal, visual, and psychological attributes while providing an optimum growing environment for maximum microalgae productivity. As a direct interface between outdoor and indoor, microalgae building enclosures take most of the climate loadings (e.g., solar radiation, temperature, prevailing wind, rain) in the first place. The microalgae enclosure changes tint and color in response to local climates and mitigates energy transfer between outside and inside while improving occupant comfort. The second level of bioclimatic design strategy is that microalgae enclosures serve as an extension of the HVAC system. Its dynamic energy attributes reduce the operation of the HVAC system. In addition, solar energy stored in the microalgae enclosure could be used for space heating and domestic hot water. Its ability to use room air with high CO_2 concentrations generated by occupants and to generate oxygen as a result of photosynthesis could be tied with the air handling unit for further improvement toward good indoor air quality. A centralized building management system can tie in with both the microalgae cultivation system and building service system. The lighting, A/C, ventilation, and cultivation are controlled and monitored. When the energy regulations from the microalgae enclosures are not sufficient to achieve occupant comfort inside the room, HVAC systems come into operation. The aeration system is operated for supplying room air. When the stationary growth state is reached, plumbing and storage tanks are required to harvest dense microalgae and refill the new nutrient-rich media. The conventional chemical and physical processes of treating wastewater are expensive and multi-process intensive. Microalgae enclosures can be combined with wastewater treatment and power production facilities. Hyperconcentrated algal cultures are highly efficient in uptaking nitrogen and phosphorus from wastewater within a short period of time. Holistic building management provides maximum control over all equipment and operation, resulting in high efficiency, energy saving, and comfort.

Depending on climatic regions, the priorities of bioclimatic design strategies vary. For hot climates, bioclimatic design focuses on the control of solar gains through the building skin. Energy stored in the building enclosures is reclaimed through recycling or heat sink mechanisms. Heat disposal could be coupled with ground, air, and water used as a heat sink or a heat exchanger using ventilative cooling or evaporative cooling techniques. The area of microalgae enclosures can be increased to absorb solar radiation and minimize overheating problems inside the building. Additional techniques could be integrated with the microalgae area to further control excessive solar stress using photovoltaic (PV) cells to filter light while reducing solar radiation and producing electricity. Geometric intervention such as self-shading can avert direct solar radiation. Surface treatments such as textured or cool surface coating reduce solar gain by reflecting the majority of solar radiation.

Microalgae architecture designs need balance between occupant satisfaction, building energy savings, and microalgae growth. To that end, there needs to be close coordination between design, engineering, and biotechnical aspects. Different climatic regions offer different light intensities and temperatures. An optimum microalgae species is selected based on performance goals and the climates (e.g., illuminance level, temperature conditions) available on the site. Most microalgae species grow well in the temperature range between 10°C and 32°C (50–90°F). Within these temperatures, photosynthetic saturation occurs at light intensities of about 200–250 μmol/m^2/s, resulting in an average utilization of solar energy at 1.3–7%.[5] Since only a fraction of the solar energy is used for photosynthesis, the remaining energy is converted to heat, and so it is important to monitor and regulate heat build-up for microalgae.

When microalgae enclosures are implemented in cold regions, bioclimatic design starts with maximizing solar heat gains through the microalgae skin and

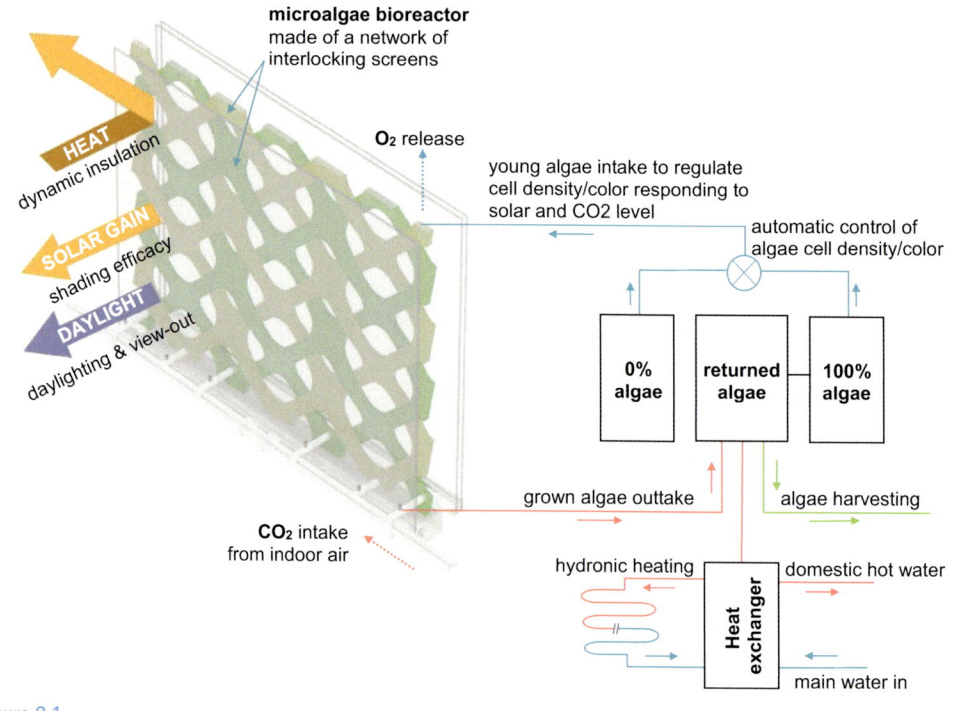

▲ Figure 9.1

Key bioclimatic design strategy for microalgae building enclosures consisting of three focus areas: (1) regulation of heat transmission, solar gain, and daylighting transmission; (2) automatic control of color and tint of microalgae enclosures responding to solar intensity and CO_2 level; and (3) energy reclamation stored in the microalgae envelope for other building service systems such as space and domestic hot water heating.

regulating heat transfer between the inside and outside. A continuous production mode allows the microalgae density to reach a steady state, and the thin density of microalgae can maximize solar gain inside the space in winter. Pumping room air into the microalgae enclosure offers dynamic insulation and reduces temperature-based heat transfer between the inside and outside, resulting in heating reduction and thermal comfort.[6] The microalgae enclosure can store solar energy in the daytime and serve as thermal mass for nighttime use. Figure 9.1 depicts microalgae enclosure bioclimatic design strategies. As a primary building envelope, heat transmission, solar gain, and daylighting are controlled in the first place. Further building energy efficiency is achieved by regulating the color and tint of microalgae depending on solar intensity and CO_2 level. The stored heat energy after photosynthesis is used for other building services such as hydronic heating and domestic hot water heating.

9.2 Energy Attributes of Microalgae Building System

The radiation transmission through a window affects both SHGC and VLT. The degree of solar transmission varies depending on solar intensity, incident angle, chemical composition, and surface treatment of windows. The sun intensity impinging on the surface diminishes as the incidental angle increases. The 90-degree incidental angle when the sun is perpendicular to the surface is subject to the maximum solar exposure. Wavelength transmission through a window varies by chemical composition of materials, and especially low-e coated glass can block infrared wavelengths while maximizing the visible light spectrum. Solar energy consists of UV (300–400 nm), visible light (400–700 nm), and infrared (700–2500 nm) regions where UV comprises 6.6% of the solar energy, visible light 44.7%, and infrared 48.7%.[7]

All microalgae enclosures need to provide energy attributes in accordance with industry standards. The energy attributes and durability of a custom system are certified through full-scale performance testing in an accredited laboratory that has been approved by the government. In addition to energy attributes, it is important to carry out long-term durability assessments under dynamic weather conditions. High-performance enclosures play an important role in enhancing energy savings by reducing heat losses through conductivity while mitigating summer solar gain and augmenting daylighting for window applications. R-value, the resistance of heat flow between the inside and outside, is one of the energy attributes for opaque building enclosures. A higher R-value corresponds to greater energy savings. The U-factor (rate of heat loss), SHGC, and VLT are important energy attributes for a window system. U-factor is the degree of energy transfer through a wall assembly, and a lower U-factor is better for heating energy load. The SHGC is a measure of the fraction of total sunlight energy that can pass through the window. A higher SHGC means more cooling load in summer. The VLT measures the fraction of visible sunlight that passes through the window. A higher VLT indicates lower artificial lighting demand. Greater heat exchange between indoor and outdoor spaces occurs through windows in the form of light and heat, which affects energy savings and occupant comfort.[8]

Dynamic U-factor: U-factor (a reciprocal of R-value) directly correlates with heating and cooling as well as occupant thermal comfort. Temperature differences between indoor and outdoor cause energy transfer, requiring HVAC systems to meet target indoor temperatures and thermal comfort. U-factor encompasses all aspects of heat transfer by conduction, convection, and radiation. Typically, the major heat transfer for a wall occurs by conduction through the solid materials and convection at the exterior and interior surface of the wall, whereas radiation is negligible. Air cavities in insulated glass units are subject to convection heat transfer. A lower U-factor (or higher R-value) is recommended practice for energy efficiency. However, depending on climatic regions and building types, static and high-efficient insulation can increase energy consumption if it exceeds the threshold of thermal resistance called "thermal inflection."[9] Dynamic insulation that provides low to high insulation values depending on climatic variations has been shown to outperform traditional static insulation.[10] For example, an opaque wall system circulated by water and air during summer effectively reduces thermal conductivity and decreases cooling load.[11] Therefore, microalgae enclosures are a great opportunity for mitigating energy flow by utilizing conditioned airflows through the microalgae system. The temperature of the microalgae system is maintained with an optimum temperature range for growing microalgae by pumping the room air. The aeration system utilizes waste energy from the return air of interior rooms, and the conditioned CO_2 (room air) flows through the microalgae enclosure. By pumping the conditioned room air, the microalgae system offers dynamic insulation mitigating temperature differences between indoor room and outdoor environments and, thereby, minimizes the space heating and cooling with minimal HVAC being used to condition the room. This process generates a smart thermal envelope to conserve building energy consumption. Enhancing the U-factor of the building enclosure as adaptable to variable climate conditions leads to substantial energy reduction.

Solar Heat Gain Coefficient (280–2500 nm): Microalgae grow fast as the sunlight becomes intense, providing dynamic SHGC depending on sunlight intensity and the time of day. The SHGC represents the solar heat gain through the window system as related to the incidental solar radiation. The SHGC is composed of the directly transmitted solar radiation and the inward-flowing portion of the absorbed solar radiation radiated to the interior of the building. The microalgae enclosure can effectively admit or block solar radiation into the interior space by changing its density and growth rate responding to solar intensity. The microalgae density in summer will remain high due to intense sunlight and active photosynthesis, providing good shading and carbon sequestration for the building. In winter, the microalgae system with the continuous cultivation mode allows a desired dilution of microalgae cells and maximizes winter solar penetration. The microalgae system offers dynamic shading efficacy to shield summer solar radiation and to maximize winter sun while effectively uptaking indoor CO_2.

Visible Light Transmittance (380 to 780 nm): The VLT measures the fraction of visible sunlight passing through the window in the direction normal to the window. Visible transmission directly affects the daylighting illumination inside of a room. A higher VLT indicates a lower artificial lighting demand. When the visible light is transmitted, it also carries heat energy that affects thermal comfort. This visible light is also important for microalgae photosynthesis to assimilate carbon and biomass cultivation. Microalgae utilize photosynthetically active radiation (PAR) and chlorophylls absorb blue light (400–500 nm) and red light (600–700 nm). While green light is dominant in PAR, microalgae reflect green light (500–570 nm).[12] Microalgae utilize up to 8–10% of solar energy for photosynthesis, the remainder of which turns into heat energy. Therefore, heat build-up requires constant monitoring and control through the room air circulation coupled with a heat exchanger as needed.

Visible-light-to-heat-gain ratio: Energy attribute testing according to industry standards is important for understanding how much the microalgae enclosure outperforms a traditional counterpart. Standardized product testing uses relevant industry standards, including American Society for Testing and Materials (ASTM) standards and the National Fenestration Rating Council (NFRC) standards. A sample of microalgae panel was tested against standards for U-factor, SHGC, and VT. Both SHGC and VLT are important energy attributes in reducing heating and artificial lighting load. The microalgae provide good shading efficacy of SHGC-0.17 (17%) at full growth. This is better than a high-performance double-pane window with triple-silver low-e coating (i.e., SB70XL). The visible light transmission was substantially reduced to VLT-0.07 (7%) at full growth. It is important to combine vision area and bioreactor to admit daylight and provide view-out when microalgae reach full growth. Testing protocols are based on ASTM C1199, C1363, E1423, E1084, E2264, and NFRC 102, 201, and 202. Figure 9.2 shows that the proposed microalgae enclosure is effective in solar shading and daylighting illumination.

Excessive Heat: Microalgae utilize a fraction of solar energy for photosynthesis, and the leftover solar energy turns into heat. Understanding the thermodynamic behavior of microalgae enclosures is important for assessing their feasibility

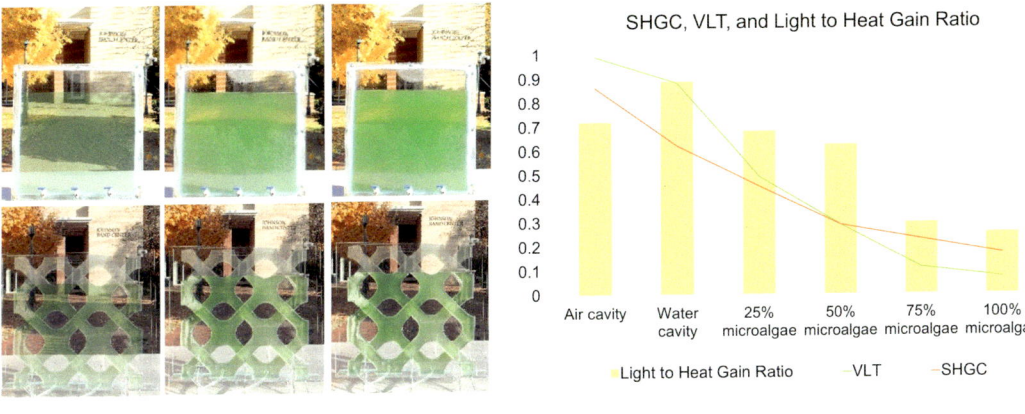

SHGC, VLT, and Light to Heat Gain Ratio

Light to Heat Gain Ratio — VLT — SHGC

▲ Figure 9.2

Pilot study of energy attributes from microalgae enclosures (left) and SHGC, VLT, and light-to-heat-gain ratio measurements of different microalgae densities (right); the measurements confirmed that the microalgae enclosure provides dynamic shading efficacy and good daylight penetration depending on microalgae density, contributing to energy efficiency.

in building applications. According to industry standards, building enclosures are designed and engineered for extreme temperature fluctuations accounting for cold and hot seasons. Unlike outdoor bioreactors, when the microalgae enclosures utilize indoor air, the heat build-up is mitigated. Three thermal cycling conditions are specified in the ASTM E2264 using an extremely cold temperature, –30°C (–22°F) combined with three hot temperatures, 49°C (120°F)—level 1, 66°C (150°F)—level 2, and 82°C (180°F)—level 3.[13] The custom microalgae window without room air circulation was subjected to 4°C (40°F) and 49°C (120°F) under the thermal cycling of –30°C (–22°F) and 82°C (180°F) when the room air was not circulated. Using an optimum range of 10–32°C, pumping room air will enable the microalgae panel to stay within the acceptable range. A heat exchanger is activated as needed. With the higher heat-specific capacity of the culture media (i.e., nutrient water) coupled with room air circulation, the microalgae panel experiences cooler temperatures in summer than a counterpart installed in outdoor environments. In winter, the culture temperature is susceptible to warmer temperatures due to conditioned air circulation recycled from room air.

9.3 Energy Efficiency and Environmental Performance

Microalgae architecture is a promising sustainable solution for building energy savings and CO_2 reduction in the built environment. The geometric configuration along with the cell concentration and color changes of microalgae responding to environments enhance building energy savings and occupant satisfaction. Using the One World Trade Center in New York City (3.5 Mft² gross floor area) as a study building, the energy simulation result indicated that the microalgae window building is expected to reduce heating, cooling, lighting, and ventilation load by an average 20% annually. Potential energy cost savings vary across states due to different climate conditions and energy prices. A counterpart building built according to the latest energy building code (ASHRAE 90.1 2019) was also analyzed for direct performance comparison.[14] Microalgae enclosures offer dynamic

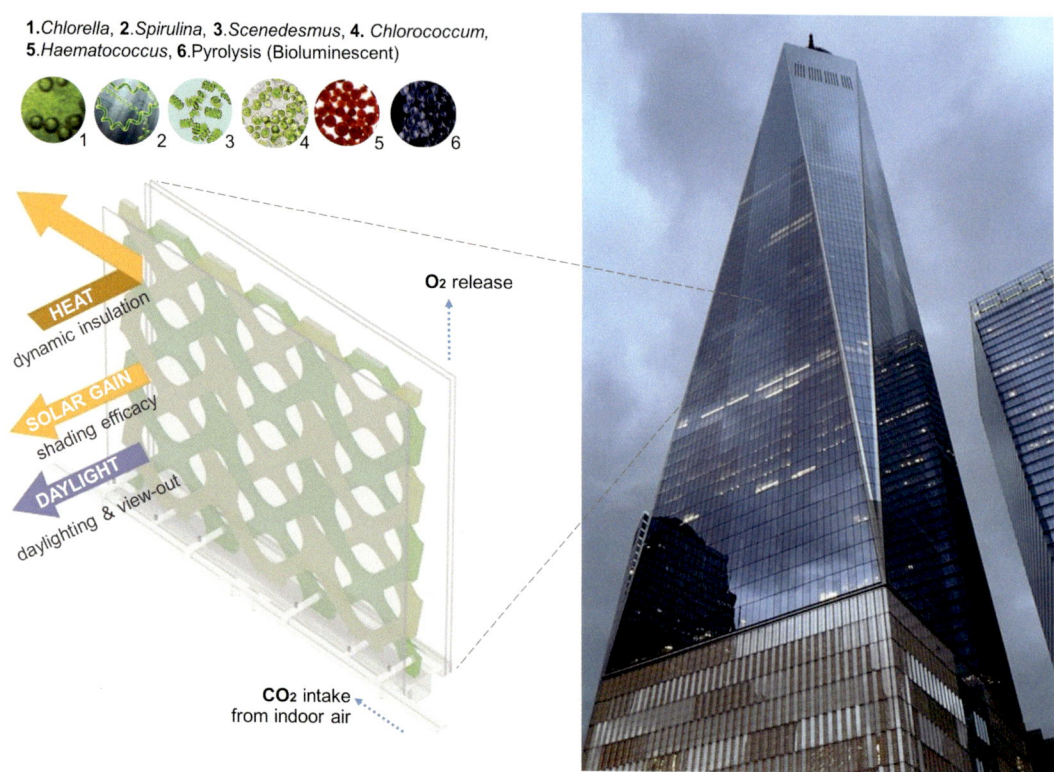

1.*Chlorella*, 2.*Spirulina*, 3.*Scenedesmus*, 4. *Chlorococcum*, 5.*Haematococcus*, 6.Pyrolysis (Bioluminescent)

HEAT
dynamic insulation

SOLAR GAIN
shading efficacy

DAYLIGHT
daylighting & view-out

O_2 release

CO_2 intake
from indoor air

▲ Figure 9.3

The One World Trade Center in New York City, a 94-story skyscraper, enclosed with microalgae windows. Computer simulation indicated that this building would reduce energy usage over 10 GWh/year with microalgae windows and save over $1 million a year in electricity costs with a seven-year ROI.

energy attributes due to their ability to change cell concentration and color using photosynthesis responding to climates. Capitalizing on the efficient photosynthetic performance of microalgae enclosures, building energy savings can be achieved by reducing heating, cooling, and artificial light demand while achieving carbon reduction and improved indoor air quality (Figure 9.3). Energy use intensity (EUI) is one of the building energy efficiency metrics that explain the level of building energy performance. It is determined by dividing annual total energy use by building. Different buildings have different EUIs, which allows cross comparisons across energy efficiency. A lower EUI equals higher building efficiency or lower energy use to operate the building.

The simulation results indicated that microalgae buildings effectively reduce EUI with good summer shading efficacy, winter solar gain, and year-round daylighting penetration. With dynamic thermal insulation by which room air is circulated, heating energy especially in cold climates is expected to be further reduced. The expected EUI reduction from microalgae is estimated to be 20–30% of the heating, cooling, and artificial lighting consumption. This energy savings also result in an average 6,000 tons of CO_2 reduction equivalent to emissions from 1,300 passenger cars. In addition to carbon reduction from energy savings, using a CO_2 sequestration rate of $5g/ft^2$, the study building can sequester 2 tons of CO_2 per day and over 7,000 tons of decarbonation annually (Figure 9.4).

Design variables for microalgae window simulations in comparison

	Microalgae building	ASHRAE code building
Climate zones in accordance with ASHRAE 90.1	15 cities representing the U.S. climate zones	15 cities representing the U.S. climate zones
Window attributes	Optimum U-factor, SHGC, VLT per climate conditions	Per ASHRAE requirements
Other simulation settings	Gross floor area: 3.5 Mft^2 Building height: 94 stories, 1268 feet tall Building orientation: facing true south Building aspect ratio: 1:1 % Window-to-wall ratio (%WWR): 40% Operation schedule: office Other enclosure requirements, lighting, HVAC and plug load requirements per ASHRAE 90.1	

Energy simulation set-up: The simulation model was a 94-story office building with 3.5 Mft^2 gross floor area, with a 200 feet by 200 feet footprint. The locations considered for the simulations were in 15 different cities representing the U.S. climate zones. For example, Miami (climate zone 1A), Charlotte (climate zone 3A), and Minneapolis (climate zone 5A) in accordance with ASHRAE 90.1. The reference model for the existing building was assumed to have a 40% window-to-wall ratio (WWR) enclosed with clear glass. Energy simulations were carried out with different design variables. Since the primary goal of the energy simulation is to understand the direct impact from a microalgae building compared to a counterpart building without microalgae enclosures, non-design factors were kept the same between the microalgae building and non-microalgae building such as building operation schedule, HVAC system, plug loads, and so on.

The microalgae building requires high-performing building enclosures to act as energy regulators and solar power producers (i.e., microalgae biomass production). A series of sensitivity analyses identified optimum energy attributes of the microalgae window. The building energy consumption also varies with different building masses and locations. Different percentages (%) of WWR also affect energy savings and biomass production. Our simulation focuses on verifying how much microalgae windows outperform in energy savings compared to a counterpart building without microalgae windows. All the design variables remain constant between the two simulation sets except the energy attributes of the windows (Table 9.1).

9.4 Full-scale Performance Mock-up

A full-scale microalgae window prototype was installed in an open studio at the school of architecture building on the UNC Charlotte campus. Figure 9.5 illustrates the performance mock-up instrumented with performance sensors for environmental data collection (e.g., PAR, VLT, SHGC, CO_2, total VOCs, air flow, water temperature, surface temperature). Field measurements are important for understanding real-world performance of the microalgae enclosures and calibrating/validating simulation data.

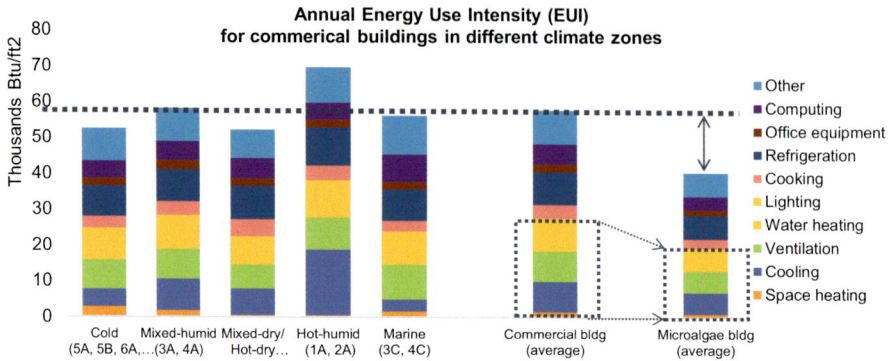

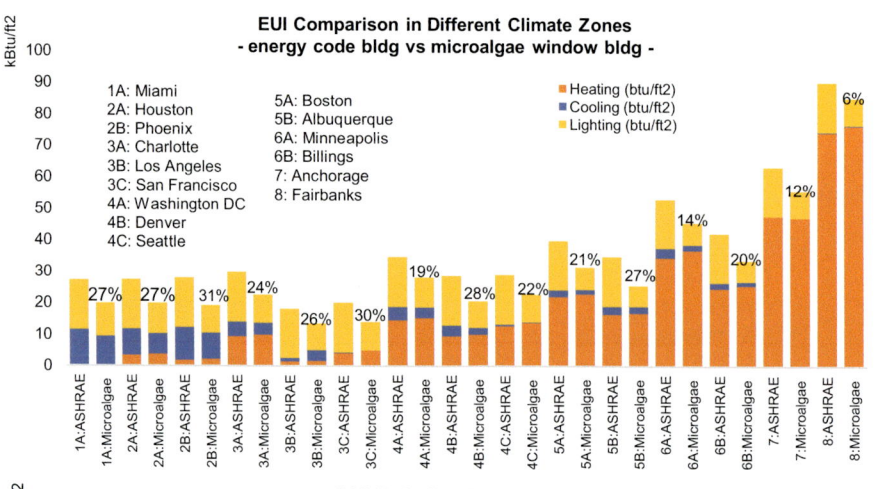

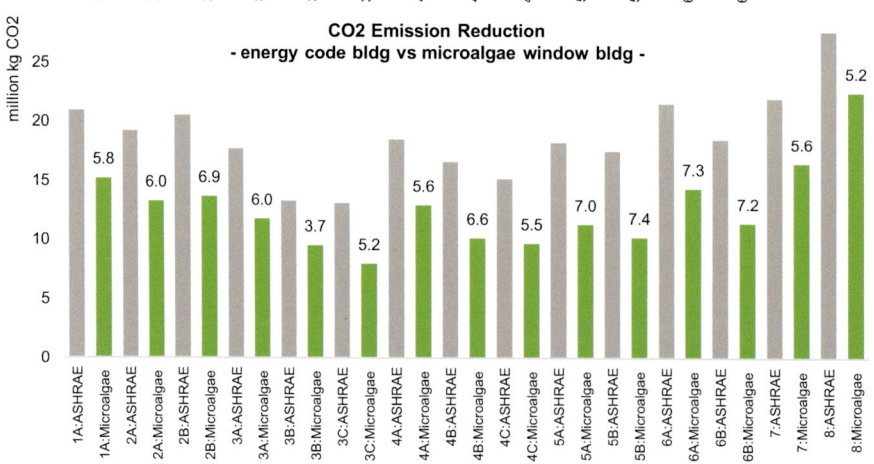

▲ Figure 9.4

Annual energy use intensity (EUI) of commercial buildings in different climate regions; average 20% energy saving can be achieved from microalgae window buildings by reducing heating, cooling, ventilation, and lighting loads (top); EUI (btu/ft^2)–level comparisons between a microalgae building and a code-complying counterpart in 15 different climate zones (middle); and CO_2 reduction potentials between a microalgae window building and a code-complying counterpart in 15 different climate zones (bottom).

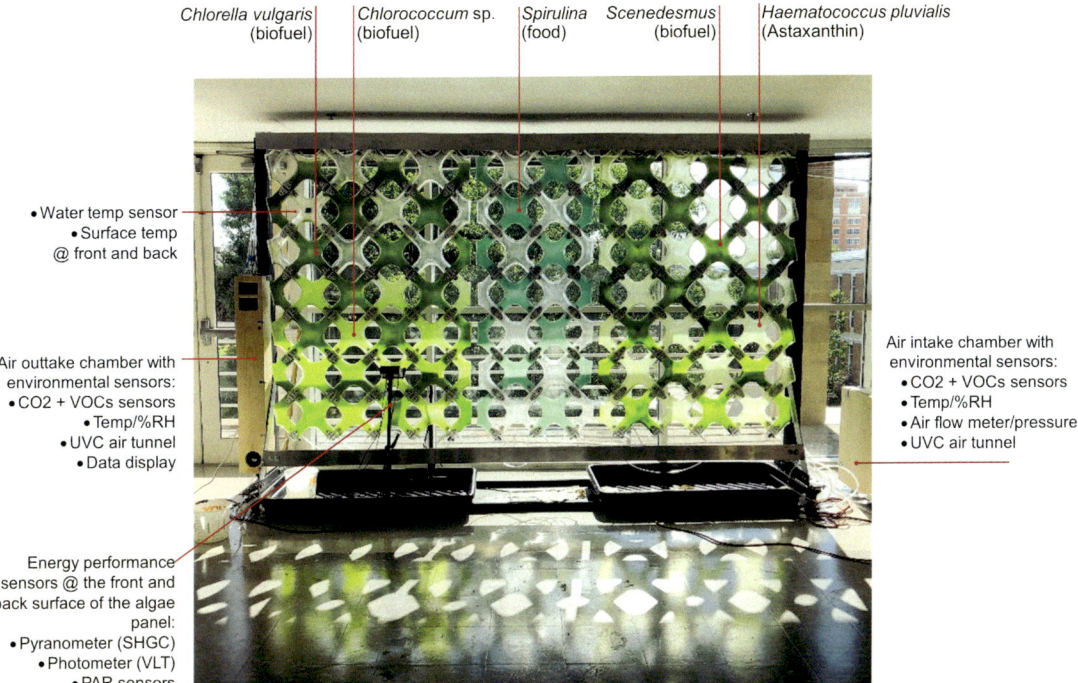

Chlorella vulgaris (biofuel) Chlorococcum sp. (biofuel) Spirulina (food) Scenedesmus (biofuel) Haematococcus pluvialis (Astaxanthin)

• Water temp sensor
• Surface temp @ front and back

Air outtake chamber with environmental sensors:
• CO2 + VOCs sensors
• Temp/%RH
• UVC air tunnel
• Data display

Air intake chamber with environmental sensors:
• CO2 + VOCs sensors
• Temp/%RH
• Air flow meter/pressure
• UVC air tunnel

Energy performance sensors @ the front and back surface of the algae panel:
• Pyranometer (SHGC)
• Photometer (VLT)
• PAR sensors

▲ Figure 9.5

Full-scale mock-up filled with five microalgae and instrumented with performance and environmental sensors.

The performance mock-up is 12 feet wide by 8 feet tall and installed behind an existing window facing west. Figure 9.6 illustrates the performance mock-up with microalgae density changes visualizing effective shading and carbon sequestration. The mock-up consists of a network of modular microalgae panels and operational systems. The modular system allows for time-efficient installation and adaptability for different window areas. The operational system includes microalgae growing apparatus (e.g., pipes, valves, pumps, storage tanks) and semiautomatic controllers to regulate intake and outtake of air, microalgae, and culture media. The mock-up was instrumented with energy performance sensors (e.g., PAR, Pyranometers, and photometer) to record data as the microalgae system responds to the environments.

The data indicated that as the density of the microalgae system gets darker, there was a gradual reduction in SHGC from a maximum of 73% to a minimum of 17%. Accordingly, visible light transmission was decreased in a similar fashion. Because the studied microalgae system incorporates a vision area for view-out and daylight, the VLT measured in the room a foot behind the microalgae system remained at 40–60%. The field measurements confirm that the microalgae window has potential for improving building energy efficiency by offering good shading efficacy while admitting daylighting that impacts heating, cooling, and artificial lighting load.

Five strains (*Chlorella*, *Chlorococcum*, *Haematococcus*, *Scenedesmus*, *Spirulina*) were cultivated in the system using a semi-continuous production mode. Their biological performance and environmental benefits (e.g., biomass production, CO_2 reduction potentials) and environmental conditions (e.g., culture temperature, surface temperature, air flow rate, pH, PAR) were monitored. By utilizing 26 harvesting valves at the bottom of the panel, grown microalgae were easily

▲ Figure 9.6

1:1 performance mock-up filled with five different microalgae strains; color and density change as biomass increases.

extracted using gravity. The intake pipe to fill in culture and media was installed at the top and also run using gravity feed.

Preliminary data collection indicated that the system provides a good growing environment for microalgae tested in general. The mock-up is able to

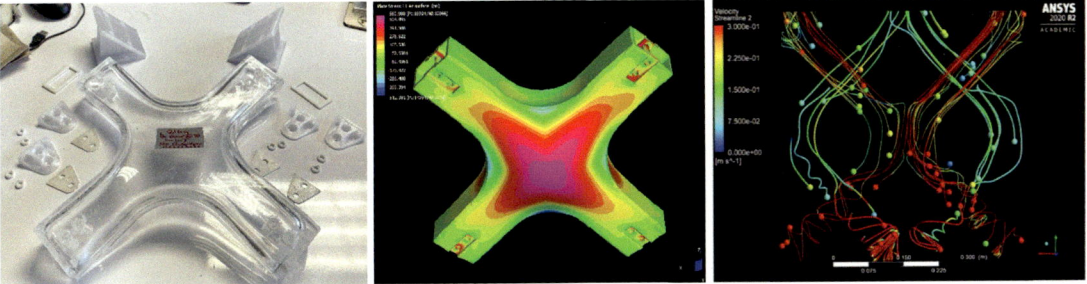

Prototyping of X-module made by rotational casting (left); principal stress of FEM structural analysis under hydrostatic pressure (middle); integrating CFD simulations to verify aerodynamic behaviors of the configuration and study an optimum air distribution mechanism to maximize microalgae growth.

sequester CO_2 generated by three occupants and provide an output of 500 g per day (~200 kg per year). The mock-up provides a feasibility study of retrofitting existing windows with a microalgae system. It also allows for the evaluation of the economic and environmental potentials of the microalgae window compared to traditional windows. The outcome of this prototyping is an enabled technology demonstration, verifying technical and practical challenges in a real-world application and evaluating economic and environmental benefits. The mock-up is intended to run for a minimum of one year to document seasonal energy efficiency and long-term performance in building applications. Technical challenges still remain, such as operation and maintenance issues in real-world application.

When it comes to selecting a fabrication technique, one of the primary requirements is low cost, quality control, and scalability of microalgae panel production. A rotocasting technique was utilized to fabricate modular units in a cost-effective and time-efficient manner. The full-scale mock-up consists of four different modular parts (Figure 9.6). Each unit was made of a quarter-inch-thick polyurethane-based resin, and structural analysis was carried out to secure a safety factor of 10 to account for long-term performance under gravity load and hydrostatic pressure. Hydrostatic pressure in a liquid is determined by $P = \rho gh$ where, P = hydrostatic pressure, ρ is liquid density, g is gravity, and h is the height of the liquid. 20 CFM (cubic foot per minute) of room air was circulated to supply CO_2 for biomass productivity. The environmental factors affecting algae growth include light, nutrients, and CO_2 levels. Bubbling air into the system is important for not only supplying CO_2 but also simultaneously providing uniform temperature, light exposure, and nutrients. Because CO_2 is essential for microalgae growth, the larger interfacial area (i.e.., higher surface area-to-volume ratio) of a bubble, with slow bubble velocity, results in a high CO_2 residency time in the growth media. Computational fluid dynamics (CFD) analysis can be used to verify aerodynamic behaviors of bubbles in the media and their velocities to minimize hydrodynamic stress that causes reduced growth (Figure 9.7). Full-scale prototype installation consists of interlocking the modular photobioreactor facing the western orientation. The microalgae mock-up provides view-out and daylighting for better occupant satisfaction. Figure 9.8 shows the installation and operation of the microalgae mock-up for performance monitoring and verification. Table 9.2 summarizes microalgae building research found in scientific publications.

▲ Figure 9.8

Color and density of the microalgae provide summer shading, passive solar heating, daylight illumination, affecting cooling, heating, and artificial lighting load.

Summary

The building sector accounts for approximately 40% of U.S. energy use and 39% of anthropogenic CO_2 emissions from building operations. Due to increasing construction activities and stringent environmental protection policies, the building industry

▼ Table 9.2

Bibliographic search examples of microalgae building research

Year	Study title	Researchers	Location	Research activities
2013	Living Skins: A New Concept of Self-Active Building Envelope Regulating Systems	Dewidar et al.	Egypt	Four case studies of smart adaptive microalgae building envelope for energy production, thermal performance, daylighting, and shading
2013	Beyond Green: Growing Algae façade	K. Kim	USA	Structural and thermal performance evaluations of microalgae façade prototypes using experiments and computation
2013	German Building to Test Algae-Filled Facade as Source of Shade and Energy (introduction)	J Landers	USA	Case study of the bio-intelligent-quotient (BIQ) house in Hamburg, Germany; 30 kWh/m^2/year biomass production and 150kWh/m^2/year heat production; $2,300–3,000/m^2 for microalgae envelopes and service system
2014	Analysis for Energy Efficiency of the Algae Façade: Focused on Closed Bioreactor System	T.R. Kim and S.H. Han	South Korea	Energy efficiency, CO2 reduction, and biomass production potentials from algal facades using computer simulations
2015	Mathematical Modeling of Energy Balance in the Photobiological Treatment Plants	Buzalo et al.	Russia	Mathematical model for wastewater treatment and energy production using PBR facades
2015	Algae Architecture (graduate thesis)	F. Qui	The Netherlands	Prior art review on microalgae facades for energy production and carbon sequestration
2015	CO$_2$ Enrichment from Flue Gas for the Cultivation of Algae—A Field Test	Wolff et al.	Germany	Pilot field test of the use of flue gas as a source of CO$_2$ for the photobioreactors; membrane flue gas technology mounted on BIQ building; 13g/m^2/day growth rate
2015	A Hybrid Facade That Combines an Algal Photobioreactor with Photovoltaics	Granata et al.	Switzerland	A combined study of biomass production and PV power production using a hybrid of flat bioreactor and PV modules by changing bioreactor material (polypropylene vs. acrylic) and reactor thickness (0.9 cm and 2.8 cm) testing for green algae and coccolithophore

(Continued)

Year	Study title	Researchers	Location	Research activities
2016	Microalgae Culture in Building-integrated Photobioreactors: Biomass Production Modelling and Energetic Analysis	Pruvost et al.	France	Theoretical study of thermal behaviors for a vertical microalgae façade system
2016	An Empirical Study Investigating the Impact of Micro-algal Technologies and Their Application within Intelligent Building Fabrics	A. Elnokaly and I. Keeling	UK	Empirical and theoretical study of shading efficiency and daylight factor in relation to cell concentration and building orientation
2016	Optimising the Bioreceptivity of Porous Glass Tiles Based on Colonization by the Alga *Chlorella vulgaris*	Ferrandiz-Mas et al.	UK	Empirical study of glass tiles that encourage colonization of *Chlorella v;* capillary porosity and water sorptivity of the glass tiles are important factor for bioreceptivity and growth
2016	Façade-integrated Photobioreactors for Building Energy Efficiency (Book chapter in *Startup Creation*)	Oncel et al.	Turkey	Case studies of microalgae-integrated building with a special emphasis on energy efficiency and sustainability role
2016	Energy-generating Glazing (Book chapter in *Smart Buildings*)	M. Casini	Italy	Case studies of bioadaptive facades integrated with algae bioreactors with their operation, performance, and energy yield
2016	Functional-layered Textiles in Architecture (Book chapter in *Fabric Structures in Architecture*)	Heinzelmann et al.	The Netherlands	Case studies of integrating microalgae culture into textile architecture for building performance enhancement
2016	Feasibility of Algae Building Technology in Sydney (Feasibility report)	Wilkinson et al.	Australia	Feasibility assessment of microalgae buildings compiled by interviewing leading professionals addressing environmental, technological, regulatory, economic, social, and land use issues

Year	Title	Author	Country	Description
2016	Energy-Efficient Buildings: A Case Study of Modified Facades Technology	N. Malik and S. Singh	India	Case study of different energy-efficient enclosures with a short section attributed to BIQ house introduction
2016	Bio-enabled Façade Systems—Managing Complexity of Life through Emergent Technologies	Decker et al.	USA	Prototyping of a microalgae system with empirical measurements of daylighting performance depending on cell concentration (Scenedesmus) and assessment of performance between the predicted and the real world
2016	Lighting and Energy Supply for Heating in Building Using Algae Power	F. Seperhri	Iran	Case study of microalgae system examples for biofuel production
2017	Exploring the Feasibility of Algae Building Technology in NSW	Wilkinson et al.	Australia	Feasibility of algae building technology in NSW based on environmental, technological, political, economic, and social factors
2017	Framework for Evaluating and Optimizing Algae Façades Using Closed-loop Simulation Analysis Integrated with BIM	Chang et al.	USA	Literature review of BIQ house and the use of system dynamic model-based BIM for a closed-loop system to optimize algal facades for best building performance
2017	Modeling Algae-powered Neighborhood through GIS and BIM Integration	Dutt et al.	USA, China	Design workflow development of geographic information system (GIS)-integrated building information modelling (BIM) for near-zero energy community research
2017	High Tech Startup Creation for Energy Efficient Built Environment	F. Pacheco-Torgal	Portugal	Sustainable technology case study for startup creation and the microalgae systems with potentials for biofuel production, wastewater treatment, bio-fixation, and bioproducts.
2017	Design of Marine Macroalgae Photobioreactor Integrated into Building to Support Seagriculture for Biorefinery and Bioeconomy	Chemodanov et al.	Israel	Empirical study of macroalgae PBR for integration into building façade measuring chemical compounds for biofuel production
2017	The Renewable Energy City within the City. The Climate Change Oriented Urban Design—Szczecin Green Island	Z. Paszkowsk and J. Golebiewski	Poland	Urban design strategies toward net zero energy architecture based on data metrics to inform decision-making; BIQ building as an example for energy production and decarbonation in cities

▼ Table 9.2

(Continued)

Year	Study title	Researchers	Location	Research activities
2017	A Review on Interaction of Innovative Building Envelope Technologies and Solar Energy Gain	Talaei et al.	Iran	Energy efficiency and architectural conformity of different building envelope technology with algae façade ranked the first as an emerging green tech
2017	Design of a Microalgae Bio-reactive Façade Reactor for Cultivation of *Chlorella vulgaris*	Ferreira et al.	Colombia	Experiments in growing *Chlorella vulgaris* for 74 days in different PBRs; assess different sized PBRs for optimization of biomass production
2017	Productivity of microalgae as biofuel for bioadaptive systems of facades	Zalata et al.	Russia	Empirical study of microalgae growth in a facade PBR under various temperature ranges and illumination conditions for biofuel production
2017	Algae and Building Façade Revisited. A Study of Façade System for Infill Design	Martokusumo et al.	Indonesia	Empirical study of temperature difference, illuminance response, and energy efficiency of PBR facades compared with brise soleil and horizontal louver facades estimated by computer simulations
2017	Algae Façade as Green Building Method: Application of Algae as a Method to Meet the Green Building Regulation	Poerbo et al.	Indonesia	Experiments/prototype of algae façade in verifying energy use reduction and improved indoor air quality in meeting green building regulations
2017	Utilization of Building Colors with the Energy-oriented Algae Façade System	HS Jo and SH Han	South Korea	Integration of LED lights for algae façade for productivity and aesthetic improvement of a building
2017	A mini Review on the Integration of Resource Recovery from Wastewater into Sustainability of the Green Building through Phycoremediation	A. Yulistyorini	UK	Summary of environmental benefits from microalgae facades by engaging in wastewater treatment, bioremediation, bioproducts and biofuels production, operational energy reduction, and resource recovery and nutrients for cultivation

Year	Title	Author	Country	Description
2018	Optimization of Microalgae Panel Bioreactor Thermal Transmission Property for Building Façade Applications	Umdu et al.	Turkey	Empirical study of steady-state U-factor measurements for a microalgae façade filled with *Nannochloropsis*
2018	Microalgae: Prospects for Greener Future Buildings	G. Elrayies	Egypt	Case study of microalgae facades and review of their environmental economic performance
2018	Exergy Efficiency of Solar Energy Conversion to Biomass of Green Macroalgae *Ulva* (Chlorophyta) in the Photobioreactor	Zollmann et al.	Israel	System optimization study using exergy for vertical membrane photobioreactors growing macroalgae *Ulva* (Chlorophyta)
2018	Symbiosis Optimization of Building Envelopes and Micro-algae Photobioreactors	M. Araji and I. Shahid	Canada	Theoretical study of biomass production, CO_2 reduction, and land use savings of microalgae façade filled with *Chlorella v* and *Dunaliella*
2018	Innovative Approaches on Building Envelope - Photobioreactor Facades	I. Kukdamar and T. Ozbalta	Turkey	Discussion of design principles of microalgae facades using BIQ house
2018	A New Approach for a Control System of an Innovative Building-Integrated Photobioreactor (Book chapter)	Kokturk et al.	Turkey	Design process of microalgae building integrated with programmable logic controller to control temperature, liquid level, pH, fluid velocity, CO_2, and nutrients
2018	Controlling Air Pollution with the Use of Bio Facades (A Solution to Control Air Pollution in Tehran)	M. Bastanfard	Iran	Comparative analysis of watery facades, algae facades, and green walls and feasibility assessment of algae facades in local conditions
2018	Algae Building Technology Energy Efficient Retrofit Potential in Sydney Housing	S. Wilkingson and P. Stoller	Australia	Feasibility for adoption of microalgae facades in Sydney in order to meet sustainability goals
2018	Biomimetic Facade Applications for a More Sustainable Future (Book chapter)	Tokuc et al.	Turkey	Case study of various biomimetic facades and BIQ house

(Continued)

Year	Study title	Researchers	Location	Research activities
2018	Probable Cause of Damage to the Panel of Microalgae Bioreactor Building Façade: Hypothetical Evaluation	M. Talaei and M. Mahdavinejad	Iran	Literature review of potential causes of damaging microalgae façade, including chemical, biological, physical, operational, and environmental factors
2019	Development of a Control System to Cover the Demand for Heat in a Building with Algae Production in a Bioenergy Façade	Kerner et al.	Germany	Measurements of heat production (algae façade) and consumption (space and water heating) of BIQ house
2019	A Critical Review on Designs and Applications of Microalgae-based Photobioreactors for Pollutants Treatment	Vo et al.	Multinational	Literature review of pollutant removal rates for various prior art PBRs filled with different microalgae strains
2019	Algae Window for Reducing Energy Consumption of Building Structures in the Mediterranean city of Tel-Aviv, Israel	Vegev et al.	Israel	Simulation study of energy consumption from microalgae building based on experiments of U-value, visible transmittance, and solar heat gain of algae windows
2019	Preliminary Results on a Novel photo-bio-screen as a Shading System in a Kindergarten: Visible Transmittance, Visual Comfort, and Energy Demand for Lighting	Pagliolico et al.	Italy	Computation of daylighting performance of microalgae window for kindergarten test classroom; visible light transmittance measurements carried out for 3 weeks during the summer.
2019	Adaptive Facades Performance Assessment, Interviews with Facade Experts	S. Attia	Belgium	Expert interviews on adaptive bioenergy facades with professionals' experience on microalgae system and BIQ house
2019	Comparison of Green Facades and Photobioreactor Facades in Energy Efficient Design	I. Kukdamar and T. Ozbalta	Turkey	Comparative study of green facades and PBR facades for structure, formation, energy efficiency, and environmental impact
2019	Feasibility of Algae Photobioreactor as Façade in the Office Building in Indonesia	N. Ardiani et al.	Indonesia	Feasibility study of tubular algae panels filled with Chlorella on a university campus building in Indonesia
2019	Algae façade: From Seaweed to City Grid (Graduate Thesis)	R. Halabi	Italy	Speculative study of an algal façade system for energy efficiency and self-sufficiency with green energy cycle

Year	Title	Author	Country	Description
2019	Utilizing Active Materials in Building Façade for Building Efficiency	Wahab et al.	Egypt	Empirical study of thermal energy stored in the system and illumination level for a residential building
2019	Defining the Problems in Integration of Microalgae Photobioreactor Systems to Architecture	F. Kerestecioglu and Y. Pekmezci	Turkey	Meta-analysis of assessing ten microalgae building projects in urban, architectural, and installation categories
2019	The Bio-adaptive Algae Contribution for Sustainable Architecture	D. Elmeligy and Z. Elhassan	Egypt, Sudan, KSA	Sustainable design strategies through implementing microalgae buildings in achieving economic, environmental, and social benefits
2019	Algae Building: Is This the New Smart Sustainable Technology?	Wilkinson et al.	Australia	Meta-analysis of prospects and challenges of microalgae buildings in improving energy efficiency and energy production
2019	Design Optimisation of Façade-integrated Photobioreactors Using CFD Simulation	Haskell et al.	Germany	CFD analysis for assessing air flow dynamics and velocity of bubbles and optimizing geometries by changing the number and length of the divider for the BIQ panel
2019	The Energy of the Green: Green Facades and Vertical Farm as Dynamic Envelope for Resilient Building	Trombadore et al.	Italy	Speculative design project of kinetic façade system out of ETFE embedded with microalgae bioreactor (*Spirulina*) for energy production
2020	Thermal and Energy Performance of Algae Bioreactive Façades: A Review	Talaei et al.	Iran, USA	Comparative study of green wall, double-skin façades, and microalgae facades with respect to energy efficiency, solar storage potentials, and environmental benefits
2020	Building Integrated Photobioreactor (Book chapter)	E. Umdu and Y. Univ	Turkey	Case study of microalgae building examples focusing environmental benefits of biofuel production, decarbonation, air quality improvement, wastewater treatment, and dynamic appearance

(Continued)

Year	Study title	Researchers	Location	Research activities
2020	Towards Efficient Green Architecture and Sustainable Facades Using Novel Brick Design	Z. Aldeek	Jordan	Prototyping of clay brick module that encourages colonization of microalgae that helps transpiration, thermal buffering, air pollutant reduction
2020	Integrating Algae Building Technology in the Built Environment: A Cost and Benefit Perspective	N. Biloria and Y. Thakkar	Australia	Comparative study of solar panel and microalgae system by analyzing net present value and return on investment (ROI); environmental benefits from the microalgae system outweigh 24 years of ROI (vs. 16 years of ROI for solar panel)
2020	Development of a Lightweight Multi-skin Sheet Photobioreactor for Future Cultivation of Phototrophic Biofilms on Facades	Scherer et al.	Germany	Prototyping and empirical study of a lightweight aerosol PBR with low water consumption for applications in building facades; biofilm strain—*C. chthonoplastes* and *T. sociatus*
2020	Carbon Sequestration in Microalgae Photobioreactors Building Integrated (Book chapter)	Oncel et al.	Turkey	Case study of microalgae facades focusing on bio-technology, design, and application in offering carbon capture and wastewater treatment
2020	Exploration of Microalgae Photobioreactor (PBR) in Tropical Climate Building Envelope	M. Hasnan and P. Zaharin	Malaysia	Speculative design study to investigate the use of microalgae facades in Malaysia to keep buildings at a comfortable temperature; energy production and thermal performance
2020	The Technical Issues Associated with Algae Building Technology	Wilkinson et al.	Australia	Literature review and semi-structure interview discussion on design, technical issues, regulation, construction and operation of algae building technology (ABT) in Australia

Year	Title	Author	Country	Description
2020	Use of Double-glazed Window as a Photobioreactor for CO_2 Removal from Air	Rezazadeh et al.	Iran	Empirical study of evaluating CO_2 reduction potentials from PBR windows tested in Tehran
2020	Robotic Extrusion of Algae-Laden Hydrogels for Large-scale Applications	Malik et al.	UK	Experimental study of 3D printing algae-laden hydrogels by robot for upscalization for environmental benefits of carbon reduction, biosensing and bioremediation, and wastewater treatment
2020	Development of a Lightweight Multi-skin Sheet Photobioreactor for Future Cultivation of Phototrophic Biofilms on Facades	Scherer et al.	Germany	Empirical study of aerosol-based biofilm PBR facades (cyanobacteria) using multi-wall construction to increase productivity and industry application
2021	Algae Utilization and Its Role in the Development of Green Cities	Chew et al.	Malaysia, China	Literature review of algae's possible contributions to city's energy sufficiency by bioelectricity, biogas, biofuel and bioproducts, decarbonation, and wastewater treatment
2021	Multi-objective Optimization of Building-integrated Microalgae Bioreactors for Energy and Daylighting Performance	Talaei et al.	Iran, USA, the Netherlands	Verification of energy-saving potentials from microalgae window compared to counterpart using multi-objective optimization by varying façade orientation, WWR, and cell concentration; EUI and UDI as performance metrics.

is now faced with challenges in reducing environmental impacts. Microalgae can be integrated with different building systems and infrastructures such as building enclosures, roofs, interior systems, furniture, urban farming, waste treatment facilities, community gardens, and so on. Bioclimatic design is important in executing microalgae systems that employ climate adaptive technology. Building enclosures are the primary location to harness natural resources because they directly interact with outdoor environments. The integration of microalgae within a building enclosure architecture can reduce heating, cooling, and artificial lighting loads inside buildings while sequestering indoor carbon and improving air quality. The design and placement of microalgae enclosures is determined by a balance between climates (e.g., sun, wind, orientation), building space, WWR, aesthetics of microalgae growth, energy savings, and occupant satisfaction. Low-performing windows are responsible for more than half of the building energy consumption and pollutant emissions. High-performance windows integrated with microalgae can offer environmental, economic, and commercial opportunities.

Notes

1   Victor Olgyay, *Design with Climate: Bioclimatic Approach to Architectural Regionalism—New and Expanded Edition* (Princeton University Press, 2015).
2   Baruch Givoni, "Man, Climate and Architecture," *Elsevier* (1969).
3   Sopa Visitsak and Jeff S. Haberl, "An Analysis of Design Strategies for Climate-controlled Residences in Selected Climates," *Proceedings of SimBuild* 1, no. 1 (2016).
4   ASHRAE Standards Committee, "Standard 55-2010, Thermal Environmental Conditions for Human Occupancy," *American Society of Heating, Refrigerating and Air Conditioning Engineers* (2020).
5   Ibid., 76.
6   Mohammed Salah-Eldin Imbabi, "A Passive–Active Dynamic Insulation System for All Climates," *International Journal of Sustainable Built Environment* 1, no. 2 (2012): 247–258.
7   Berkeley Lab, "Cool Roofs," Heat Island Group, accessed June 27, 2021, https://heatisland.lbl.gov/coolscience/cool-roofs
8   Helenice Maria Sacht, Luís Bragança, Manuela Almeida, and Rosana Caram, "Specification of Glazings for Façades Based on Spectrophotometric Characterization of Transmittance," *Sustainability* 13, no. 10 (2021): 5437.
9   M. Dabbagh and M. Krarti, "Evaluation of the Performance for a Dynamic Insulation System Suitable for Switchable Building Envelope," *Energy and Buildings* 222 (2020): 110025.
10 Ibid., 2.
11 M. Jothilakshmi, A. Mohan, and M. Tholkapiyan, "Performance Evaluation and R-value Analysis for Thermally Insulated PCM Wall with Fluted Sheets," *Materials Today: Proceedings* 22 (2020): 912–918.
12 Kristin Collier Valle, Marianne Nymark, Inga Aamot, Kasper Hancke, Per Winge, Kjersti Andresen, Geir Johnsen, Tore Brembu, and Atle M. Bones, "System Responses to Equal Doses of Photosynthetically Usable Radiation of Blue, Green, and Red Light in the Marine Diatom Phaeodactylum tricornutum." *PLoS ONE* 9, no. 12 (2014): e114211.
13 ASTM Standards Committee, "ASTM E2264-05 Standard Practice for Determining the Effects of Temperature Cycling on Fenestration Products" (2013).
14 ASHRAE Standards Committee, "ANSI/ASHRAE/IES Standard 90.1-2019: Energy Standard for Buildings Except Low-Rise Residential Buildings" (2019).

Chapter 10

Microalgae Façade Design

10.1 Design and Engineering of Microalgae Enclosures

Microalgae enclosures serve as a primary building skin protecting the interior space from external environments. Design and engineering of microalgae enclosures requires a delicate balance between different design criteria such as structural integrity, environmental performance, air tightness, waterproofing, durability, user comfort and safety, and aesthetics. First, all components of macroalgae enclosures must work together to withstand external structural loads such that they provide safe indoor environments against external structural and environmental forces. The American Society of Civil Engineers' (ASCE) standard 7 defines loads as "forces or other actions that result from the weight of all building materials, occupants and their possessions, environmental effects, differential movement, and restrained dimensional changes."[1] Structural design loads acting on building enclosures are mainly generated by three types: gravity load, lateral load, and environmental load. Gravity load can be further divided into dead load such as self-weight of enclosure materials and live load such as maintenance load if applicable. Lateral loads mainly refer to wind loads and seismic loads. Wind load acts perpendicular to the enclosure surface causing deflection and internal structural stress. Seismic load which applies in-plane of enclosures requires more stringent conditions for enclosures to withstand. Microalgae enclosures are also subject to external environmental loadings from temperature, humidity, solar radiation, and barometric pressure due to the altitude of the site location. As direct interface regulating energy flows between indoor and outdoor environments, microalgae enclosures work closely with the heating, ventilation, and air conditioning (HVAC) system by controlling heat transfer, solar radiation, air infiltration, and daylight illumination. The microalgae enclosure also stores solar energy for possible energy reclamation (e.g., space and domestic hot water heating), reduces CO_2, and generates oxygen as an extension of the HVAC system. It also plays an important role in enhancing user comfort at low building operation cost.

1 Technological Performance Requirements

- Structural Integrity: Microalgae enclosures serve as a primary building skin that separates the conditioned indoor environment from the outdoor environment. Fundamental structural integrity refers to the adequate strength and stiffness of

DOI: 10.4324/9780367814410-13

microalgae building enclosure systems. Microalgae systems should limit their deflection by span over 175 under lateral design load when tested according to the American Society of Testing and Materials (ASTM) standard E330. No material failures should occur under various structural loads and load combinations in accordance with ASCE 7.

- Energy attributes: Microalgae systems should provide adequate R-value when tested according to the American Architectural Manufacturers Association (AAMA) 1503 Thermal Testing U-factor. The solar heat gain coefficient (SHGC) of microalgae systems provide dynamic shading efficacy and year-round daylight penetration when tested in accordance with National Fenestration Rating Council (NFRC) 200, 300. Energy attributes should exceed the American Society of Heating, Refrigerating and Air-Conditioning Engineers (ASHRAE) 92.1 requirements.

- Water Penetration Resistance: Microalgae enclosures shall not exhibit evidence of water leakage when tested according to the ASTM E331 at a differential pressure of 575 MPa and according to AAMA 501.1 under a dynamic pressure of 575 MPa.

- Air Infiltration: Air infiltration is unwanted air migration between outdoor and indoor spaces. Microalgae enclosures shall have permanent resistance to air leakage through the system of not more than 0.1 cfm/ft^2 of fixed wall area when tested according to ASTM E283 at a minimum uniform static air pressure differential of 300 Pa.

- Longevity: The longevity of the system can be evaluated by carrying out thermal cycling tests according to AAMA 501.5 Test Method for Thermal Cycling of Exterior Walls. This test method utilizes convective hot air to create comparable real-world conditions.

(2) Biological Performance Requirements

- The growth of microalgae is determined by the photosynthesis rate and the specific solar wavelength used in the photosynthesis. Microalgae, similar to terrestrial plants, utilize photosynthetically active radiation (PAR) in the range of 400–700 nm, so the microalgae enclosure should be able to provide optimum PAR range. The system should provide optimum photosynthetic photon flux density (PPFD) ranges of 50–400 $\mu mol/m^2 s$. Unused wavelength radiation will turn into heat, and the system needs to provide an optimum temperature range for maximum microalgae production.

- Another biotechnical requirement is optimum growing temperatures. Different strains have their own optimum temperature for growth. An optimum growing temperature range is 10–32°C (50–90°C) in microalgae bioreactors.[2] The solar energy stored in the microalgae enclosure can be reclaimed to supplement other building services (e.g., geothermal heat pump, space heating, domestic hot water). An active heat exchanger should be available to regulate extreme hot and/or cold conditions and maintain the optimum temperature. Bubbling room air is essential to alleviate extreme temperature responses with reduced energy consumption.

- Bubbling air is important to evenly mix CO_2, light, and nutrients in the media. CO_2 bubble size and residence time in the media affect the cells' CO_2 fixation efficiency. Slower bubble aeration and smaller diameter promote CO_2 diffusion. Smaller bubble sizes have a greater interfacial area (area in contact with media), resulting in higher residency time for microalgae growth. Bubble sizes between 1 and 7 mm diameter with 0.05 m/s of bubble velocity are recommended (Table 10.1).

Design and engineering process and performance requirements for microalgae enclosures

Microalgae enclosure design phases	Target	Performance metrics	Control mechanism
Bioclimatic design	Climate-responsive geometry and orientation Aesthetics	Shapes, regulatory, performance, maintenance, and longevity	Evidence-based design
Engineering development	Structure performance Energy performance Water tightness Air tightness Longevity	Strength and stiffness per ASCE 7 U-factor/SHGC/VLT per ASHRAE 90.1 No leakage 0.1 cfm/ft^2 per ASHRAE Fundamental >10 years	Material selection, Joint design, Field craftsmanship
Biotechnical development	Illumination requirements Temperature Aeration and mixing pH requirements	50–400 µmol/m^{2}s (100–200 µmol/m^{2}s) 50–90°F (70–80°F); 10–32°C (21–27°C) <0.05 m/s, 2ACH, 1–7 mm bubble diameter Neutral; some strains can withstand higher pH and salinity	PAR sensors, Temp sensors/heat exchanger, Mixing and aeration system
Microalgae system operations	Capital savings Operational savings Wastewater treatment Flue gas treatment Biofouling maintenance Cultivation mode Visual, acoustic, olfactory performance	Return on investment (ROI) $/year % nutrient uptake, biomass growth % CO_2, SO_2, and NO uptake, biomass growth Frequency Biomass growth Comfort level	Building management system coupled with automated monitoring and control system
End of life cycle	High-value product Biofuel	Tons of dry weight Gallons of biofuel	Post-processing system

Cost-saving Potentials: The primary catalyst for widespread implementation of microalgae building applications will be the cost savings from both capital and operational expenditures. The shading efficacy and daylight penetration response to solar intensity are expected to reduce cooling and artificial lighting loads. The utilization of room air to supply CO_2 is expected to improve the U-factor, thus reducing heating and cooling loads. The carbon uptake improves indoor air quality. The bio intelligent quotient (BIQ) house in Hamburg, Germany is a multistory apartment building incorporating a 2.5 m × 0.7 m × 0.018 m flat panel with a growing capacity of 24 liters covering 129 panels for a 185 m^2 façade area (135 m^2 as actual bioreactor-filled area in 129 reactors × 0.55 m × 1.90 m).[3] The initial analysis showed an energy conversion rate of 40% for heat and 8% for biomass. However, the final conversion efficiency for heat energy was 21%, resulting in total heat of 22,500 kWh/annual. The final conversion efficiency for biomass production was

4.4% instead of the 8% original prediction based on the microalgae enclosure yield of 600 kg dry mass per year (12 g/m^2/day or 1.1 g/ft^2/day). Depending on growing conditions and strain types, 8–10% photosynthetic efficiency can be achieved. The BIQ house cultivated *Chlorella vulgaris* obtained from a local stream with a temperature tolerance within 8°C and 38°C. Further economic improvement was realized through recycling the thermal energy stored in the bioreactor. Excessive heat build-up is a negative by-product for microalgae enclosure applications especially during summer. However, when integrated with building service systems, it can be reclaimed to meet the demand for other utilities such as heat and hot water. For the BIQ house, the excessive heat during summer is stored underground with a probe that is tied to the district heating. Carbon storage generated by the building is an added value. Finally, harvested bioproducts from grown microalgae can incentivize stakeholders to adopt microalgae enclosures.

Operation: There are three operation modes for industrial production: (1) batch cultivation, (2) continuous cultivation, and (3) semi-continuous cultivation. With batch cultivation microalgae are grown in a confined batch without discharge for a certain period of time. It is relatively cost-effective and low maintenance and can protect microalgae from contamination because no additional water/media is introduced. To improve low biomass productivity, batch modes intake nutrient supplementation which results in a growth boost. Continuous production operation modes continuously discharge grown microalgae and add nutrient-rich new media. It is more costly and requires higher maintenance. Because the culture is constantly fed with new media, the chance of introducing contaminants is also greater than with the other two modes. The semi-continuous operation applies the advantages of the batch and continuous methods to yield maximum growth with reduced contamination and maintenance. In semi-continuous modes, withdrawal of grown microalgae and the supply of new media occur at a certain period/interval depending on the strain's growth rate. All cultivation modes require the post-cultivation processes of dewatering, drying, and downstream processing, which involves high energy and resource input. The BIQ house was operated in both continuous and batch production modes. The harvested biomass was shipped to subcontracts for the production of high-value bioproducts such as amid acids and lipids.

Maintenance: Upfront cost and long-term maintenance are of crucial importance for real-world applications. Contamination during cultivation is one of the challenges in mass production and industrial processes. To this end, the microalgae enclosure incorporates preventative measures such as sterilization systems that use high temperature, UV-C, and chemical treatment against biological pollutants.[4] Centralized control systems can play a significant role in monitoring and regulating optimum environment and biological contaminants with regard to temperature, dissolved oxygen, pH, pumps, aeration, sterilization, and performance sensors. During maintenance for local areas, the microalgae enclosure can be designed in a way to be compartmentalized with local mechanical systems such that the entire system does not need to be shut down for local maintenance. Biofouling is another maintenance issue where microalgae cells adhere to the inner surface of microalgae windows which hinders light penetration and leads to aesthetic dissatisfaction. Potential solutions could incorporate mountable and movable parts for easy cleaning of the interior surface. The BIQ house was monitored for two years and provided valuable lessons in technical and process optimization as well as appropriate material selection for bioreactor applications.

Longevity: The microalgae enclosure is the outermost building skin that experiences harsh temperatures, air movement, and relative humidity across the day and season. Environmental loadings such as humidity, air, and temperature (HAM) cause expansion and contraction of system components. Every material has a specific expansion of coefficient that causes changes in size with variations in material temperatures. Microalgae enclosures should be designed to accommodate expansion and contraction, be free from dynamic environmental loadings, and to minimize any internal stresses. In addition to material longevity under external environments, the materials should be compatible with microalgae culture. A valuable lesson from the BIQ house is that aluminum spacers in contact with media corrode and become a root cause of clogging distribution pipes. Glass and polymers are typical bioreactor materials assembled with plastic fittings. Bioreactors are operated with a heat exchanger as well as a pump and distribution system. Monitoring and maintenance for the operational system is critical in maximizing cultivation. Aesthetics of the system along with acoustic and olfactory qualities are also important aspects for wide public acceptance.

Environmental stewardship: The microalgae enclosure contributes to a carbon-neutral building and occupant health and well-being by sequestering CO_2 and bioremediating contaminants. Microalgae cultivation does not need arable land, agricultural water, and fertilizers. Microalgae have a faster growth rate compared to terrestrial plants. Population doubling time is less than a day for most microalgae species and a shorter time is required for a final appearance. Substantial research and pilot studies have been executed with industrial-scale deployment by integrating microalgae systems for waste processing plants such as wastewater treatment and power plants where nutrients are supplied to grow microalgae and operational costs are saved. Economic benefits are achieved by uptaking nutrients, inorganic carbon, SO_2, and NO from wastewater and gas flue sources and harvesting renewable biomass for biofuel and/or high-value bioproducts. The effect of high solar irradiance on culture productivity was studied using *Arthrospira platensis* and *H. pluvialis* growing in a roof-installed interior tubular bioreactor. For a 450 L bioreactor, Fresnel lenses are installed to intensify light illumination above 6 mmol photon m^{-2} s^{-1}. Better light utilization results in exponential cell growth and higher astaxanthin production.[5]

Green walls can be a competitor of microalgae enclosures. Green walls offer multiple environmental benefits including biodiversity, urban heat island mitigation, and social and psychosocial well-being. Plants phytoremediate air pollutants through both uptake (absorption) and deposition (adsorption). The role of plants in indoor airborne pollutant removal is well documented in laboratory studies under controlled conditions. However, mixed results on their effectiveness in improving indoor air quality in real-world applications have been shown. In addition, green walls take a longer time for surface coverage and often remain with scattered growth and surface deterioration,[6] Microalgae, on the other hand, have a short cultivation period and reach high density in a short time. Plant selection for the green wall is highly climate specific, and it is often hard to predict the behavior of local species,[7] whereas microalgae are season independent and exceptionally tolerant to extreme environmental conditions. Green walls consume substantial water and nutrients,[8] whereas microalgae consume less freshwater and can use rainwater and nutrients from wastewater/power plants. Table 10.2 summarizes the comparison between microalgae building enclosures and green walls.

Comparison between microalgae enclosures and green walls

Microalgae enclosures	Green walls
Short cultivation period	Longer cultivation period
Year-round cultivation	Season dependent
Automatic homogenous distribution over façade	Extra maintenance for homogenous appearance
Less contamination risks	Protection against pest
No direct influence on adverse insects	Direct influence on biodiversity

10.2 Microalgae Enclosure Examples

This section discusses five design typologies of microalgae enclosures synthesized with building science and biotechnical performance requirements discussed in the previous section. The development of each typology involves the determination of the geometrical properties and material techniques of key components. The inherent material characteristics improve structural integrity, constructability, durability, and other practical issues that can further support a symbiotic relationship between built environments, occupants, and nature. Figure 10.1 illustrates the five typologies of the microalgae curtain-wall application.

Another design goal is to decrease upfront capital and operational cost thorough augmenting multi-functionalities from the microalgae enclosures. Multiple benefits are important to justify the upfront investment and influence public acceptance on the microalgae enclosure through economic incentives and environmental stewardship, including high-value products, wastewater treatment, flue gas uptake, CO_2 sequestration, carbon credits and subsidies, and real estate values. Microalgae buildings can be part of the societal infrastructure for producing food and fuels while phytoremediating pollutants generated from anthropogenic activities.

Additional operational cost savings could be possible with the use of nutrients from wastewater treatment or the use of recycled water from rainwater collection. In this case, pretreatment by settlement or sterilization for adverse bacteria in wastewater may be required before microalgae is grown in wastewater. If additional pretreatment units are required on-site, the microalgae system may impose an extra economic burden. Instead, concentrated microalgae can be transferred and subjected to separate wastewater treatment and treated water can be used as the primary media. Governmental financial incentives, carbon credits, and high-value products will play an important role in attracting stakeholders.

(1) Divided Typology©

The Divided microalgae enclosure system utilizes a flat bioreactor suspended within a window assembly surrounded by metal frames. The system is a factory-assembled unit with a size of 5 feet wide by 13 feet tall that encloses the full floor-to-floor height. The Divided microalgae enclosure incorporates a photobioreactor

▲ Figure 10.1

Patent-pending microalgae curtain walls: the Divided (a), the Inflated (b), the Stranded (c), the Suspended (d), and the Woven (e).

area for growing microalgae and vision area for allowing view-out and daylighting transmission. The flat bioreactor is made up of a network of micro flat bioreactors and micro bioreactors that are separated for view-out yet connected to each other with watertight connectors. The placement of flat bioreactors encourages microalgae growth while the vision area allows occupants to connect with outside and daylighting penetration. The bioreactor of the Divided system is made of thin glazing sheets (e.g., bioplastic, UV-resistant polymer, glass), where a nutrient-rich liquid and culture are contained. Maximum cultivation needs uniform daylight penetration across the bioreactor. Flat systems like the Divided are less susceptible to dark zones due to their constant depth and because they have a larger illuminated surface area to unit volume than a tubular counterpart.

The Divided bioreactor is suspended between glass panes, and the media cavity is approximately 6 mm to 12 mm thick depending on overall building dimensions and engineering of biotechnical/performance requirements. Its geometry and layout of bioreactors in elevation reflects a balance between solar exposure for microalgae growth and view-out for occupants. All the edges of the bioreactor provide a continuous seal against water leakage. The top and bottom of the system is connected with growing apparatus and operation systems. Carbon dioxide (room air) is pumped through check valves at the bottom into a network of bioreactors. Young microalgae are gravity-fed from the top of the system, and grown microalgae are withdrawn from the bottom using gravity. Oxygen generated by the system is collected at the top and transported to a centralized mechanical room for good indoor air quality.

The operation system consists of PVC pipes, check valves, manifolds, pumps, storage tanks, and so on tied with an automatic monitoring and controlling system that can be set to a semi-continuous production mode. All the pipes and valves are housed in a glazing frame where maintenance access can be performed from the outside or inside. The vision area offers visually unobstructed view-out and leftover area after bioreactors are populated within the Divided system. The geometry and combination of bioreactors and vision area provide the multiple functionalities of view-out, daylighting, building energy efficiency, thermal comfort, and acoustic insulation. Figures 10.1.1–10.1.4 shows an overview of a Divided microalgae building enclosure system.

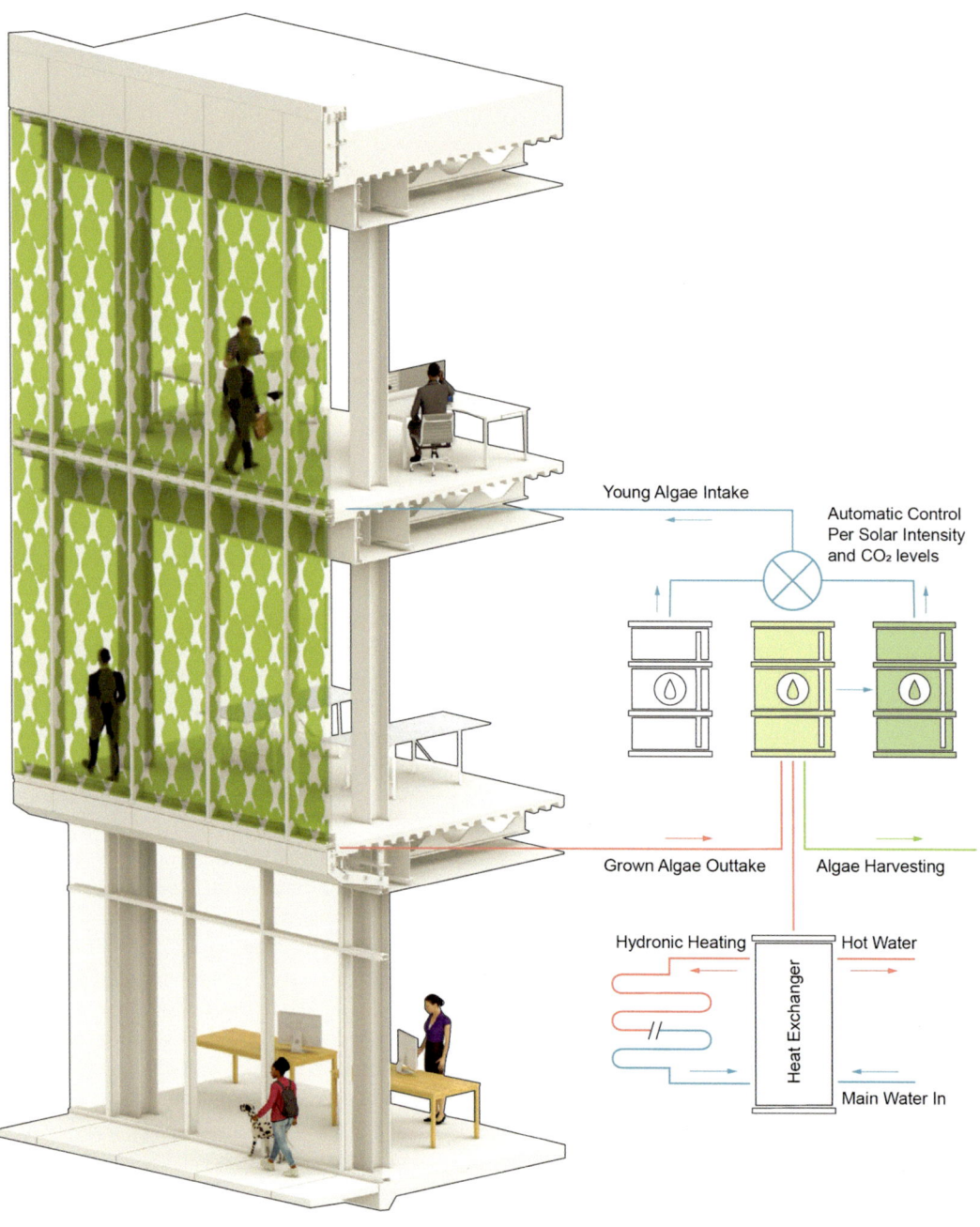

Young Algae Intake

Automatic Control
Per Solar Intensity
and CO₂ levels

Grown Algae Outtake

Algae Harvesting

Hydronic Heating

Hot Water

Heat Exchanger

Main Water In

▲ Figure 10.1.1

The Divided curtain wall consists of a network of flat bioreactors and clear vision areas; the automatic monitoring and control system further enhances energy efficiency, indoor air quality, and occupant satisfaction; the solar energy stored in the system can be reclaimed for space and water heating while the integration of a heat exchanger mitigates excessive temperature ranges in the culture especially during summer.

Unitized Curtainwall System

① Insulated Glass Unit (IGU)
② Divided bioreactor
③ Mullion
④ Transom
⑤ Interior glazing panel

▲ Figure 10.1.2

The Divided curtain wall is composed of insulated glass unit (IGU), bioreactors, and metal frames; prefabrication of the unit allows speedy installation and high quality control.

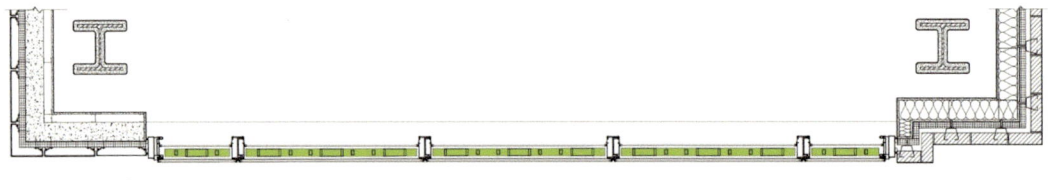

▲ Figure 10.1.3

The flat bioreactor of the Divided offers a maximum illuminated surface-to-volume ratio and sunlight penetration while the vision area allows for view-out and daylight transmission.

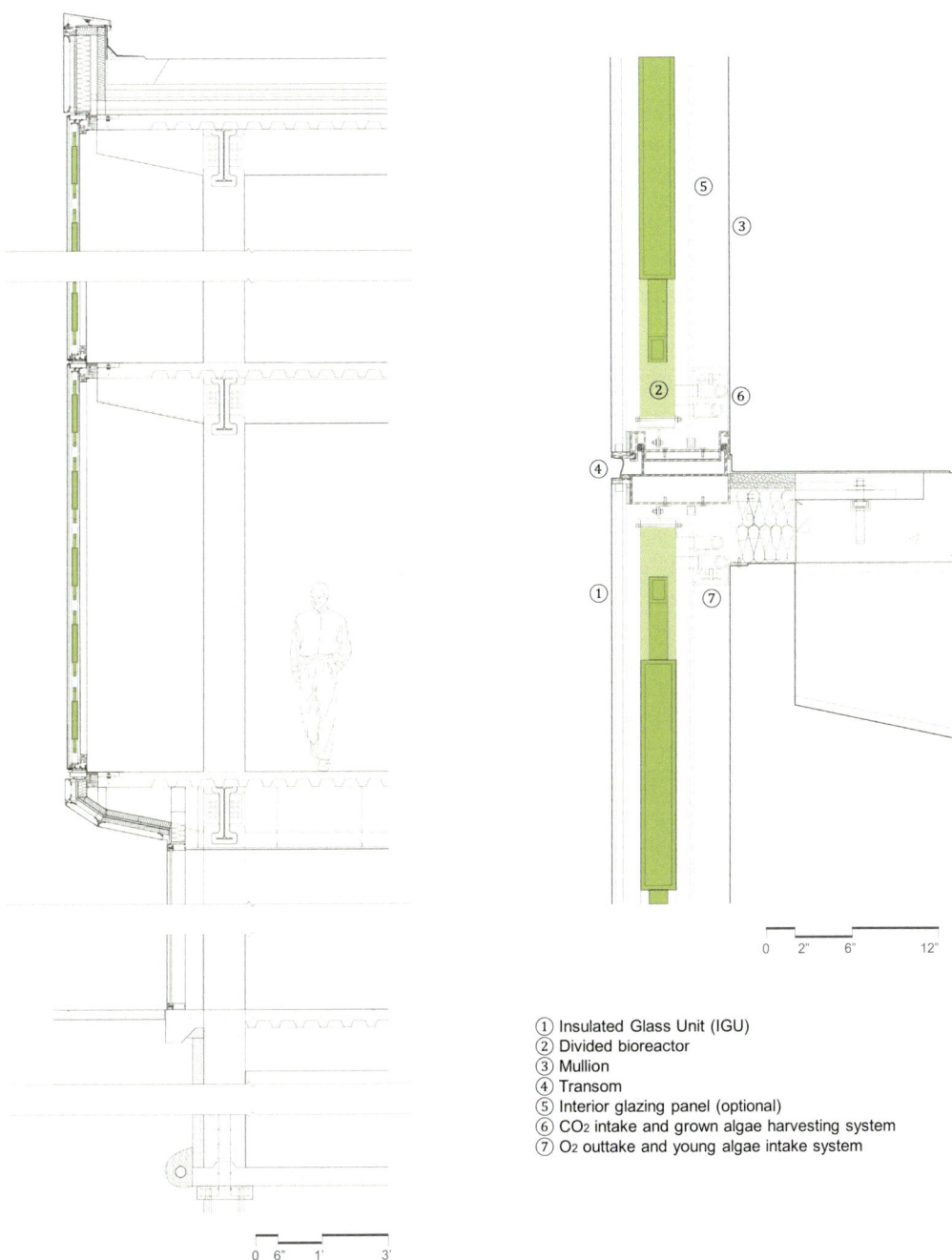

1. Insulated Glass Unit (IGU)
2. Divided bioreactor
3. Mullion
4. Transom
5. Interior glazing panel (optional)
6. CO_2 intake and grown algae harvesting system
7. O_2 outtake and young algae intake system

▲ Figure 10.1.4

Section details of the Divided system with growing apparatus installed at the top and bottom of the system, providing easy access for operation and maintenance.

(2) Inflated Typology©

Buildings integrated with ETFE (ethylene tetrafluoroethylene) have gained popularity because the ETFE foil provides good structural, thermal, acoustic, and solar performance as well as durability. It also has energy production potential when integrated with organic photovoltaic (OPV) elements. The ETFE pillow keeps its shape by constantly pumping air into the cavity with inner air pressures between 250 Pa and 1000 Pa, creating a cushion appearance. ETFE has high optical clarity, transmitting most of the visible light range for photosynthesis and is resistant to excessive temperatures. At 1% of the weight of glass, the inflated panel allows for spanning a large space and reduces the use of metal frames to support the panel. The ETFE pillow typically consists of a multi-air cavity created by inner membranes, which improves the U-factor and energy transfer.

As the name implies, the Inflated microalgae enclosure incorporates a micro bioreactor on the outermost surface of the ETFE cushion by adding ETFE bioreactors heat welded to the primary cushion. The Inflated bioreactor is a series of 12 feet long by 1 feet wide and 1-inch deep enclosures, aiming to have high surface-to-volume ratio for maximum solar exposure. Extremely hot or freezing temperatures in the culture over the year need close monitoring and temperature control. Room air circulation coupled with an active air conditioning unit is introduced to limit extreme temperature swings. The room air is pumped from the bottom of the system, and oxygen generated from photosynthesis is collected from the top. Likewise, grown microalgae will be withdrawn from the bottom and culture media will be added from the top using gravity. Due to space constraints, the manifold connection is utilized to operate air and media using one fixture. The distribution system consists of PVC pipes, valves, and environmental sensors for transporting air, media, nutrients, and cultures between the bioreactor and centralized mechanical systems (e.g., pumps, storage tanks).

The control systems measure the media temperature, air flow rate, optical density (i.e., turbidity), and pH level and monitor growing conditions. Automated monitoring and control systems optimize system operation for maximum algae growth and energy savings. The placement of micro bioreactors on the outermost surface of the ETFE pillow allows maximum solar exposure, thermal insulation, and acoustic separation. One criticism of the ETFE pillow is a drumming noise from rain, but the multiple micro bioreactors are possibly able to suppress the noise from rain droplets while encouraging microalgae growth. The distribution components are housed along the metal frames that support the ETFE pillow and can be maintained from the outside via access panels. Figures 10.2.1–10.2.4 shows an overview of an Inflated microalgae building enclosure system.

(3) Stranded Typology©

The Stranded microalgae building enclosure system grows microalgae within a vertical framing while keeping the vision glazing visually unobstructed. The vertical bioreactor is made of extruded polymer. In response to structural requirements and microalgae growing environments under solar exposure, the extrusion is deformed in two axes along with in-plane and out-of-plane relative to vertical glazing. The

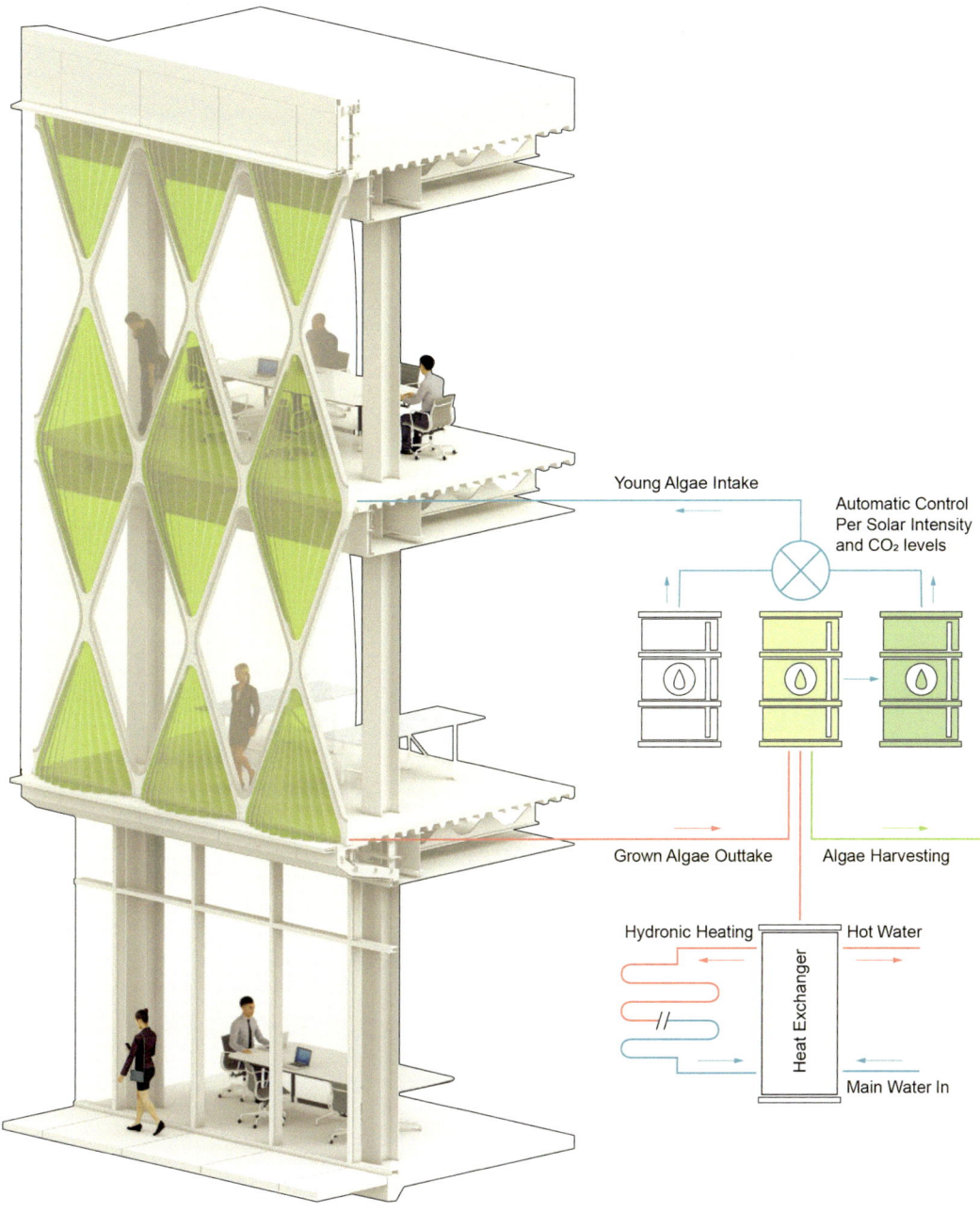

Young Algae Intake

Automatic Control
Per Solar Intensity
and CO_2 levels

Grown Algae Outtake   Algae Harvesting

Hydronic Heating   Hot Water

Heat Exchanger

Main Water In

▲ Figure 10.2.1

The Inflated curtain wall incorporates a micro bioreactor on the outermost surface of the ETFE cladding; the system offers good thermal insulation, shading efficacy, acoustic simulation, daylighting penetration, and view-out; the growing apparatus is connected at the top for controlling young algae intake and oxygen outtake and at the bottom for CO_2 intake and grown algae outtake.

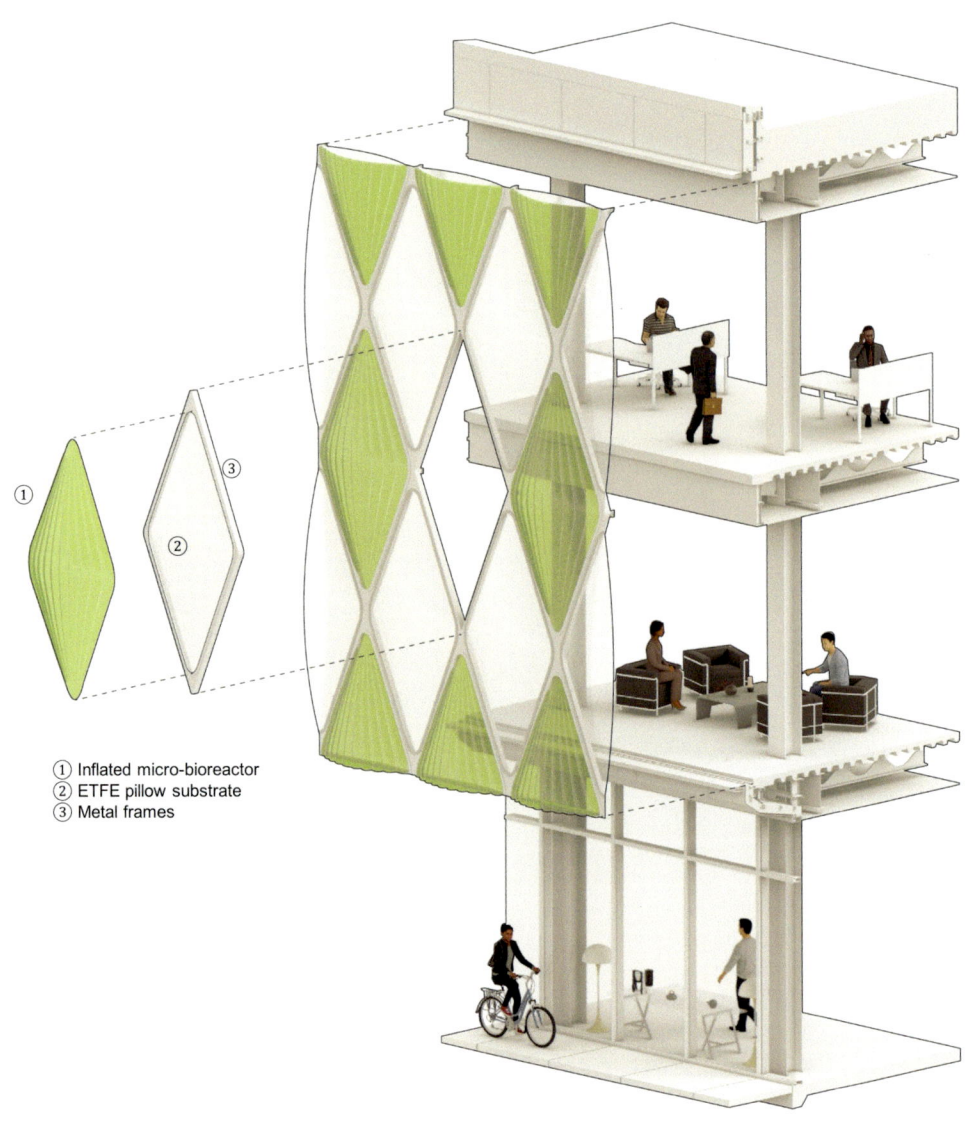

① Inflated micro-bioreactor
② ETFE pillow substrate
③ Metal frames

▲ Figure 10.2.2

The Inflated bioreactor offers a high illuminated surface-to-volume ratio with lightweight construction, maximizing solar exposure and the microalgae productivity rate.

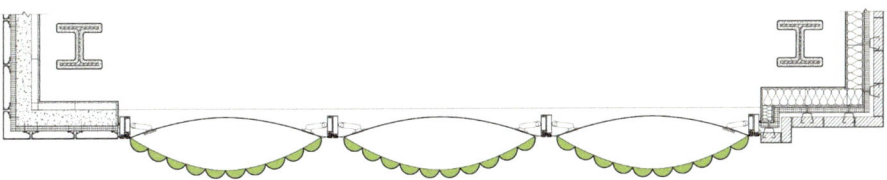

The building façade enclosed with the Inflated system offers a balance between maximum algae growth, occupant view-out, and daylight penetration.

top and bottom of the deformed extrusion is deeper and connected with horizontal metal frames that are anchored to concrete slabs. The depth of the deformed extrusion flattens toward the middle to provide better sunlight penetration (depth up to 4 inches is appropriate for light penetration).

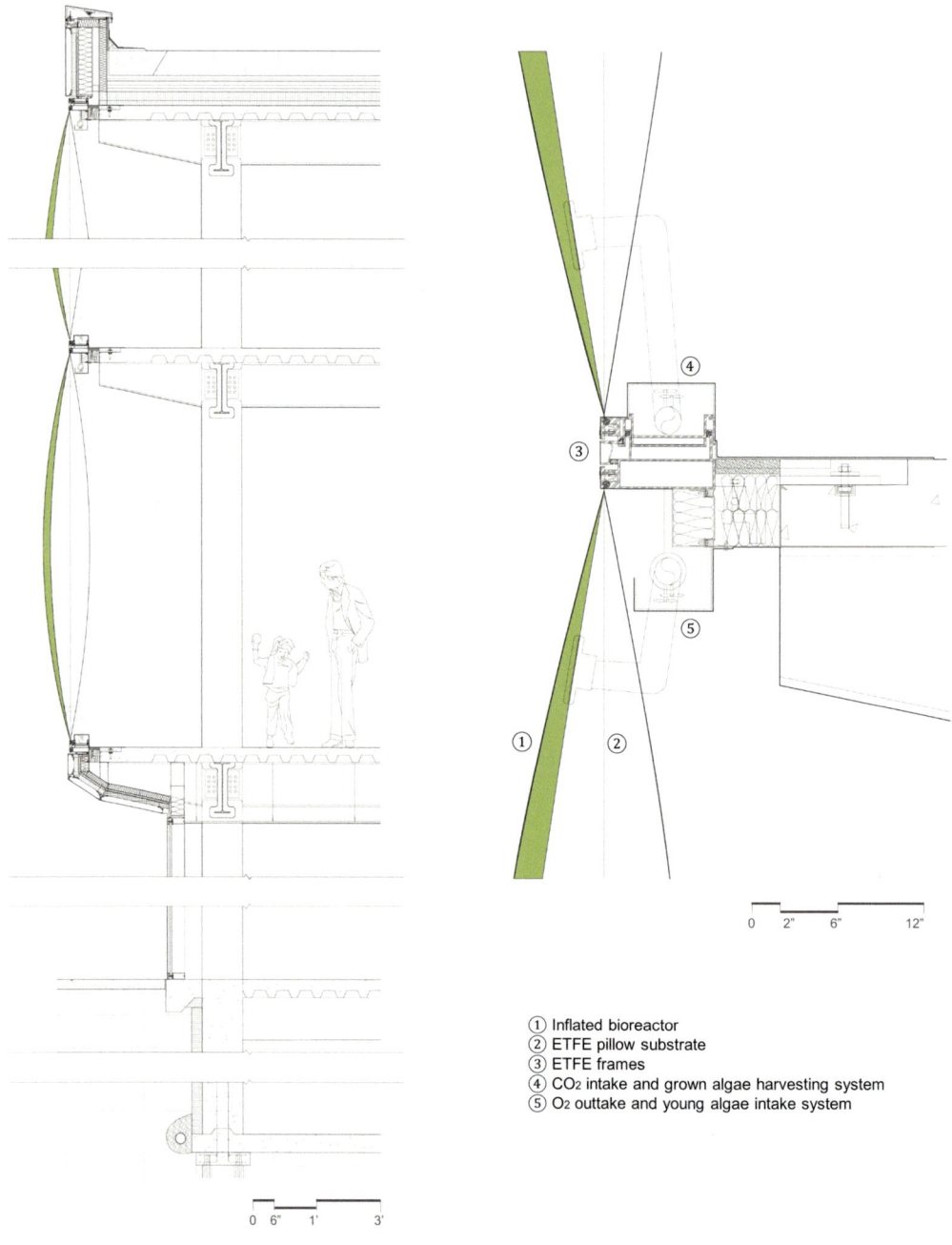

▲ Figure 10.2.4

Section details of the Inflated system with growing apparatus installed at the top and bottom of the system, providing easy access for operation and maintenance.

Legend:

① Inflated bioreactor
② ETFE pillow substrate
③ ETFE frames
④ CO_2 intake and grown algae harvesting system
⑤ O_2 outtake and young algae intake system

The top of the deformed extrusion needs to be deeper to provide structural rigidity against a later load such as wind pressure. The sinusoidal geometry of the bioreactor in elevation adds dynamic aesthetics. The viewing glazing bounded by the Stranded bioreactor is made of clear glass and/or an ETFE pillow (vision system). The Stranded bioreactor grows microalgae and, in the meantime, acts as a structural framing system to support the glazing/ETFE vision pillow. The glazing/ETFE vision pillows between the Stranded bioreactor provide view-out, daylight transmittance, waterproofing, airtightness, an acoustic buffer, thermal insulation, and natural ventilation when necessary.

Similar to other typologies, the microalgae growing devices of the Stranded system consist of distribution and mechanical systems. The distribution system is directly connected at the top (for young algae intake and oxygen outtake) and bottom (for CO_2 intake and grown algae outtake) of the Stranded bioreactor and linked to the pumps, storage tank, and control/monitoring systems. The distribution systems consist of PVC pipes, valves, and sensors to transport air, media, nutrients, and cultures. Aeration helps maintain mixing of the culture with homogeneous illumination, temperature, pH, nutrients, and CO_2. Excessive aeration causes shear stress and deters productivity while increasing energy consumption. The distribution system is contained in horizontal metal frames which can be accessed for maintenance either from the inside or the outside through access panels. Figures 10.3.1–10.3.4 show an overview of a Stranded microalgae building enclosure system.

(4) Suspended Typology©

The Suspended microalgae enclosure consists of a network of interlocking modular units encapsulated between glass panels. Individual X-shaped modules are connected to each other to generate an aggregate bioreactor. The modular construction allows for partial or full coverage of a window area, and various configurations could be devised for functional needs and aesthetics. They could be screen types or louver/fin types and regulate energy transfer between indoor and outdoor spaces by controlling solar reduction, daylighting, view-out, microalgae growth, and aesthetics. Each X-module is around 1 feet by 1 feet and 1-inch thick (around 1 L), made of UV-resistant polymers (e.g., polycarbonate, biopolymer) or glass material.

The connection of X-modules incorporates double defense lines of anti-leak details. The X shape of the module allows for interlocking multiple layers of bioreactors, and different microalgae can be grown separately in each layer. The top of the suspended system is connected to intake pipes for supplying young algae and outtake pipes for collecting oxygen generated from photosynthesis. The bottom of the suspended system is connected to intake pipes for CO_2 aeration and outtake pipes for grown algae withdrawal. These distribution components are concealed within horizontal glazing frames and maintenance access is possible either from the outside or inside. A centralized system automatically controls inflow and outflow of media, air, and cultures for maximum productivity. Temperature, air flow rate, pH, and optical density are monitored to maximize production. Long-term creep of the polymer under sustained hydrostatic pressure needs attention.

Controlling the environmental growth conditions is important for microalgae survivability and maximum productivity. Key environmental conditions include

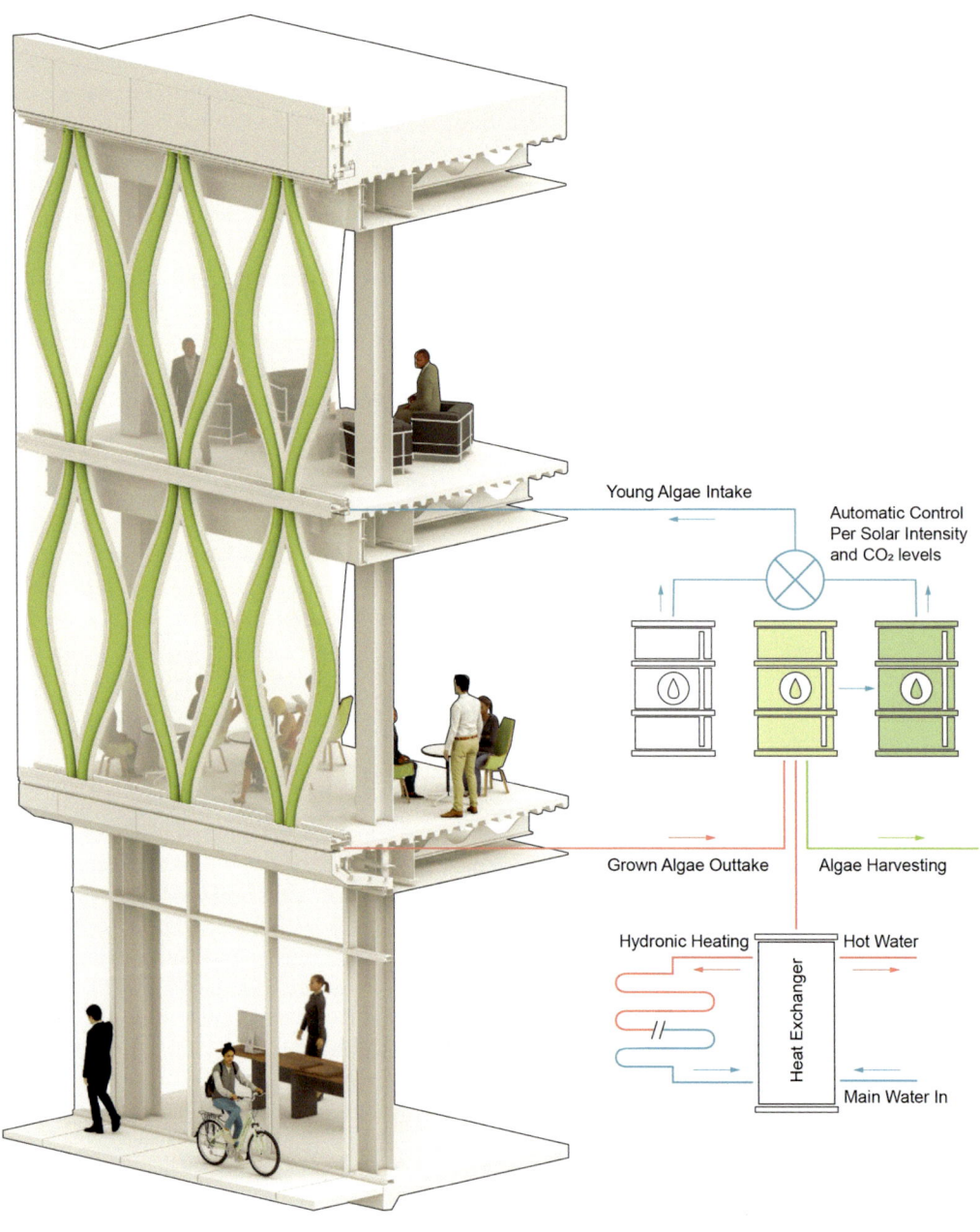

Labels within figure:
Young Algae Intake
Automatic Control Per Solar Intensity and CO₂ levels
Grown Algae Outtake
Algae Harvesting
Hydronic Heating
Hot Water
Heat Exchanger
Main Water In

▲ Figure 10.3.1

The Stranded system grows microalgae within a vertical framing system that is deformed in two axes along with in-plane and out-of-plane relative to the vertical glazing in response to structural requirements and microalgae growing environments.

illumination strategy, temperature swings, aeration and mixing, pH, and dissolved O_2-to-CO_2 ratio. Controlling the irradiance level from the sun is challenging, and, therefore, microalgae enclosures need optimization in system design and operation protocol. The density and color of microalgae respond to sun intensity, and a

① Stranded bioreactor
② Glazing infill panel
③ Horizontal transom

▲ Figure 10.3.2

The Stranded system consists of vision glazing and bioreactor framing; the top and bottom of the frame is deeper toward the concrete slab for structural rigidity and is flatter toward the middle.

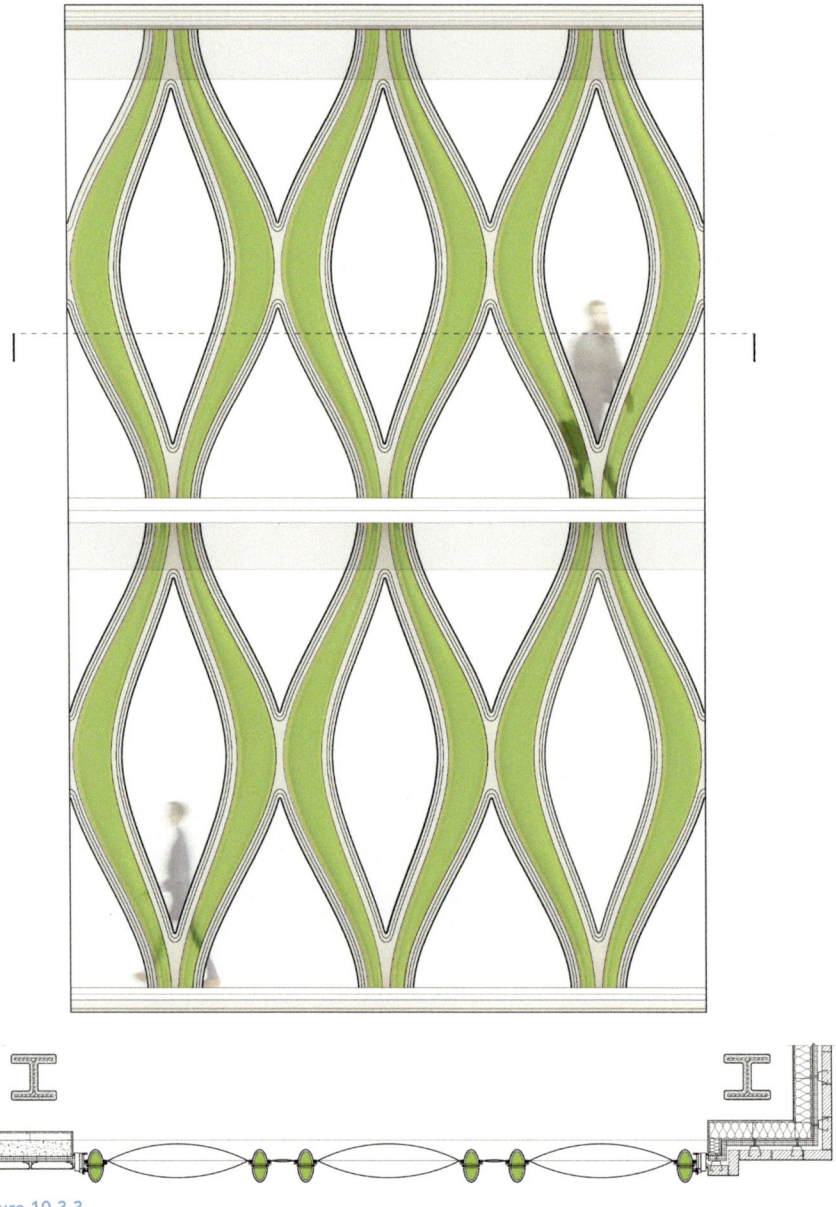

▲ Figure 10.3.3

The sinusoidal geometry of the bioreactor in elevation adds dynamic aesthetics; the vision glazing is bounded by the bioreactor framing for view-out and daylight penetration.

maximum of 8–10% of PAR is used for photosynthesis. The culture temperature can be regulated with circulating interior air, which helps maintain optimum conditions at a constant temperature. Active heat exchangers can be operated depending on extreme temperatures. Figures 10.4.1–104.4 show an overview of a Suspended microalgae building enclosure system.

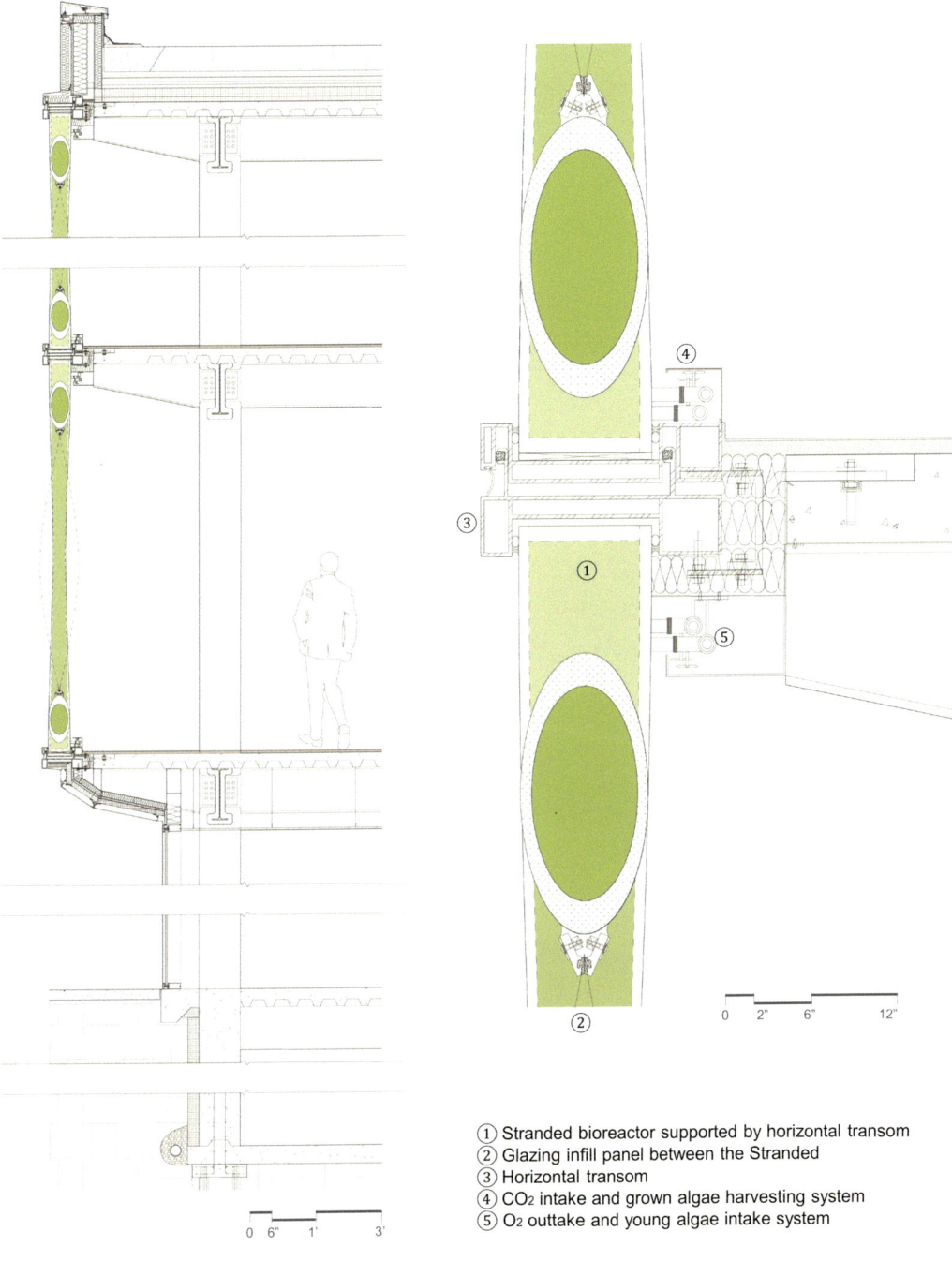

① Stranded bioreactor supported by horizontal transom
② Glazing infill panel between the Stranded
③ Horizontal transom
④ CO_2 intake and grown algae harvesting system
⑤ O_2 outtake and young algae intake system

▲ Figure 10.3.4

Section details of the Stranded system with growing apparatus installed at the top and bottom of the system, providing easy access for operation and maintenance.

(5) Woven Typology©

The Woven system provides a continuous watertight bioreactor. The opacity of Woven weft and warp tubing is a resulting balance of the solar exposure for maximum microalgae growth, access to view-out, and daylighting potentials while

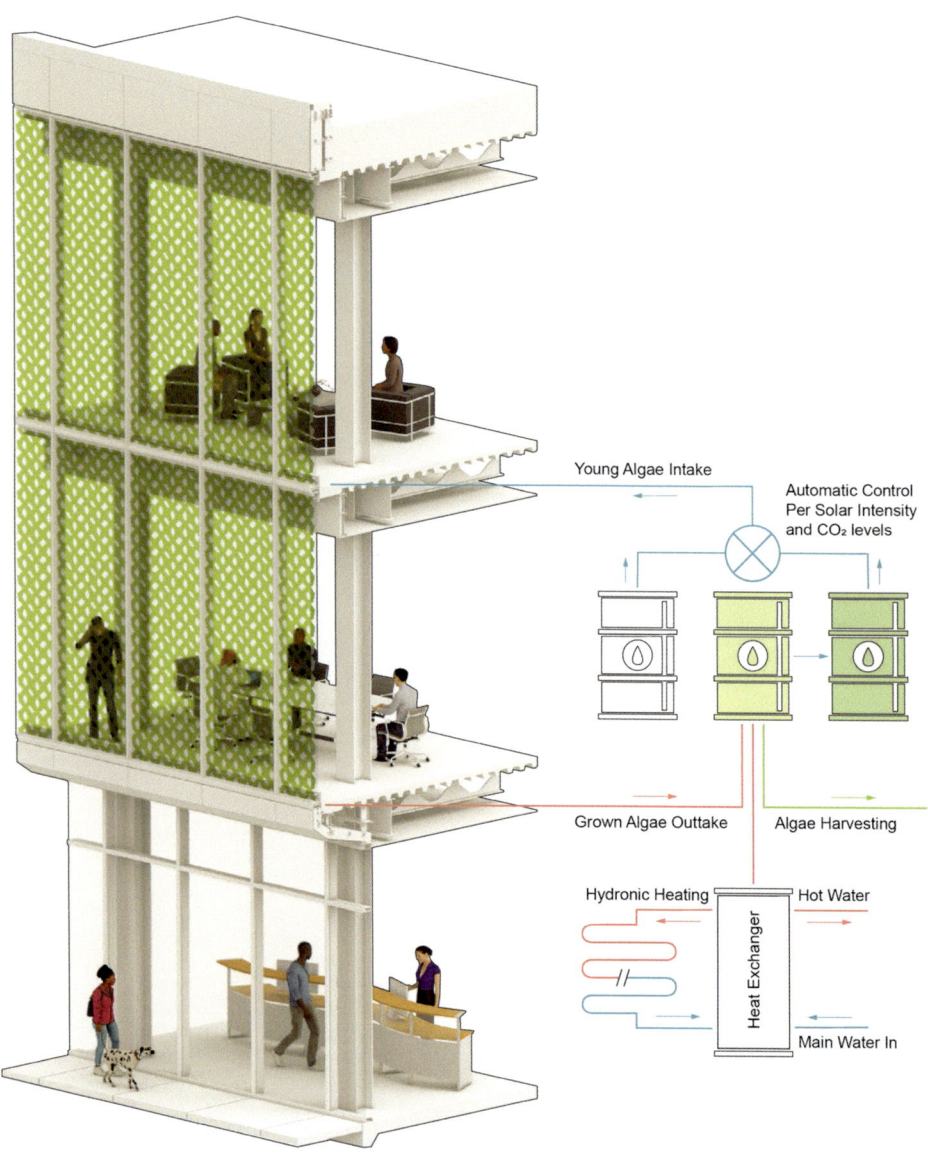

Young Algae Intake

Automatic Control Per Solar Intensity and CO_2 levels

Grown Algae Outtake   Algae Harvesting

Hydronic Heating   Hot Water

Heat Exchanger

Main Water In

▲ Figure 10.4.1

The Suspended system incorporates an interlocking bioreactor made of X-shaped modular units; an active monitoring and control system enables the system to regulate cell density and colors depending on solar intensity and CO_2 levels; the system can be tied with a centralized building management system for additional energy savings, improved indoor air quality, and occupant satisfaction.

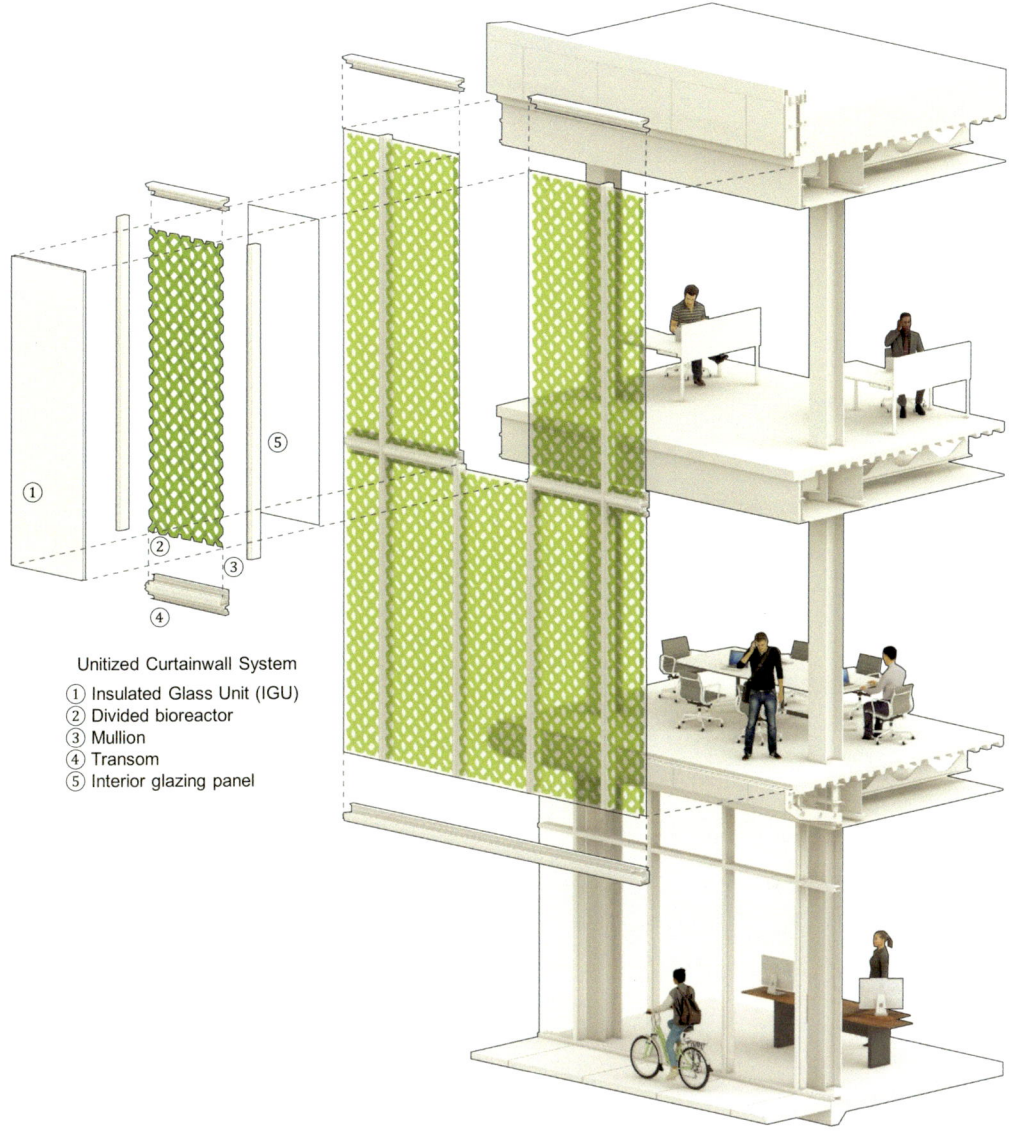

Unitized Curtainwall System
1. Insulated Glass Unit (IGU)
2. Divided bioreactor
3. Mullion
4. Transom
5. Interior glazing panel

▲ Figure 10.4.2

The Suspended system is a factory-installed prefabricated unit, and the bioreactor is encapsulated between glass panels for easy maintenance.

filtering thermal and visual environments. The Woven photobioreactor is made of continuous flexible tubing, and Woven knots provide the geometric stability for bioreactor tubing especially when it is suspended in the window cavity. The Woven photobioreactor can be hung in the air cavity of the window assembly or be cast within the glazing layer. The diameter of the tubing can be 6 mm–12 mm and is expected to have good solar exposure for maximum microalgae growth.

The microalgae growing apparatus is connected to both ends of the photo-bioreactor tubing located at the top and the bottom of the system. The bottom of

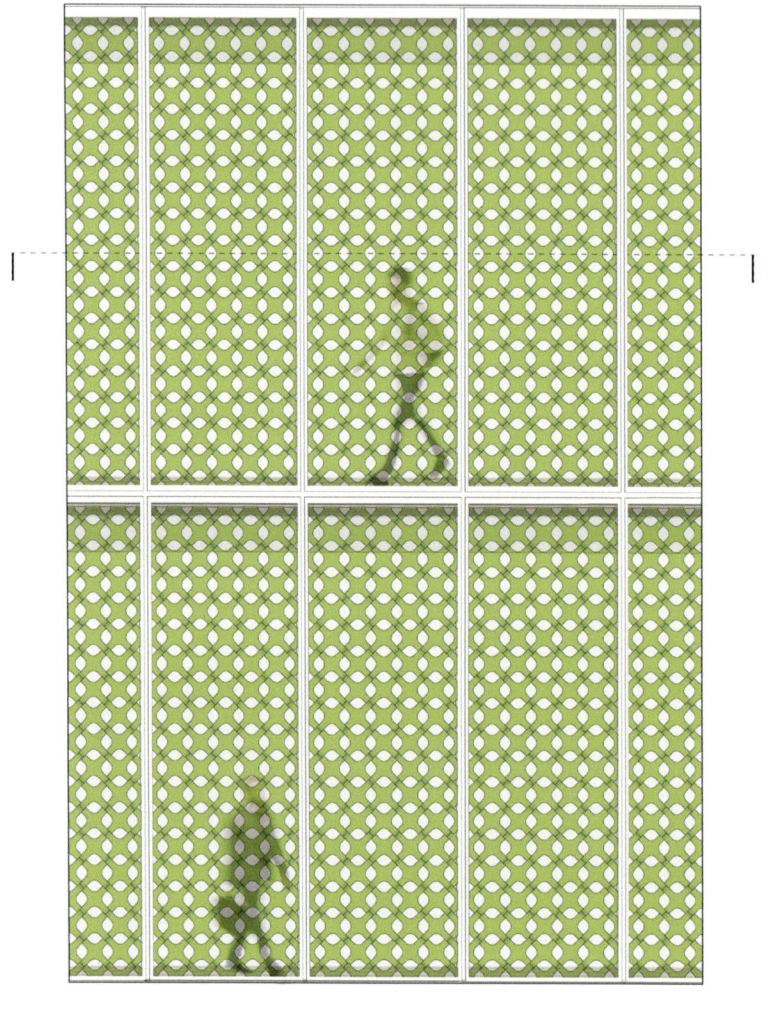

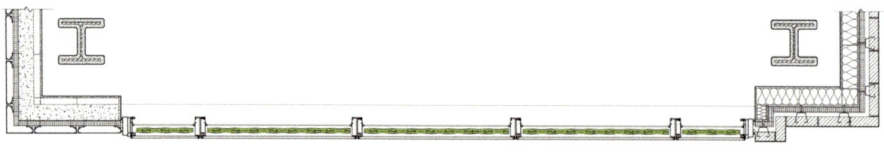

▲ Figure 10.4.3

The modular construction of the Suspended curtain wall allows for partial or full coverage of a window area.

the Woven bioreactor is connected with a manifold fitting to continuously supply room air and periodically discharge grown algae. Likewise, the top of the Woven bioreactor has manifold fittings to constantly collect oxygen and fill in young algae at a certain interval.

The microalgae-growing apparatus consists of distribution and mechanical systems. The distribution systems consist of PVC pipes, valves, and sensors and distribute air, water, nutrients, and cultures between the photobioreactor and the

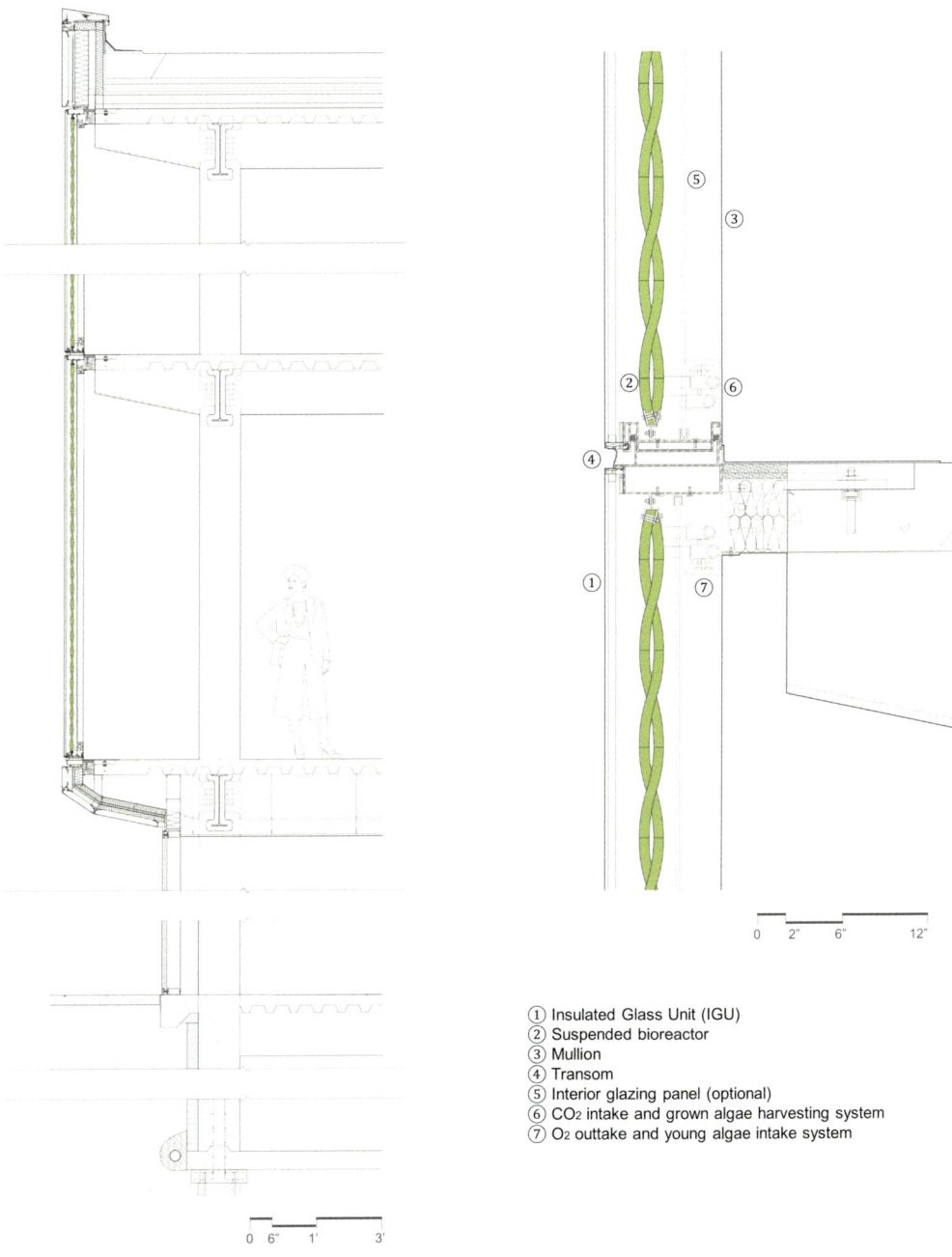

1. Insulated Glass Unit (IGU)
2. Suspended bioreactor
3. Mullion
4. Transom
5. Interior glazing panel (optional)
6. CO_2 intake and grown algae harvesting system
7. O_2 outtake and young algae intake system

▲ Figure 10.4.4

Section details of the Suspended system with growing apparatus installed at the top and bottom of the system, providing easy access for operation and maintenance.

mechanical systems. The mechanical systems consist of pumps, storage tanks, and a monitor/control system (e.g., temperature, flow rate, pH, and turbidity). A harvesting system can be housed in a mechanical room for processing on-site, or grown microalgae can be trucked away to third-party biofuel producers. Figures 10.5.1–10.5.4 show an overview of a Woven microalgae building enclosure system.

10.3 Post-Cultivation Processing

Microalgae have been identified as a promising renewable energy source. However, the harvesting process is considered as the challenging step in the production of microalgae biofuel because it is hard to separate the tiny cells (typically 3–30 μm) of microalgae from water. Economic and sustainable harvesting techniques are critical in the scale-up of mass cultivation. There are many strains living in natural habitats, and only a fraction of them have been analyzed and studied for biofuel production. Microalgal harvesting techniques involve mechanical, chemical, and biological methods. The biological approach is the least expensive, but the mechanical method is the most reliable and commonly used approach in the field.[9] Chemical and biological methods are available for coagulation/flocculation to thicken the cell density. The physical method utilizes gravity sedimentation and flotation. It is cost-effective but time-consuming. Combined approaches of different techniques are more cost-effective than a single approach.[10]

After cell growth reaches the stationary phase, the culture will be collected for harvesting. Various harvesting techniques have been researched, but no universal method is available. Harvesting is processed through the two steps of thickening the culture into a slurry (about 2–7% of total solids) and dewatering the slurry to a cake (20–25% of total solids).[11] The thickening process reduces the overall biomass volume to be processed in the dewatering step. Chemical-and/or biological-based coagulation/flocculation are an initial step for improving the efficiency of the harvesting process. Chemical coagulation/flocculation is a simple and fast method that doesn't rely on external energy. However, chemical coagulation is expensive and the culture media is difficult to recycle due to potential toxicity.[12] Bioflocculation functions by increasing pH levels or introducing microorganisms that promote flocculation. It is relatively inexpensive, and not toxic, but it changes cell compositions with a possibility of microbiological contamination.[13] Gravity sedimentation is not practical in industry applications because settling rates are very low and time-consuming. Therefore, coagulation is necessary for faster settling prior to gravity sedimentation.[14] Flotation is the opposite of sedimentation where bubbling gases lift and separate cells from the media. Chemical flocculation is still required for process efficiency.[15] Electrical-based coagulation uses an electrical field to separate cells. It is environmentally friendly and toxic free due to no requirement for a chemical coagulant, but it has challenges in high energy requirements and equipment cost for large-scale application.[16]

Once microalgae slurry is prepared, a further dewatering process is required for an efficient drying process. The filtration method dewaters algal slurry through a membrane with fine pores. It is a low-cost, low-energy process, but the periodic cleaning requirements for clogged pores increases operational cost with additional cost associated with membrane replacement.[17] Centrifugation is a widely used method for the

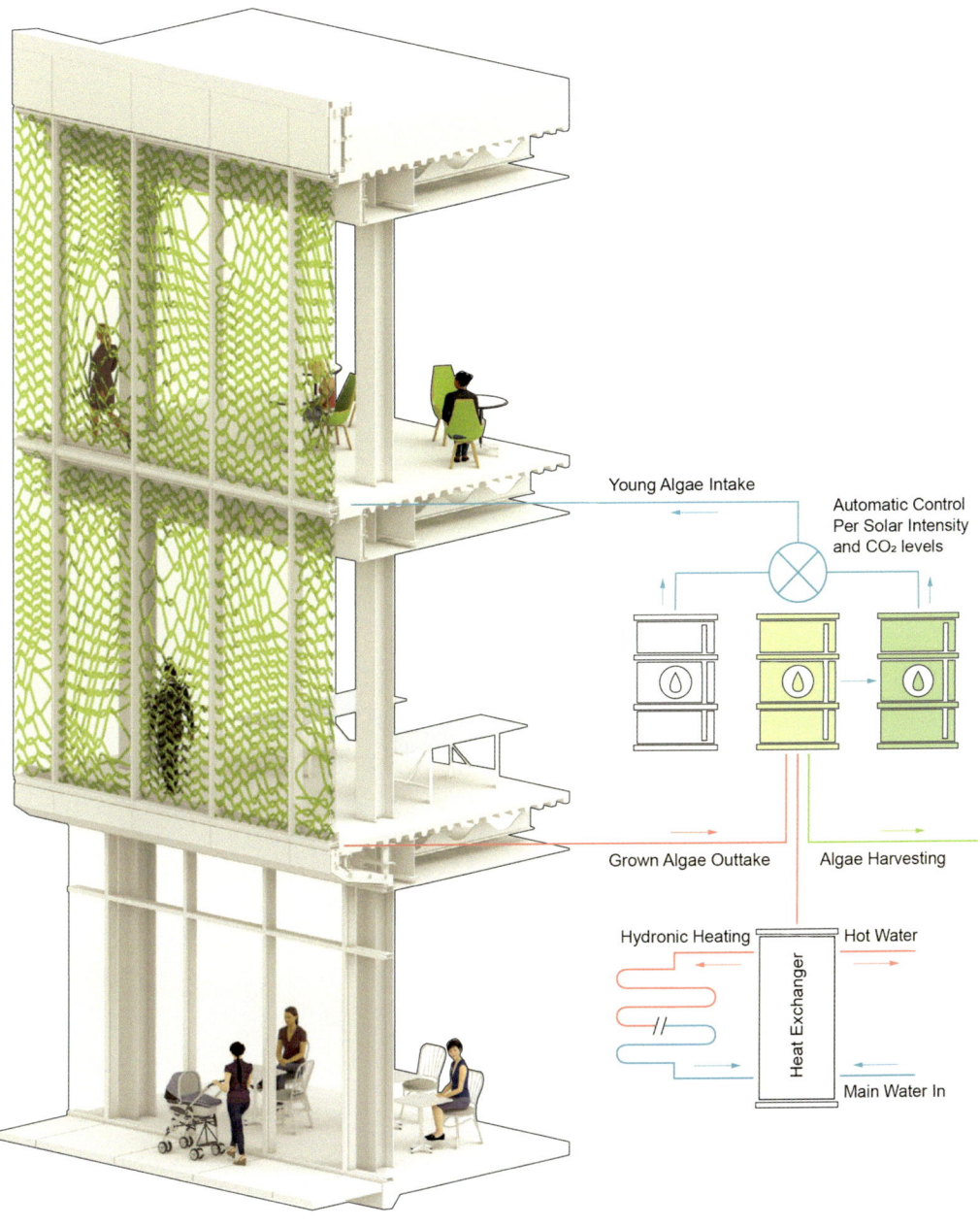

Young Algae Intake

Automatic Control
Per Solar Intensity
and CO_2 levels

Grown Algae Outtake

Algae Harvesting

Hydronic Heating

Hot Water

Heat Exchanger

Main Water In

▲ Figure 10.5.1

The Woven curtain wall utilizes a flexible tubing as a bioreactor that can be suspended in the air cavity between glass panes; both the top and bottom of the Woven bioreactor is connected with a manifold fitting to continuously circulate air while periodically fill in young algae and discharge grown algae at a certain cultivation interval.

dewatering process and is commercially available in industry. Due to high energy input and high operational cost, this technique is more popular for high-value bioproducts in pharmaceutical and nutraceutical applications.[18] One disadvantage is that high centrifugal and shear forces may cause cell damage.[19] To improve overall efficiency and cost

Unitized Curtainwall System
1 Insulated Glass Unit (IGU)
2 Divided bioreactor
3 Mullion
4 Transom
5 Interior glazing panel

▲ Figure 10.5.2

The Woven curtain wall consists of an IGU, woven bioreactor, and metal frames; the woven photobioreactor can be hung in the air cavity between glass layers or be cast within the glazing layer.

savings, the use of a combination of different harvesting techniques is recommended. Table 10.3 summarizes the pros and cons of harvesting techniques. Figure 10.6 shows microalgae harvesting techniques for microalgae biofuel.

Fast growth rates and high lipid and carbohydrate content make microalgae attractive feedstock for biodiesel and bioethanol production, respectively. Microalgae-based biodiesel is biodegradable and has lower environmental impacts than fossil diesel. First-generation biodiesel sources were edible food crops (e.g., soybean, rape-seed, sunflower, palm oil) followed by second-generation biodiesel from nonedible

▲ Figure 10.5.3

The density of Woven weft and warp tubing is a balance between the solar exposure for microalgae growth, access to outside views, and daylighting penetration.

feedstock (e.g., Jatropha trees). Both generation crops have high environmental impacts from high consumption of fertilizers and pesticides impacting water resources, soil degradation, and loss of wild habitat. Although Jatropha can grow in non-arable land, most biofuel feedstock affects food cultivation because it competes with arable land and freshwater. In response to the need for more sustainable biofuel, microalgae have drawn worldwide attention. Assuming a U.S. consumption of 180 billion gallons of fuel (approximately 130 billion gasoline and 50 billion diesel) and 5,000 gallons

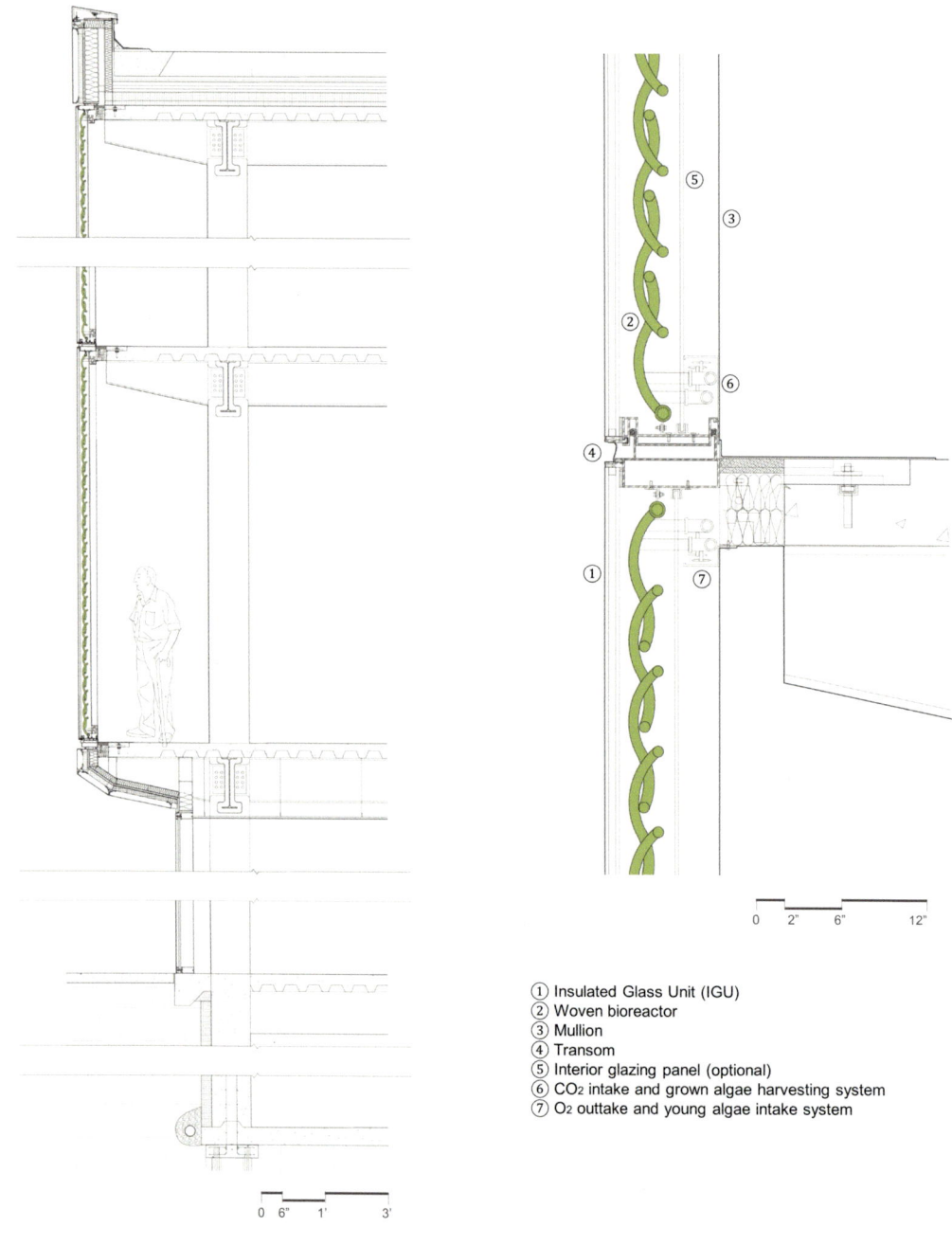

① Insulated Glass Unit (IGU)
② Woven bioreactor
③ Mullion
④ Transom
⑤ Interior glazing panel (optional)
⑥ CO_2 intake and grown algae harvesting system
⑦ O_2 outtake and young algae intake system

▲ Figure 10.5.4

Section details of the Woven system with growing apparatus installed at the top and bottom of the system, providing easy access for operation and maintenance.

Pros and cons of various harvesting techniques (reproduced from Barros et al., 2015)[20]

Harvesting techniques	Pros	Cons
Chemical coagulation/flocculation	No energy input requirements Simple and fast process	Potentially toxic and expensive process Limit of recycling culture media
Bioflocculation	Inexpensive process Recyclability of culture media Nontoxic to biomass	Changes in cell composition Possibility of biomass contamination
Gravity sedimentation	Low-cost, slow process	Time consuming Possibility of biomass deterioration
Flotation	Low-cost, fast process Low space requirements Possibility for large-scale applications	Need of chemical flocculants Not good for marine microalgae applications
Electrical coagulation	Wide variety of microalgae applications No need of chemical flocculants	High energy intensity and high equipment cost Lack of dissemination
Filtration	High recovery efficiency Shear sensitive strain applications	High operational cost from biofouling and clogging High maintenance cost for membrane replacement Periodic membrane maintenance
Centrifugation	Fast and high recovery efficiency Suitable for the majority of strains	High energy-intensive and expensive process Suitable for high-value bio-product applications Possibility of cell damage from high shear forces

per acre for microalgae productivity according to research data,[21] a 36 million ft^2 area similar to an area the size of the state of North Carolina can supply the U.S. annual traffic fuel demand. After the lipid is extracted, microalgae still contain high contents of carbohydrate and protein that make them an ideal source for bioethanol production.[22] Carbohydrates and protein are used for organic carbon sources for fermentation, and bioethanol production shows high conversion efficiency.[23] The harvesting process for this application can be more cost-effective because wet algae can be used for hydrolysis and fermentation stage. Figure 10.7 shows biodiesel productivity comparisons between first-, second-, and third-generation fuel stock.

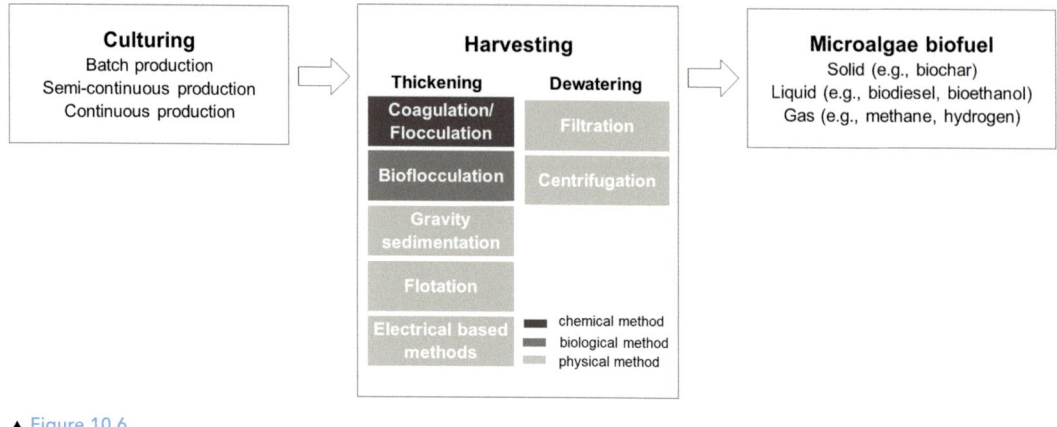

▲ Figure 10.6

Microalgae harvesting techniques for microalgae biofuel (reproduced from Barros et al., 2015).[24]

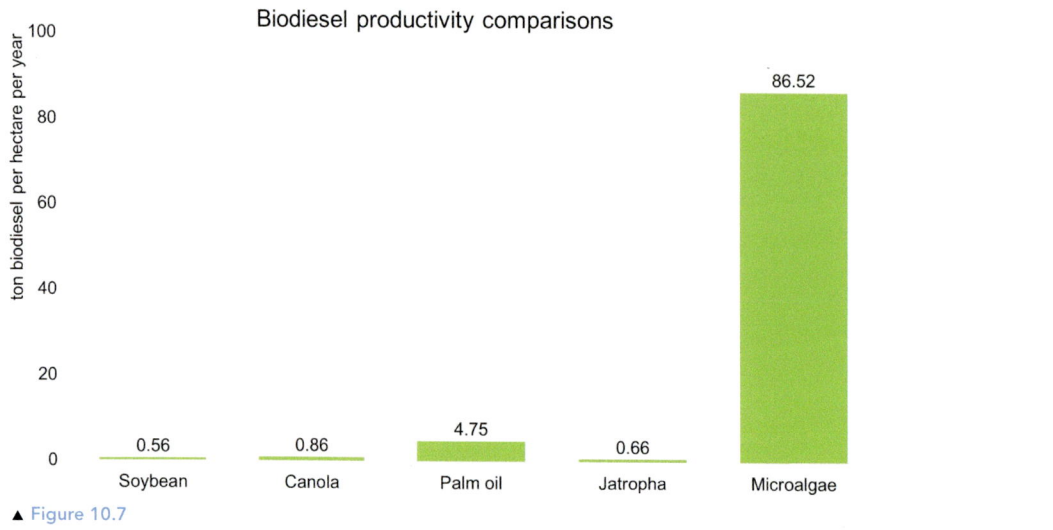

▲ Figure 10.7

Biodiesel productivity comparison between first-generation biofuel crop (soybean, canola, palm oil), second-generation crop (Jatropha), and third-generation microalgae fuel stock.

Summary

Global imperatives on mitigating climate changes have contributed to the growth of regenerative, revolutionary solutions to outcompete anthropogenic activities. The integration of microalgae systems in the built environments offers the promising, new sustainable options. Microalgae enclosures can have a wide variety of configurations and material selections. Key performance goals during the design phase need holistic understanding of engineering and biotechnical requirements including structural integrity, energy attributes, water tightness, air tightness, growth environments of temperatures, solar intensity, aeration, and cultivation, maintenance, and harvesting techniques. Their multiple environmental and social benefits include high energy efficiency, improved air quality, carbon capture, and renewable energy production. As an interface between outdoor and indoor environments, the microalgae enclosure allows year-round daylight illumination and direct view-out and further enhances user satisfaction. Density and color changes responding to solar intensity provide suitable shade for the building and occupant thermal comfort. Effective carbon sequestration and good air quality enhance user health and well-being. For actual real-world effect, it is important to increase widespread public acceptance by inspiring users, communities, and stakeholders with notable architectural quality.

Notes

1   ASCE, "Minimum Design Loads for Buildings and Other Structures," *American Society of Civil Engineers*, 2013.
2   Ibid., 851.
3   Jan Wurm and Martin Pauli, "SolarLeaf: The World's First Bioreactive façade," *Arq: Architectural Research Quarterly* 20, no. 1 (2016): 73.
4   S.Ş. Öncel, A. Köse, and D.Ş. Öncel, "Façade Integrated Photobioreactors for Building Energy Efficiency," In *Start-Up Creation* (Woodhead Publishing, 2016), 237–299.
5   J. Masojídek, M. Sergejevová, K. Rottnerová, V. Jirka, J. Korečko, J. Kopecký, I. Zaťková, G. Torzillo, and D. Štys, "A Two-stage Solar Photobioreactor for Cultivation of Microalgae Based on Solar Concentrators," *Journal of Applied Phycology* 21, no. 1 (2009): 55–63.
6   Gabriel Perez, Lidia Rincon, Anna Vila, Josep M. Gonzalez, and Luisa F. Cabeza, "Green Vertical Systems for Buildings as Passive Systems for Energy Savings," *Applied energy* 88, no. 12 (2011): 4854–4859.
7   Ibid., 4854.
8   Clara Gerhardt and Brenda Vale, "Comparison of Resource Use and Environmental Performance of Green Walls with Façade Greenings and Extensive Green Roofs," in *Sustainable Building Conference (SB10), Wellington, New Zealand*, 2010.
9   Ana I. Barros, Ana L. Gonçalves, Manuel Simões, and José CM Pires, "Harvesting Techniques Applied to Microalgae: A Review," *Renewable and Sustainable Energy Reviews* 41 (2015): 1489–1500.
10 Ibid., 1490.
11 Ibid., 1490
12 Ibid., 1493.
13 Ibid., 1494.
14 Ibid., 1496
15 Ibid., 1496
16 Ibid., 1497
17 Ibid., 1498
18 Ibid., 1498.
19 Ibid., 1498.
20 Ibid., 1491.
21 Carla S. Jones and Stephen P. Mayfield, "Chapter Nine. Aquatic Versatility for Biofuels: Cyanobacteria, Diatoms, and Algae," in *Our Energy Future* (University of California Press, 2016), 115–128.
22 Razif Harun, Michael K. Danquah, and Gareth M. Forde, "Microalgal Biomass as a Fermentation Feedstock for Bioethanol Production," *Journal of Chemical Technology & Biotechnology* 85, no. 2 (2010): 199–203.
23 Ibid., 203.
24 Ana I. Barros, Ana L. Gonçalves, Manuel Simões, and José CM Pires, "Harvesting Techniques Applied to Microalgae: A Review," *Renewable and Sustainable Energy Reviews* 41 (2015): 1490.

Part IV | Microalgae Building Enclosure Applications

Chapter 11

Microalgae Low-rise Buildings

11.1 Microalgae Enclosures for Residential Energy Efficiency

Residential and commercial buildings account for about 40% of the total U.S. energy consumption and pollutant emissions and 76% of electricity use.[1] High energy end-users in buildings are space heating and cooling, lighting, water heating, and ventilation, consuming more than half of the building energy use. Electricity consumption in the building sector has increased in the past decades from 20% of the U.S. electricity consumption in 1950 and 40% in the 1970s to 76% in 2012.[2] An average primary energy use intensity (EUI) of residential and commercial stocks is around 120 kBtu/ft^2/year and 200 kBtu/ft^2/year, respectively,[3] and stringent energy codes and policies have been implemented to significantly reduce energy use and associated greenhouse gas (GHG) emissions. Majority of commercial buildings are low-rise buildings and are responsible for higher energy use. U.S. residential buildings consume one-fifth of the total energy use. American households spend around $1,900 on average annually with houses located in cold regions such as the Northeast and Upper Midwest paying 28% more.[4] More than half of the buildings in the United States were built prior to 1980 when building energy codes were not fully implemented. Significant energy waste has been attributed to poor building enclosure construction and inefficient HVAC (heating, ventilation, and air conditioning) systems. Table 11.1 summarizes U.S. building energy consumption.

An average of 1.7 million housing units were constructed annually between 2000 and 2009. Due to affluent lifestyles, the floor space per capita for new construction is getting bigger and outpacing the numbers of new construction. Heat loss and gain occur with various architectural systems such as exterior walls, windows, roof, and slab on grade. Internal load and air infiltration rate also determine internal thermal loads. Residential buildings are typically skin load dominant where building envelopes play an important role in regulating energy flow between the inside and the outside. There are other design factors affecting energy consumption, including the geometry of the building, energy attributes of opaque walls and windows, window-to-wall ratio, and microclimate control such as shading, trees, and landscape.

U.S. households account for 20% of the U.S. energy consumption and 38% of electricity consumption.[5] Space cooling and appliances are the larger end-users of residential electricity, although water heating accounts for 30% of electricity use.[6] High-performance building enclosures typically increase upfront cost due to

DOI: 10.4324/9780367814410-14

Comparisons of building characteristics and energy consumption of commercial and residential buildings

	Total number of buildings (millions)	Total floor area (billion f²)	Total energy use (trillion Btu)	Total energy bill (billion dollars)
Low-rise commercial buildings	5.37 (97%)	68 (78%)	4,959 (71%)	108 (72%)
High-rise commercial buildings	0.19 (3%)	19 (22%)	2,004 (29%)	42 (28%)
Total commercial buildings	5.56 (100%)	87(100%)	6,963 (100%)	149 (100%)
Total residential buildings	118	237	9,114	219

additional material and construction cost. However, it can pay off with operational energy savings. In addition to energy savings, building enclosures affect indoor air quality. Continuous insulation with thermal breaks mitigates condensation and potential mold risks. Indoor air quality can be affected by other factors such as off-gassing interior materials, molds/bacteria due to leaks, or lack of ventilation. Ventilation is critical in making indoor space healthier. For nonresidential settings, ventilation equipment should supply sufficient outdoor air, and the energy should be saved through heat/energy recovery ventilation systems.

There is a positive correlation between residential building energy codes and residential energy consumption. Energy regulations in general require more efficient use of energy, protect water and air quality, and regulate wastes and pollution. The energy crisis in the 1970s sparked the development of building energy codes in the United States. At the federal level, the 1975 Energy Policy and Conservation Act made an amendment in 1978 to include building energy codes.[7] At the state level, starting with California's Warren-Alquist Act in 1974, Tittle 24 of the California Code of Regulations was enacted in 1978 to regulate residential and nonresidential buildings. The first AHSRAE 90 (American Society of Heating, Refrigerating and Air Conditioning Engineers) energy code was established in 1975, and a number of states adopted it in the 1980s.[8] It wasn't until 1992 that the Energy Policy Act included a provision to review and/or revise residential energy codes to meet the national level.[9] Therefore, many residential buildings built prior to 1980 were not energy efficient and are in noncompliance of building energy codes. Building code enforcement has significant effects in improving energy efficiency, resulting in reducing per capita electricity use by 2–5% in 2006.[10]

Thirty-five percent of all U.S. households live in rental housing. Among the rental housing, 30 million households, over 60% of all rental housing, live in the multifamily sector. About 20 million households pay more than 50% of their income for renting a house.[11] Energy cost savings through energy efficiency and on-site energy production can improve living affordability.[12] There are financial incentives available for improving energy efficiency such as the U.S. Department of Energy's (DOE's) Better Building Initiatives.[13] The program aims for building energy efficiency, waste reduction, water savings, and clean energy production. Through the Home Performance with ENERGY STAR® (HPwES) program, nearly 1 million homes improved energy and comfort, resulting in $500/year energy savings and over $35 million/year savings on energy costs and CO_2 reduction by 200,000 metric tons/year.[14] Energy interventions for residential housing play an important role in mitigating pollutant emissions and reducing energy bills. For quicker economic payment, it is important to integrate climate-responsive design strategies with energy-efficient active systems and renewable energy production. Typically, architectural features and structural systems last longer than active building service systems.

The recent Paris Agreement aims for GHG reduction around the world. Greater awareness of the climate emergency and economic returns encourage more public agreement to improve building energy efficiency. Both mandatory and voluntary implementations are important to tackle the climate crisis. New York City mandates carbon neutrality by 2050 and calls for building energy efficiency improvements of 23% above 2012 levels by 2030. It sets GHG limits for all buildings larger than 25,000 ft^2 and targets city GHG reductions of 40% by 2025 and 50% by 2030. It also requires the NYC Housing Authority properties to meet GHG reductions of 40% by 2030 and sets emissions reporting requirements.[15] In New York City, all roof space above new buildings, building expansions, and structural roof cover must incorporate green roofs, solar, or both. At the voluntary level, the Passive House (PH) standard is one of the high-performance building standards with strict design criteria such as super insulation and air tightness and post-construction testing. Started in Germany in the 1990s, there are over 65,000 PH buildings certified around the globe.[16] The performance requirements are 15 kWh/m^2 (19 kBtu/ft^2) of each heating and cooling demand focus with maximum 60 kWh/m^2/year of renewable primary energy demand (heating, hot water, and domestic electricity use).[17] Strategies to meet the energy requirements include (1) high insulative building enclosures, (2) energy-efficient windows, (3) thermal breaks, (4) airtightness, and (5) ventilation heat recovery. The strategies for (1) to (4) are related to high-performance building enclosures and (5) is related to energy-efficient HVAC systems. High-performance building envelopes and windows can result in downsizing heating and cooling equipment.

Microalgae enclosures can serve as an alternative building system to provide operational cost savings and occupant health and well-being. They offer good summer shading efficacy by increasing density and color responding to solar intensity, thus reducing cooling load. They offer maximum winter solar gain because their growth rate in winter would be slower and less dense, thus reducing heating demand. Microalgae enclosures can achieve daily, seasonal density targets by withdrawing grown microalgae and filling in new media or vice versa. They can also contribute to CO_2 capture and increase their biomass for potential economic return. In addition, a microalgae-based closed-loop system can be implemented for an off-grid residential community that is able to process wastewater treatment and clean energy production on-site without relying on city grids. Community-generated wastewater and flue gas can supply nutrients for microalgae growth, and the biomass produced can provide biofuel energy for community use. The recovered water from wastewater treatment can be used for irrigation and aquaculture production.[18] A closed-loop building information model can help estimate holistic food–water–energy feedback from a microalgae-integrated building

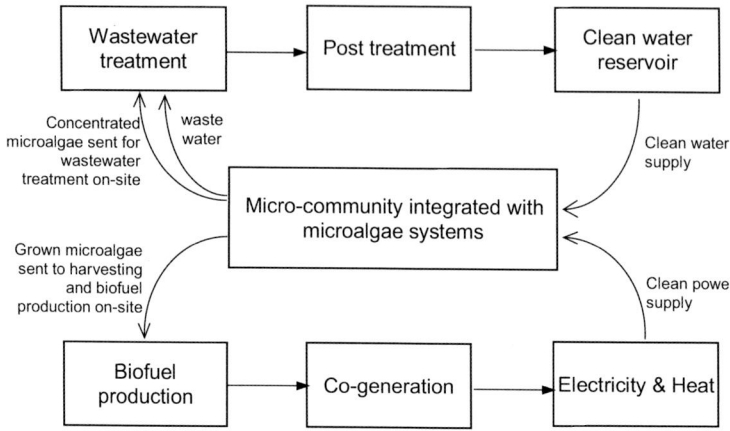

▲ Figure 11.1

Micro community integrated with microalgae systems restores wastes from buildings and converts them into valuable resources for operating a micro community; they can achieve off-grid power and water independency while decarbonating air pollution and processing wastewater.

enclosure.[19] Implementation of microalgae in community environments still poses technological, economic, environmental, social, and regulatory challenges, including long-term performance in energy efficiency and effective CO_2 sequestration, thermal insulation, acoustic insulation, maintenance, and harvesting of microalgae culture. Application hurdles include high upfront cost and negative environments such as potential toxins and odor produced by harmful algae (Figure 11.1).[20]

11.2 Microalgae Enclosures for Children's Well-being and School Performance

The World Health Organization (WHO) estimates that nine out of ten urban dwellers are exposed to air quality levels that do not meet WHO standards for healthy air.[21] Data show a correlation between air pollution and adverse health effects, including lung function, asthma attacks, cardiac hospital admissions, and even premature death.[22] People inhale 15,000 l/day (or 6–10 l/min) of air along with air pollutants indoors.[23] Indoor environments present their own challenges, such as off-gassing of building materials becoming trapped indoors. Typical indoor air pollutants include PM, NO_2, CO, VOCs, and so on.[24] Common sources of these indoor air pollutants include fuel-burning appliances, tobacco products, building materials and furnishings, cleaning products, HVAC devices, excess moisture, and outdoor sources such as radon, pesticides, and air pollution.[25] In addition, the worldwide pandemic increased the amount of time spent indoors by most people as well as their use of sanitizing and cleaning products, making these sources of indoor pollution more relevant.

Building energy codes require that building enclosures are built to be airtight to minimize heat flow between inside and outside. Ventilation equipment is undersized to save ventilation energy by reducing outdoor air exchange rates. In the early 1970s, building ventilation standards called for about 15 cubic feet per minute (cfm) of outside air for each building occupant, but this standard was reduced to 5 cfm after the 1973 oil embargo and tightened national energy conservation measures. After decades of indoor air quality–related complications, the ASHRAE standard was updated in the late 1980s to 15 cfm per occupant. However, the current ASHRAE standard 2019 calls for 5 cfm (office) and 10 cfm (classroom) of outdoor air flow per occupant, depending on the interior space to save building operational energy. This hermitically sealed construction, combined with inadequate fresh air exchanges, results in more harmful levels of indoor air quality.

Multiple studies show a correlation between indoor air quality and occupant health and performance. In the late 1990s, Fisk and Rosenfeld estimated that reducing sick building syndrome (SBS) in the United States could yield $10–$20 billion in increased productivity.[26] Another study found that increasing ventilation rates for office workers would result in productivity gains that would yield at least 10 times higher returns than energy and maintenance costs.[27] In classroom environments, a high CO_2 concentration contributes to poor academic performance,[28] and good indoor air quality significantly narrows test score gaps between low- and high-income students.[29] While CO_2 is not often considered to be a pollutant, it can indicate low ventilation rates and the presence of harmful pollutants.[30] California has declared a mandate for monitoring CO_2 for all K-12 schools, and other states are expected to adopt similar practices soon. Monitoring CO_2 is a way to identify ventilation deficiencies and combat indoor viruses.[31]

Air pollution has detrimental health impacts for children, and a positive correlation between air pollution and child academic performance has been shown. Primary factors affecting academic performance are

(i) school absenteeism due to illness caused by pollution; (ii) attention problems in school due to illness caused by pollution; (iii) fatigue when doing homework due to illness caused by pollution; and (iv) a direct negative effect of pollution on brain development.[32]

Air pollution increases children's emergency room visits from asthma, causes acute bronchitis and asthma attacks among children each year, and causes missed school days each year.[33] Children's math test scores increased when outdoor pollutants (e.g., PM_{10}, $PM_{2.5}$, and NO_2) were decreased.[34] On the other hand, there is a positive association between exposure to green surroundings and children's educational attainment. Attention Restoration Theory by Kaplan and Berman (2010) and Stress Recovery Theory by Ulrich et al. (1991) provide a theoretical framework for demonstrating nature's influence on cognitive performance and lower stress levels, healing the physical, mental, and cognitive development of children. The third-grade students surrounded by greenness showed higher academic outcomes in math and English on standardized tests,[35] and high school students with direct views to greenness showed higher academic scores and graduation rates.[36]

Children are often disconnected from natural systems due to sparse green space available in densified cities. Integrating microalgae systems with building materials can increase opportunities to interact and learn with nature without the need for extra land. Like an indoor potted plant, the need for pumping room air into the microalgae system further encourages students to interact, creating chances for physical exercise. Culturing microalgae and operation of the system can be part of the curriculum. Microalgae reach high density in a short time. Rapid changes of appearance and density may attract children's attention and encourage continuous explorations. This interaction time may help provide a break from studying and restore and mitigate mental fatigue. In return, children can enhance their study focus and boost academic outcomes. Interaction with green features in virtual environments shows a positive influence in lowering stress and mental distress.[37] Therefore, it is expected that microalgae systems will aid the well-being of urban children and enhance academic performance.

11.3 Key System Development Approaches

The "best-fit" static building enclosure is the industry standard. High-performance envelopes that are responsive to climates and local conditions are one of the key design strategies toward energy-efficient, healthy built environments. With multifunctional benefits for the building and the occupant, wide deployment of the microalgae system can be a promising solution for the climate crisis. Performance objectives for microalgae envelopes include maximum solar exposure (i.e., CO_2 sequestration utilizing microalgae's photosynthesis), shading efficacy (i.e., cooling load reduction), daylighting penetration (i.e., lighting load reduction), and view-out (i.e., user satisfaction). For large-scale deployment, it is important to reduce capital and operational costs while bringing in added values such as energy savings, power production, decarbonation, and bioremediations. Symbiotic benefits among microalgae systems, buildings, and occupants can offset upfront costs by saving operational expenses.

To meet mutual needs, geometric configurations of the microalgae system address both architectural constraints (e.g., space, aesthetics, constructability) and bioreactor environments for biomass productivity. Microalgae systems rely on continuous operation year-round. A robust system (i.e., longevity) and easy management is ideal. The pneumatic intake system constantly feeds the microalgae system with indoor and/or outdoor CO_2. Each air tube is connected with a check valve to prevent backflow and the air sparger at the bottom of the system creates CO_2 bubbles. CO_2 bubble size and residency time in the culture affect the CO_2 fixation efficiency of microalgae cells. Slower bubbling speed with smaller diameters promote CO_2 uptake. Higher CO_2 sequestration efficiency means good air quality and improved occupant health and well-being. Geometric configuration and materials of bioreactor, evaporation of medium, and strain types such as motile versus immotile should be considered to mitigate the degree of biofouling and maintenance activities. The microalgae system changes its tint and color responding to solar intensity and CO_2 concentration levels generated by occupants. The biochromatic quality from the microalgae system encourages user interactions between solar intensity, air quality, and daylight conditions.

Maintaining an optimum light range is another important design criteria for the system. The microalgae system in the vertical facade application is less susceptible to extreme overheating compared to horizontal bioreactors due to the tilted sunlight angle and reduced sunlight intensity impinging on the vertical surface. However, an active heat exchanger is required to adjust extreme temperatures. The use of recycled room air reduces or eliminates dependency on an active heat exchanger. A unique differentiator of the building application compared to outdoor application is that a microalgae system within the building utilizes conditioned room air and provides optimum temperature ranges to grow microalgae. The solar energy stored in the microalgae system can also be used to operate other building service systems, such as a hot water boiler or heat exchanger, potentially saving more building energy consumption.

The bioreactor will be preassembled in a factory as a building product. Site installation requires the assembly and connection of the growing apparatus and supporting equipment. The microalgae system can vary in size, shape, and material components, from a small-size panel to a mega unit comprised of various bioreactor and window components. Modularity of the system promotes standardization of design and yet allows for customization of the final assembly. The system should provide flexibility in assembly and installation depending on whether it is a residential or commercial application. A centralized monitoring/control system coupled with the building management system is recommended for further performance enhancement, better user interaction, and managing practical issues such as periodic maintenance, harvesting, and the longevity of the system.

Summary

The majority of U.S. buildings are low-rise buildings, contributing to more than half of the building energy consumption and associated GHG emissions. Residential buildings account for one-fifth of U.S. energy usage and 38% of electricity consumption. Buildings should be adaptable to climates and augment benefits from the natural resources surrounding them. Microalgae buildings can improve energy efficiency through the provision of shade in summer and solar gain in winter, which is tied with the HVAC system for further efficiency enhancement and occupant satisfaction. The microalgae enclosures will benefit from continuous interaction with users, and, in return, they provide opportunities to experience and learn with nature. Children connected with green features show higher physical and psychological well-being and school performance. Thermal break and adequate ventilation through microalgae systems improve energy savings and indoor air quality. Micro communities can utilize microalgae systems to support their energy demands through renewable energy systems. Microalgae systems can provide clean energy production and recycled water from wastewater treatment. Indoor air quality affects occupant health and well-being and school/workplace performance. Financial incentives will make a wide acceptance through carbon credits and the production of high-value bioproducts or biofuel.

Notes

1 U.S. Department of Energy (DOE), "An assessment of Energy Technologies and Research Opportunities," *Quadrennial Technology Review. United States Department of Energy* (2015).

2 Ibid., 145.

3 Ibid., 146.

4 U.S. Energy Information Administration (EIA), "Residential Energy Consumption Survey," accessed June 30, 2021, https://www.eia.gov/consumption/residential/data/2015/

5   Ibid.

6   Ibid.

7   Anin Aroonruengsawat, Maximilian Auffhammer, and Alan H. Sanstad, "The Impact of State Level Building Codes on Residential Electricity Consumption," *The Energy Journal* 33, no. 1 (2012).

8   Ibid., 33.

9   Ibid., 33

10 Ibid., 50.

11 DOE's Better Buildings Program, "Partnering for the Future: Leadership, Innovation, and Proven Solutions," accessed June 30, 2021, https://betterbuildingssolutioncenter. energy.gov/sites/default/files/attachments/DOE_BBI_2021_Progress_Report.pdf

12 Ibid.

13 Ibid., 37.

14 Ibid., 37.

15 "NYC Climate Mobilization Act," *Mayor's Office of Sustainability*, PowerPoint presentation.

16 Dick Clarke and Andy Marlow, "Mind the Gaps: Passive House from the Inside," *Sanctuary: Modern Green Homes* 45 (2018): 62–66.

17 Passive House Institute, "Passive House Requirements," accessed June 30, 2021, https://passivehouse.com/02_informations/02_passive-house-requirements/02_ passive-house-requirements.htm

18 Luo Shipeng and Daniel Castro-Lacouture, "Holistic Modeling of Microalgae for Powering Residential Communities," *Energy Procedia* 88 (2016): 788–793.

19 Soowon Chang, Daniel Castro-Lacouture, Florina Dutt, and Perry Pei-Ju Yang, "Framework for Evaluating and Optimizing Algae Façades Using Closed-loop Simulation Analysis Integrated with BIM," *Energy Procedia* 143 (2017): 237–244.

20 S.J. Wilkinson, Paul Stoller, Peter Ralph, and Brenton Hamdorf, "Feasibility of Algae Building Technology in Sydney," *Feasibility of Algae Building Technology in Sydney* (2016).

21 World Health Organization, "WHO Releases Country Estimates on Air Pollution Exposure and Health Impact," last modified September 27, 2016, https://www.who.int/ news/item/27-09-2016-who-releases-country-estimates-on-air-pollution-exposure-and-health-impact

22 Dean E. Schraufnagel, John R. Balmes, Clayton T. Cowl, Sara De Matteis, Soon-Hee Jung, Kevin Mortimer, Rogelio Perez-Padilla, et al., "Air Pollution and Noncommunicable Diseases: A Review by the Forum of International Respiratory Societies' Environmental Committee, Part 2: Air Pollution and Organ Systems," *Chest* 155, no. 2 (2019): 417–426.

23 R.A. Wood, M.D. Burchett, R.A. Orwell, J. Tarran, and F. Torpy, "Plant/Soil Capacities to Remove Harmful Substances from Polluted Indoor Air," Plants and Environmental Quality Group, Centre for Ecotoxicology, UTS, Australia (2002) (Bezugsquelle: www. plants-for-people.de)

24 U.S. Environmental Protection Agency, "Report to Congress on Indoor Air Quality, volume II: Assessment and Control of Indoor Air Pollution," *Technical Report EPA/400/1-89/001C* (1989).

25 Ibid., 2-1 and 2–6.

26 William J. Fisk and Arthur H. Rosenfeld, "Estimates of Improved Productivity and Health from Better Indoor Environments," *Indoor Air* 7, no. 3 (1997): 158–172.

27 R. Djukanovic, Pawel Wargocki, and Povl Ole Fanger. "Cost-benefit Analysis of Improved Air Quality in an Office Building," *Proceedings of Indoor Air 2002* 1 (2002): 808–813.

28 A.N. Myhrvold, E. Olsen, and O. Lauridsen, "Indoor Environment in Schools—Pupils Health and Performance in Regard to CO_2 Concentrations," *Indoor Air* 96, no. 4 (1996): 369–371.

29 Jacqueline S. Zweig, John C. Ham, and Edward L. Avol, "Air Pollution and Academic Performance: Evidence from California Schools," *National Institute of Environmental Health Sciences* (2009): 1–35.

30 Xin Zhang, Xi Chen, and Xiaobo Zhang, "The Impact of Exposure to Air Pollution on Cognitive Performance," *Proceedings of the National Academy of Sciences* 115, no. 37 (2018): 9193–9197.

31 California Legislative Information, "AB-841 Energy: Transportation Electrification: Energy Efficiency Programs: School Energy Efficiency Stimulus Program," last modified October 2, 2020, https://leginfo.legislature.ca.gov/faces/billTextClient.xhtml?bill_id=201920200AB841

32 Zewig, et al., "Air Pollution and Academic Performance: Evidence from California Schools" (2009): 2.

33 Travis Madsen and Elizabeth Ouzts, "Air Pollution and Public Health in North Carolina," 26.

34 Ibid., 24.

35 Chih-Da Wu, Eileen McNeely, J.G. Cedeño-Laurent, Wen-Chi Pan, Gary Adamkiewicz, Francesca Dominici, Shih-Chun Candice Lung, Huey-Jen Su, and John D. Spengler, "Linking Student Performance in Massachusetts Elementary Schools with the 'Greenness' of School Surroundings Using Remote Sensing," *PloS ONE* 9, no. 10 (2014): e108548.

36 Rodney H. Matsuoka, "Student Performance and High School Landscapes: Examining the Links," *Landscape and Urban Planning* 97, no. 4 (2010): 273–282.

37 Bin Jiang, Dongying Li, Linda Larsen, and William C. Sullivan, "A Dose-response Curve Describing the Relationship between Urban Tree Cover Density and Self-reported Stress Recovery," *Environment and Behavior* 48, no. 4 (2016): 607–629.

Chapter 12

Microalgae Tall Buildings

12.1 Microalgae Curtain Wall

Two out of three people of the world's population will live in cities by 2030. The growing urban population will demand more urbanization and new construction of tall buildings. The number of high-rise commercial buildings comprises less than 5% of the total buildings, but they account for more total floor area with high embodied energy and energy use intensity (EUI). While tall buildings are often considered an economically and socially sustainable solution, their environmental stewardship is critical for mitigating climate change. The building industry around the world is adopting more stringent building regulations and voluntary actions toward net zero energy architecture practice. The three key design strategies toward net zero energy towers are as follows: (1) Bioclimatic design to fully utilize available resources on-site (solar energy, wind, rain, etc.), (2) holistic integration of energy-efficient building systems and components to minimize operational building energy, and (3) on-site energy production to offset building energy use. Considering the tall building enclosure is a primary interface between indoor and outdoor spaces to manage energy flows and renewable energy production, a microalgae curtain wall could engage all key design strategies. It regulates heat transfer from the inside to the outside, solar heat gain from the outside to the inside, daylight illumination and glare controls, and maximum harnessing of solar energy. A microalgae curtain wall, as the name implies, is hung from primary building structures such as the concrete slab and perimeter beam. It can be integrated with other building materials such as glass, stone, metal, concrete, ceramic, and so on surrounded by metal frames.

The curtain wall, one of modern tall building envelope constructions, offers multi-performance capabilities and unique aesthetic features. There are a growing number of all glazed curtain wall skyscrapers in metropolitan cities such as New York, Hong Kong, Shanghai, and Dubai. Although the design, procurement, production, and installation of a curtain wall is particular to each project, the global curtain wall market in general is increasing from $32 billion in 2019 and expected to reach $66 billion in 2027,[1] accounting for approximately 30% of the overall global building enclosure market.[2] Contemporary skyscraper enclosures typically adopt a curtain wall system primarily due to its faster construction and good quality control. The prefabricated and factory-assembled system allows speedy installation without the need for many laborers and elaborate scaffolding on-site. Contemporary curtain wall practice adheres to systematic quality assurance and

DOI: 10.4324/9780367814410-15

quality control (QA/QC) protocols according to industry standard, ensuring targeted performance and workmanship.

From the environmental stewardship perspective, the glass windows of U.S. buildings play a crucial role in conserving heating, cooling, and artificial lighting load. In response to the climate changes from anthropogenic CO_2 generation, more stringent building regulations and voluntary actions toward net zero energy architecture have been enforced around the world. The European Union (EU) projected a "zero energy goal" for all new buildings by 2020 while the United Kingdom is committed to cut carbon emissions by 80% by 2050. The 2030 Challenge in the United States is targeted to attain net zero energy practice by 2030. Twenty-nine states of the U.S. Climate Alliance have committed to net zero or nearly net zero greenhouse gas (GHG) emissions and a 100% renewable energy target by no later than 2050. Considering building enclosure is the primary barrier between indoor and outdoor spaces where multidirectional energy transfer occurs, energy attributes and environmental stewardship of curtain walls are pertinent to the heating, cooling, and lighting of a building.

The new development of microalgae curtain walls aims to materialize the symbiotic interaction between innovative building enclosures, biological systems, and users. The microalgae curtain wall typically consists of a network of bioreactors, vision insulated glass unit (IGU), spandrel panel (i.e., glass, stone, or metal), metal frames, and bracketry system. Unlike traditional walls that rest on the ground (e.g., concrete wall of Pantheon), it is a non-loadbearing enclosure without transferring gravity load to each curtain wall system. The microalgae curtain wall provides structural integrity, water/air tightness, thermal insulation, daylighting, ventilation, and low maintenance while enabling high quality control and speedy construction for limited construction site in cities. Depending on the project conditions, the microalgae curtain wall could be hung from the primary structure, typically a concrete slab or bottom supported by the bracket embedded in a concrete slab. Potential configurations of microalgae curtain walls are discussed in Chapter 10 based on the synthesis of innovative building science and biotechnical performance.

The development of each microalgae curtain wall type integrates the geometrical properties and material performance with the building service system to be effective in building energy consumption. Because of watertight construction, proven material techniques can be utilized such as glazing plate (Divided system), ETFE (ethylene tetrafluoroethylene) membrane (Inflated system), extruded tube (Stranded system), cast modules (Suspended system), and flexible tubing/cast bioresin (Woven system) as discussed in the previous chapter. Check valves and manifold connections can minimize the chances of water leakage from connections between different materials and components. Integrated parametric modeling and performance assessment workflow assist in a holistic understanding of system design and performance requirements along with some practical issues of aesthetics, fabricability, installation, operation, and maintenance.

12.2 Vertical Landscape

Cities are pushing to be more compact and vertical to accommodate population increases. Densification isolates urban dwellers from green space, reducing opportunities to interact with nature. Given the limited green space on ground, microalgae technologies are endowed with potential natural systems or biophilic elements on vertical surfaces such as building envelopes, balconies, and sky gardens. The integration of microalgae systems provides environmental advantages. By-products that are not useful for the building and infrastructure but can be used for microalgae cultivation are CO_2, wastewater, flue gas, and so on. The vertical microalgae system helps CO_2 sequestration, wastewater treatment, and flue gas reduction while the biomass can be converted into clean biofuel for building supplies. Economic benefits can be realized through energy bill savings and high-value biomass production. Proximity to the microalgae system can improve cognitive performance and reduce stress, enhancing the psychological and physical well-being of occupants.

There are many practical issues conducive to successful integration of nature within tall towers, including the reduction of upfront and operational costs, short return-on-investment periods, and other practical issues such as operation, maintenance, and the harvesting of the microalgae culture for high-value production. Vertical microalgae systems can serve as a secondary biological membrane set out from the primary window using a maintenance catwalk. Operation and maintenance happen in the outdoor space, limiting disturbance to users. Collected rainwater can be used for cultivation and saves fresh water use. Microalgae systems installed on intermediate floors serve as sky gardens and communal space. Microalgae sky gardens can improve microclimates and outdoor/indoor air quality. The system runs in a closed loop where nutrients from wastewater and CO_2 from city emissions are utilized.

In addition to environmental benefits from microalgae systems in tall building applications, scientific research indicates that there is an inborn human affiliation with nature and that contact with nature contributes to psychological and physical well-being, called biophilia. Biophilia stems from the Greek roots meaning "love of life," coined by the social psychologist Erich Fromm (1964) and populated in the 1980s by the biologist Edward Wilson. Biophilic architecture took shape in the 1990s and addressed the growing issue of the disconnect with nature in an urbanizing world. Scientific data also support the benefits of biophilic architecture ranging from health benefits in health care environments to workers' productivity in office environments and to improved academic performance in school environments. While biophilic design is regarded as a positive approach, biophobia theory shows fears or negative attitudes toward nature in part due to decreased contact with nature.[3] Biophobia is the association of hazards found in nature such as insects, animals, heights, and even the discomfort of being outside of mechanically conditioned spaces. Interaction with a microalgae system increases environmental education and may enhance positive attitudes toward nature and a willingness to protect environments.

Microalgae use light energy to convert water and CO_2 into organic compounds such as lipids, carbohydrates, and proteins. Photosynthetic efficiency (PE) is defined by the light energy stored in microalgae over the total available incident light. Owing to high PE and full canopy absorption, microalgae can grow much faster than terrestrial plants, with some species doubling biomass 1 to 3 times in one day.[4] A PE of up to 40% from microalgae can be attained under optimal environmental conditions compared to 1–8% of terrestrial plants.[5] This quick growth rate favors microalgae as an optimal candidate for reducing carbon concentrations and anthropogenic emissions. Microalgae are known to biofixate CO_2 due to their remarkable sequestration rates by uptaking 1.8 kg of CO_2 per 1 kg biomass.

12.3 Microalgae System's Effect on User Responses

Indoor environmental quality (IEQ) concerns indoor air quality and thermal and visual comfort for occupants and energy savings of space. Good IEQ positively impacts occupant comfort, health, well-being, and productivity. Indoor air quality is determined by CO_2 from occupant respiration and volatile organic compounds (VOC) off-gassed from building materials. These air pollutants accumulate in indoor air due to a lack of or reduced ventilation rates. According to an industry rule of thumb, an average operational expense for a commercial building is $3/ft^2 for energy cost, $30/ft^2 for rental cost, and $300/ft^2 for occupant salaries. Therefore, it is important to have a workspace with good IEQ to promote occupant positive responses and productivity.

Active interaction with nature provides positive outcomes of well-being, but passive interaction (e.g., virtual views through TV and virtual reality) can also have a positive effect on mood.[6] The influence of nature on occupant well-being and productivity has been well documented in office, hospital, and educational settings. Indoor greenery experiences support better mental health for university students working from home during a pandemic[7]; indoor plants contribute to positive mental well-being for residents, especially those living in reduced space and light availability due to health confinement[8]; and studying in a library room with green indoor features results in higher cognitive performance compared to a control room without the same green features.[9] View-out of nature for high-rise residents contributes to positive mental well-being too.[10] Indoor green features provide visual interaction, and specific plants reduce CO_2 levels through photosynthesis. Making connections to nature in both active and passive interaction either indoors or outdoors can mitigate mental stress and improve cognitive performance.

Although nature contributes to positive effects on health and happiness, many people in high-rise buildings have limited access to nature. Plants affect psychological restoration and stress mediation. Mangone et al. found that office workers working in a room where there were plants reported 10% more thermal comfort than in a control setting without plants.[11] Qin et al. indicated that plants increased human comfort and satisfaction in an office environment using subjective questionnaires and physiological measurements. In addition, most people prefer window views, and green spaces through windows show a positive association with the improvement of human health and well-being.[12] Viewing green spaces from tall buildings also positively affects the "physiological and psychological" well-being of residents.[13] Therefore, for tall buildings where access to green space is limited, microalgae systems can play an important role in fostering a more positive relationship between people and microalgae, while promoting a sense of relaxation and restoration. The microalgae absorb the carbon dioxide in the air, allowing them to grow and release oxygen. Through this interactive process such as a manual air pump, the public in turn learns more about microalgae and their benefits to the community and the environment. Their green color will also provide psychological comfort and physical adjustment.

Summary

In contemporary high-rise buildings in urban settings, the popularity of glazed towers is on the rise due to sleek aesthetics and view-out. However, tall tower densification results in less connection with nature on the ground. Urban densification causes the loss of green space, and microalgae integrated within the built environment could help restore that connection with nature. Microalgae systems can be incorporated as part of building skins, vertical landscapes, or sky gardens. An innovative outcome from microalgae tall buildings is to establish nature-based dialogue between architecture space and inherent human affiliation. They also provide opportunities to integrate new forms of nature in our built environment with access to daylight, fresh air, clean water, and natural systems. When it serves as a tall building skin, the microalgae curtain wall provides biophilic aesthetics, good daylight transmission, and adaptive sunshade with adequate thermal and structural performance. Dynamic thermal insulation effectively keeps the heat out in the summer and the cold out in the winter. Microalgae's innate and superior photosynthesis improves air quality through O_2 production and CO_2 absorption. Microalgae grown from tall buildings have the potential to be converted into renewable fuel stocks such as biomass or biofuel.

Notes

1 Grand View Research, "Aluminum Curtain Wall Market Size: Industry Report, 2020–2027," accessed July 1, 2021, https://www.grandviewresearch.com/industry-analysis/aluminum-curtain-wall-market

2 QY Research, Inc., "Global Building Envelope Market Expected to Reach US $182510 Mn by 2025," Newswire, December 23, 2019, https://www.newswire.com/news/global-building-envelope-market-expected-to-reach-us-182510-mn-by-2025-21061466.

3 Weizhe Zhang, Eben Goodale, and Jin Chen, "How Contact with Nature Affects Children's Biophilia, Biophobia and Conservation Attitude in China," *Biological Conservation* 177 (2014): 109–116.

4 Carla S. Jones and Stephen P. Mayfield, *Our Energy Future: Introduction to Renewable Energy and Biofuels* (University of California Press, 2016).

5 Ana P. Carvalho, Susana O. Silva, José M. Baptista, and F. Xavier Malcata, "Light Requirements in Microalgal Photobioreactors: An Overview of Biophotonic Aspects," *Applied Microbiology and Biotechnology* 89, no. 5 (2011): 1275–1288.

6 Lee Bak Yeo, "Psychological and Physiological Benefits of Plants in the Indoor Environment: A Mini And In-Depth Review," *International Journal of Built Environment and Sustainability* 8, no. 1 (2021): 57–67.

7 Angel M. Dzhambov, Peter Lercher, Matthew H.E.M. Browning, Drozdstoy Stoyanov, Nadezhda Petrova, Stoyan Novakov, and Donka D. Dimitrova, "Does Greenery Experienced Indoors and Outdoors Provide an Escape and Support Mental Health during the COVID-19 Quarantine?" *Environmental Research* (2020): 110420.

8 Luis Pérez-Urrestarazu, Maria P. Kaltsidi, Panayiotis A. Nektarios, Georgios Markakis, Vivian Loges, Katia Perini, and Rafael Fernández-Cañero, "Particularities of Having Plants at Home during the Confinement Due to the COVID-19 Pandemic," *Urban Forestry & Urban Greening* (2020): 126919.

9 Nicole van den Bogerd, S. Coosje Dijkstra, Sander L. Koole, Jacob C. Seidell, and Jolanda Maas, "Greening the Room: A Quasi-experimental Study on the Presence

of Potted Plants in Study Rooms on Mood, Cognitive Performance, and Perceived Environmental Quality among University Students," *Journal of Environmental Psychology* 73 (2021): 101557.

10 Mohamed Elsadek, Binyi Liu, and Junfang Xie, "Window View and Relaxation: Viewing Green Space from a High-rise Estate Improves Urban Dwellers' Wellbeing," *Urban Forestry & Urban Greening* 55 (2020): 126846.

11 Giancarlo Mangone, S.R. Kurvers, and P.G. Luscuere, "Constructing Thermal Comfort: Investigating the Effect of Vegetation on Indoor Thermal Comfort through a Four Season Thermal Comfort Quasi-experiment," *Building and Environment* 81 (2014): 410–426.

12 Roger S. Ulrich, "View through a Window May Influence Recovery from Surgery," *Science* 224, no. 4647 (1984): 420–421.

13 Mohamed Elsadek, Binyi Liu, and Junfang Xie, "Window View and Relaxation: Viewing Green Space from a High-rise Estate Improves Urban Dwellers' Wellbeing," *Urban Forestry & Urban Greening* 55 (2020): 126846.

Chapter 13

Microalgae Retrofitting Buildings

13.1 Tall Building Enclosure Retrofitting

Rapid urbanization and projected environmental stress exacerbate the sustainability of built environments around the world. Urbanization of cities involves pavement, buildings, roads, parking lots, and so on, which creates an extent of impervious surfaces over the cities without allowing the absorption of water through land. Heavy rainfall, floods, and hurricanes create storm water run-off, causing an overloading of water treatment facilities. The impairment of sewer and water catch basin pollutes receiving waters and endangers public health. Hurricane Harvey in Texas in August 2017, for example, was catastrophic in destroying and flooding the built environment and displacing thousands of people exposed to infectious diseases. While water is available in nearly every corner of the planet, usable water is a scarcity in many regions of the world. Population growth has a large effect on the scarcity of water. One-third of the world's population lives under conditions of relative water scarcity, and about 450 million people live under conditions of severe water stress.[1]

Furthermore, many urban centers face serious urban air quality problems mainly due to the increasing urban populations, number of buildings, and cars on the road. Urban air quality seriously affects public health and security as well as ecosystems. Major causes of air pollution in the city are attributed to large traffic emissions such as NOx (nitrogen oxides), SO_2 (sulfur dioxide), CO (carbon monoxide), PM (particulate matter), and VOCs (volatile organic compounds). The Clean Air Act (CAA, 1963) and Air Quality Act (AQA, 1967) have been implemented to regulate emissions since the 1960s. Challenges still exist in identifying the health impacts of human and ecosystem associated pollutant levels. Protecting low-income and minority communities from environmental injustice is important because they tend to be disproportionately exposed to higher pollution.[2] Outdoor air pollutants also negatively affect indoor air quality impacting occupant health. It also complicates the durability and maintenance of building materials.

Urban building retrofitting is the most cost-effective, energy-efficient strategy and provides excellent opportunities to enhance economic benefits, ecological sustainability, and occupants' well-being. Building energy efficiency retrofits are estimated to mitigate 600 million metric tons of CO_2 annually which is equivalent to approximately 10% of total U.S. emissions.[3] Retrofitting projects in urban areas represents a large percentage of the overall construction market, which can create job opportunities. An average of 3.3 million accumulative jobs can be created

 DOI: 10.4324/9780367814410-16

from energy-efficient retrofits.[4] Furthermore, building envelope retrofitting also offers increased marketability and provides a good public image. In addition to the economic benefits, retrofitting increases the energy efficiency of a building, thus reducing the building's energy consumption, operational costs, and environmental impact over the course of the building's lifecycle. Other benefits from retrofitting the building enclosure include a healthier indoor environment quality (IEQ) and increased occupant satisfaction and work productivity.

Thorough planning and addressing challenges associated with building envelope retrofitting are important. Renovation and replacement cause upfront cost and disturbance to the users/urban residents if construction time is prolonged. Building-site access and effective replacement of old building systems need to be resolved. Jurisdiction and state agencies regulate issues related to protection of interior and exterior environments, public health and safety, temporary sidewalk closures, rerouting of pedestrian flow, demolition removal, delivery and storage of building materials and equipment on-site, and location of temporary hoisting systems. Each retrofitting construction has different conditions that need to accommodate and facilitate minimal disruption to existing tenants while protecting the public's health and safety as well as construction workers' safety.

Envelope retrofitting requires holistic evaluation of the economic suitability in the areas of investment and return-on-investment (ROI) period from energy savings in building operations. Ecological and performative benefits from energy-efficient retrofitting will be compared with as-built base condition with respect to energy savings, improved occupant satisfaction, indoor air quality, and other added benefits such as pollutant reductions and increased productivity. A series of sustainable technologies for building envelope retrofitting are available. Key performance attributes are (1) heat transmission (U-factor) or thermal resistance (R-value) without minimum thermal bridge, (2) solar heat gain coefficient (SHGC), (3) visible light transmittance (VLT), (4) air infiltration, and (5) sustainable material composition with recyclability and no VOC content. In addition to energy savings and pollutant reduction, energy-efficient retrofit plays an important role in improving workplace performance as well as asset value. Worker productivity from a building envelope retrofit is increased due to work dedication and less sick leave. The value of the building at sale or rental is also increased due to building envelope improvement.

In particular, low-performing windows are responsible for major energy consumption and pollutant emissions. The energy-efficient window market in 2015 was approximately $6 billion for replacement and renovation and $4 billion for new construction.[5] In 2021, the global window market has grown to $10 billion for replacement and renovation and $6 billion for new construction with a CAGR (compound annual growth rate) of 8.6% and 8.3%, respectively.[6] Energy-efficient window retrofitting accounts for over 60% of the global window market share.[7] There are an estimated 3.8 million commercial buildings (51 billion square feet of floor space) and 65 million households (130 billion square feet of floor space) built before 1980 in the United States when building energy codes were not fully implemented. Ten- to fifteen-year-old windows are recommended for replacement, and the environmental and economic benefits are the major driving force for the increasing retrofitting market.

Legislative adoption is also a driver for improving building energy efficiency. New York City, for example, aims to curtail CO_2 emissions by 80% by 2050, requiring that older buildings larger than 25,000 ft^2 undergo deep retrofitting to meet

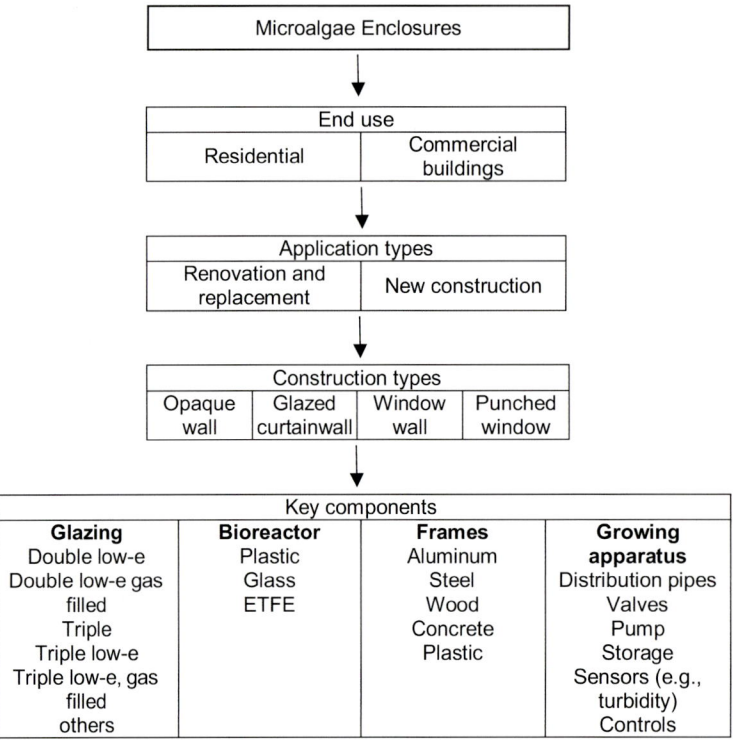

▲ Figure 13.1

Microalgae enclosure application overview.

the benchmark energy efficiency target. Out of a million total buildings in New York City, 50,000 account for 60% of the total energy use, and it is essential to make tall buildings highly energy efficient. On the West Coast, the California Energy Efficiency Strategic Plan set a goal that 50% of commercial buildings be retrofitted to net zero energy buildings by 2030. These cities are also imposing a high energy cost that motivates energy efficiency goals for owners (Figure 13.1).

13.2 Low-rise Enclosure Retrofitting

Low-performance windows with high thermal transmission (U-factor) allow energy to escape from the inside to the outside. They also cause condensation due to thermal bridging and low insulative values. Existing technologies serve to improve thermal insulation such as a low-e coated insulated glass unit (IGU). The number of air cavities within the IGU and the number/location of low-e coating contribute to thermal insulation. The low-e coating type also affects U-factor depending on hard and soft low-e coating with various coating materials and fabrication techniques. Adding inert gas to the IGU air cavity is also a commonly used method in practice. The three inert gases available in the building industry include argon, krypton, and xenon. Another intervention is to turn the air cavity into a vacuum cavity by drawing out the air in the cavity or to fill the cavity with aerogel. Both vacuum and aerogel are good insulators and discourage heat transfer by convection in the air cavity. The inert gases, vacuum, and aerogel technologies incur additional costs. Their long-term performance should be taken into consideration because infilled gases or vacuum cavities tend to be compromised over the course of operation.

Microalgae windows offer a dynamic U-factor that can improve heating and cooling as well as occupant thermal comfort. The room air is pumped into the microalgae window, and an optimum temperature range for growing microalgae is maintained. The interior space has warmer temperatures than the outside in winter and cooler temperatures than the outside in summer. By pumping in the conditioned room air, the microalgae window offers dynamic insulation which can mitigate energy flow between indoor and outdoor spaces. This process generates a smart thermal envelope that also provides better thermal comfort while saving cooling and heating load. Its effective role in mitigating energy transfer can reduce the size of HVAC systems and overall energy use. With its dynamic thermal attributes, the microalgae window offers a climate responsive U-factor. Static, highly efficient insulation exceeds threshold thermal resistance called thermal inflection and increases energy consumption.[8] Dynamic insulation that changes insulation values depending on climates outperforms traditional static insulation.[9] Active circulation of conditioned air in the culture media effectively lowers heat transmission and improves thermal comfort.[10] Microalgae window retrofitting provides opportunities in enhancing energy attributes and occupant satisfaction.

Another benefit of microalgae window retrofitting is to control solar heat gain and daylighting performance. Their superior photosynthesis produces high growth rate, and their cell concentration offers good shading efficacy responding to solar intensity. The microalgae window contains culture media made of water and nutrients. The microalgae window utilizes up to 8–10% of solar energy for photosynthesis, the remaining of which turns into heat energy. Although the majority of solar energy becomes heat within the system, the culture media (mainly composed of water and nutrients) coupled with room air circulation effectively dissipates the heat absorbed by the system. The solar energy stored in the system can be reclaimed for space or domestic hot water heating. With a smart control system, the tint and color can change in real time for further energy improvement and user interaction. Common techniques of controlling solar heat gain in window systems include low-e coating, air cavity, tint/color of glass, and addition of shading devices. The glass coating types, colors or dichroic treatment of glass, interlayers, and other components affect solar gain. External or internal shades integrated within the IGU cavity mitigate solar heat gain too. Electrochromic, photochromic, and thermochromic glazing or coatings are also available for active solar controls by changing opacity through electric current, heat, and light intensity. With similar logic, the microalgae window is biochromic glass in that it changes its density from photosynthesis and, coupled with an automatic control system, balances solar gain, daylighting, and user interaction.

Daily 1–5g of biomass/ft^2 of building enclosure is feasible productivity for microalgae retrofitting applications when optimum growing environments and operation modes are implemented. A 100 ft^2 (10 feet × 10 feet) microalgae enclosure retrofitting can sequester CO_2 generated by three office users. In other words, approximately 30–40 ft^2 microalgae retrofitting is enough to offset one individual's CO_2 production. Using the maximum growth rate, a medium-size office building with 100 feet × 100 feet × 5 stories (65-feet tall) retrofitted with microalgae envelopes can sequester 17–85 metric tons of CO_2, 10–50 metric tons of dry biomass, and 1,400–7,000 gallons of biofuel. Using the commercial rate of carbon removal in the range of $500 to $1,690 per ton of CO_2,[11] the cost savings according to this case study could be up to $145,000 per year. The conventional chemical and physical process of treating wastewater is expensive and multi-process intensive. Hyperconcentrated algal cultures are highly efficient in uptaking nitrogen and phosphorus in the wastewater within a short period of time. Combined with the wastewater treatment process, the algal culture removal of phosphorus and nitrogen has an average range of 80–100% efficiency. The system can be installed to treat wastewater from other manufacturing industries such as brewery wastewater, domestic wastewater, textile wastewater, pharmaceutical waste streams, slaughterhouse waste, heavy metal-containing wastewater, palm oil mill effluents, and agro-industrial wastewater.[12] Microalgae-based wastewater treatment is cost-effective and yields higher biomass with ample nutrients (Table 13.1).

▼ Table 13.1

Overall performance and benefits from microalgae envelope retrofitting

Architectural performance	Climate-responsive design Dynamic aesthetics; animated skin by biological growth.
Construction types	Opaque wall, curtain wall, window wall, punched window supported by metal frames
Material types	UV-resistant clear materials (e.g., glass, Plexiglas, polycarbonate, and ETFE)
Control system and sensors	Medium temperature sensor, light meter, pH sensor, optical clarity sensor (i.e., turbidity sensor), flow meter
Unit cost	$200–300/ft^2, including materials and installation
Energy performance	Dynamic U-factor –0.1–0.3 Btu/ft^2-h-F Dynamic SHGC –0.17–0.62 Dynamic VLT —0.07–0.88
Technological performance	Strength and stiffness w/a safety factor of 5–10 under self-weight, hydrostatic pressure, wind, impact, and maintenance load Acoustic Sound Transmission Class ≥35 Air tightness—0.1 cfm/ft^2 Water tightness—no water leakage
Operation and maintenance	Monitoring/controlling (control logic and hardware systems/sensors) Cultivation (batch production, semi-continuous production, and continuous production) Cleaning Harvesting (dewatering, drying) Post-processing (bioproducts, biofuel)

(Continued)

Environmental benefits	Building energy efficiency, pollutant reduction		
	Potential daily productivity ranges	1 g/ft^2/day	5 g/ft^2/day
	Biomass production	9 metric tons/year	47 metric tons/year
	Carbon sequestration	17 metric tons/year ($8,500–29,000/year)	85 metric tons/year ($43,000–145,000/year)
	Biofuel production	1,400 gallons/year	7,000 gallons/year
	Wastewater treatment	80–100% removal of phosphorous and nitrogen	
Economic benefits	Energy savings up to 30% in heating, cooling, lighting, and ventilation load		
	High-value bioproducts for use in nutraceutical and pharmaceutical applications		
	Use of the building waste (e.g., nutrients and inorganic carbon)		
	Biofuel production		
Social benefits	Harmony with nature		
	Increased health and well-being (e.g., Stress Recovery Theory)		
	Productivity gain, improved school performance (e.g., Attention Restoration Theory)		
Challenges	Design guidelines unavailable		
	Insufficient track record of green performance and longevity in real-world applications		
	Lack of private and public investment for scale-up		
	Multi-trade engagement and lack of infrastructure for post-cultivation processing		
	Long ROI (i.e., difficult in economic return)		
	Health and safety issues caused by potential microalgae contamination as well as toxins and VOCs generated by some strains		

Summary

All existing and new buildings will become old and inefficient unless built with net zero energy and/or net zero carbon modifications. Therefore, building energy-efficient retrofitting plays an important role in addressing environmental, economic, and social aspects. Microalgae enclosure retrofitting can serve as an alternative energy-efficient retrofitting and help combat a climate emergency caused by global warming due to excessive anthropogenic activities. The optimum system operation of the microalgae enclosure can reduce heating, cooling, and artificial lighting loads inside of the building, sequester carbon, and treat wastewater or other contaminants. Microalgae system, HVAC, and building management systems should work closely together for holistic efficient performance. Aesthetic enhancement through microalgae envelopes can be a potential driver for widespread public adoption. The biological integration of the microalgae window provides appealing aesthetics and emotional comfort. The carbon reduction capability further benefits occupant health and well-being, making the building more appealing to occupants, industry professionals, and stakeholders.

Notes

1   Charles J. Vörösmarty and Dork Sahagian, "Anthropogenic Disturbance of the Terrestrial Water Cycle," *Bioscience* 50, no. 9 (2000): 753–765.
2   National Research Council, *Air Quality Management in the United States* (National Academies Press, 2004), 5.
3   Rockefeller Foundation and DB Climate Change Advisors, "United States Building Energy Efficiency Retrofits: Market Sizing and Financing Models," last modified in 2012, http://www.dbcca.com/research
4   Ibid., 7.
5   MarketsandMarkets, "Energy-efficient Windows Market by Glazing Type (Double, Double Low-e, Triple, and Triple Low-e), Component (Glass, Frame, and Hardware), Application (Replacement & Renovation and New Construction), End-use Sector, and Region—Global Forecast to 2026," 62.
6   Ibid., 62.
7   Ibid., 62.
8   M. Dabbagh and M. Krarti, "Evaluation of the Performance for a Dynamic Insulation System Suitable for Switchable Building Envelope," *Energy and Buildings* 222 (2020): 110025.
9   Ibid., 2.
10 M. Jothilakshmi, A. Mohan, and M. Tholkapiyan, "Performance Evaluation and R-value Analysis for Thermally Insulated PCM Wall with Fluted Sheets," *Materials Today: Proceedings* 22 (2020): 912–918.
11 Kit Wayne Chew, Kuan Shiong Khoo, Hui Thung Foo, Shir Reen Chia, Rashmi Walvekar, and Siew Shee Lim, "Algae Utilization and Its Role in the Developments of Green Cities," *Chemosphere* (2020): 129322.
12 Ibid., 7.

Conclusions

Microalgae buildings in general offer environmental benefits, economic opportunities, and societal impacts. However, lack of solid evidence in practical questions when installed in the real world makes the technology debatable in the main areas of upfront/operational cost, green performance, longevity, operation, and maintenance.

Over the decade, substantial research has demonstrated the promise of microalgae systems with innovative applications from small installation to architecture and city scales. Despite increasing worldwide attention and R&D activities, the number of built examples has been limited to speculative design projects, laboratory-scale experiments, or urban installations with short-term periods of technology demonstrations.

When it comes to actual realization of the technology, two main criteria need to be answered—cost and longevity. The high upfront cost of the microalgae building represents a significant challenge in widespread actualization. The microalgae system in building applications needs to be built with a much higher caliber to meet multiple technical and biotechnical requirements, increasing engineering requirements, and system cost. It also needs additional hardware and software for automatic monitoring and controls, adding additional upfront cost. The self-weight is not outrageously heavy, but it will add extra weight onto new or existing structures that may need reinforcement.

At the technical level, optimum performance and long-term durability are primary concerns. The system needs to be multifunctional for microalgae, occupants, and buildings. The ability to withstand various structural and environmental loadings is an essential attribute of a building enclosure system. Maximum biomass productivity leads to economic returns as high-value bioproducts. Providing good growing environments by regulating temperatures, solar intensity, carbon dioxide (CO_2) aeration, pH, and salinity is critical. Summer shading, winter solar heating, and year-round daylighting are key to offsetting operational cost by energy cost savings. Carbon sequestration leads to improved indoor air quality and can be monetized as a carbon credit. Interaction with nature and good indoor environments promote occupant health and well-being.

At the maintenance and operation stage, excessive heat build-up in the culture requires additional cost for a heat exchanger. A majority of the sunlight is converted to heat energy in the media. The use of a heat exchanger to temper overheating increases operational cost. It is important to reclaim the solar energy stored in the system for other uses such as space and water heating. Microalgae

DOI: 10.4324/9780367814410-17

must also be protected from external contaminants. Potential toxin and volatile organic compounds (VOC) emissions from microalgae have been reported and require protective measures. Nutrients contribute to substantial cultivation cost. Operational cost can be saved by uptaking nutrients from wastewater. When the microalgae system is tied with wastewater treatment, it is recommended to send concentrated microalgae to a wastewater facility to avoid contamination.

Biofouling diminishes light penetration as well as aesthetics, so periodic cleaning is one of the main maintenance obligations during cultivation. Most of the microalgae are immotile and have a tendency to stick to the inner surface of the bioreactor. When the configuration of the bioreactor does not receive aeration bubbles, cell sedimentation occurs at the bottom of the reactor which may impact architectural appearance.

When cultivation is done at the architecture scale, working with existing infrastructure is necessary for bioproducts and biofuel production. Integration with existing infrastructure is important for the harvesting process. Due to low utilization and building code requirements, it is not economical to set up a post-cultivation system in buildings. Microalgae-based biofuel is much more expensive than other forms of renewable energy. More private and public investment is critical in cost-effective upscale industrialization.

Index

Note: **Bold** page numbers refer to tables and *italic* page numbers refer to figures.

residential energy consumption 4;
sick building syndrome 40, **41**, 227;
thermoelectric power 42–43
urban: agricultural systems 109; air quality
problems 238; algae canopy 77–78;
algae culture 67–68; anthropogenic
emissions 10; building retrofitting 238;
infrastructure 57; population 3, 232
urban heat island (UHI) effect 9–10
urban influx 3
urban intervention: AlgaeClad 82–84; Algae
Dome 81–82; AlgaeGarden 76–77;
Algae Photobioreactor Parking Canopy
78–79; Algaevator 79–81; ALGA(e)
zebo 77; Ecopods 75–76; Flower Street
Bioreactor 75; Helix BioReactor Perth 75;
Urban Algae Canopy 77–78
urbanization 238
UrbanLab 91–92
U.S. Climate Alliance 12–13, **14–16**, 233
U.S. Department of Energy's (DOE) 31, 224
U.S. Environmental Protection Agency (EPA)
19, 40–41
U.S. Global Change Research Program's
2018 report 9
U.S. Green Building Council Leadership
(USGB-C) 13

vertical landscape 233–235
visible-light-to-heat-gain ratio 168
visible light transmittance (VLT) 166, 168
volatile organic compounds (VOC) 38,
41–42, 235, 246
voluntary commitment 12

waste-free circular economy 6
wastewater treatment 26–27, 31, 38, 43, 65
water: building industry 43; irrigation
43; public supply 43; quality 42–47;
thermoelectric power 42–43
wavelength transmission 167
Wilson, Edward 234
window-to-wall ratio (WWR) 171
World Health Organization (WHO) 38, 226;
nitrogen dioxide air quality guidelines 39
World Meteorological Organization reports
6, 9
World Trade Organization (WTO) 40
Woven Typology© 208–212, *213–216*

X SEA TY 58–64

zero energy architectural practices 5, 12,
232, 233
zero energy goal 233